Deep Learning for Radar and Communications Automatic Target Recognition

To my family, Irshita Majumder, Tanishee Majumder, and Dr. Moumita Sarker; Mrs. Mamata Majumder, and Mr. Ranajit Majumder; Doctors Kalipada Majumder, Ramaprasad Majumder, and Khokan Majumder; Mrs. Ila Sarker, Mr. Ananda Kanti Sarker, Dr. Moushumee Dey and Mrs. Bithi Kona Paul—for your love, support, and encouragement.

—Uttam K. Majumder

To my family Vojtech, Matej, Ondrej, and Jitka for overseeing the book progress

—Erik P. Blasch

To my family, wife Penny Garren, daughter Anna Grace Garren, daughter Mary Elizabeth Garren, father Dr. Kenneth Garren, mother Sheila Garren, brother Prof. Steven Garren, sister Dr. Kristine Garren Snow—for your enduring love and support—with special celebration of my father's exceptional service and upcoming retirement as President of the University of Lynchburg for almost two decades!

—David A. Garren

For a listing of recent titles in the
Artech House Radar Series,
turn to the back of this book.

Deep Learning for Radar and Communications Automatic Target Recognition

Uttam K. Majumder
Erik P. Blasch
David A. Garren

ARTECH
HOUSE
BOSTON | LONDON
artechhouse.com

Library of Congress Cataloging-in-Publication Data
A catalog record for this book is available from the U.S. Library of Congress.

British Library Cataloguing in Publication Data
A catalog record for this book is available from the British Library.

ISBN-13: 978-1-63081--637-7

Cover design by John Gomes

685 Canton Street
Norwood, MA 02062

10 9 8 7 6 5 4 3 2

Contents

Foreword

This timely book provides the context and technical background to participate in the exciting renaissance of synthetic aperture radar (SAR) automatic target recognition (ATR) fueled by the revolutionary advances in machine learning (ML), particularly deep learning (DL). This book, however, does not begin with a treatise in DL. As a self-contained reference, it provides a historic view of SAR ATR; an academic view where fundamentals in linear algebra, multivariate calculus, and probability are covered, a comprehensive review of machine learning as a jumping-off point for DL, and an application view where these tools are applied to both SAR and radio frequency (RF) signal based systems.

This book introduces the second coming of SAR ATR. From a historic perspective it also documents the first coming of SAR ATR, which happened in the mid-1990s when Defense Advanced Research Projects Agency (DARPA) and Air Force Research Laboratory (AFRL) invested in model-based SAR ATR as part of the Moving and Stationary Target Acquisition and Recognition (MSTAR) program and its predecessors. This technology, along with its sister method, template-based ATR technology, defined the state of the art that has persisted to the present time. Its staying power stemmed from the fact that it was based on strong electromagnetic, signal processing, and statistical decision theory principles. In fact, one could argue that it was an optimal approach given its statistical assumptions of conditional independence.

However, the power of DL unleashed the ability to find patterns and dependencies in the data that are not limited by the conditional independence assumptions of the current generation of SAR ATR algorithms. These more expressive features provide the potential for significant improvement in performance. This book provides the requisite material needed to hop on board this exciting research field applied to the very important RF application areas. Unlike the model-based approaches of the past that are, in principle, based on linear systems theory and are very modular, the DL approaches are just emerging, are highly nonlinear, and use multiple architectures with multiple objectives. In essence, there are a myriad of approaches that are emerging. Fortunately, this book provides a comprehensive treatment of these emerging architectures and gives multiple application examples.

Although the fields of artificial intelligence and machine learning (AI/ML) have been researched since the late 1950s, the field of DL is just emerging. There are many more challenges ahead as we don't yet understand how the algorithms work—we have no fundamental theory, many applications are data poor, and current generation DL approaches require large amounts of data, particularly when considering the variability of targets, environment, and sensors. This need for large amounts of data motivates the use of synthetic data to cover the variety of conditions; however, there is a gap between synthetic data and operational sensors. In addition, DL approaches have been primarily applied to closed set problems where the targets in the test set are well represented in the training set. For many application areas and specifically in ATR, the algorithm will see objects in operations that were not in the training set, which is known as the *open set problem*. As well, many networks are not well calibrated in probability and give point estimates rather that full posteriors. The need for accurate confidence predictions is very important in ATR applications where a wrong decision can have dire consequences.

Finally, this book gives you very powerful DL approaches to tackle the important challenges inherent in the ATR problem. However, it reminds you to not forget the principles of previous ATR approaches based on statistical, physical, and signal processing fundamentals. The marriage of these fundamentals with the emerging DL tools will be required to solve the difficult challenges moving forward. I know you will enjoy and learn much from this book. Armed with the knowledge gained, you will be well equipped to attack the ATR problem with renewed vigor and rigor. We are counting on you!

Edmund G. Zelnio
Fellow, Air Force Research Laboratory (AFRL)
Director, Autonomous Technology Research Center, AFRL
Former Chief and Technical Advisor, ATR Technology Division, AFRL Sensors Directorate

Preface

Deep Learning for Radar and Communications Automatic Target Recognition presents a comprehensive illustration of modern artificial intelligence/machine learning (AI/ML) technology for radio frequency (RF) data exploitation. While numerous textbooks focus on AI/ML technology for non-RF data such as video images, audio speech, and spoken text, there is no such book for data in the RF spectrum. Hence, there is a need for an RF machine learning (ML) book for the research community that captures state-of-the-art AI/ML and deep learning (DL) algorithms and future challenges. Our goals with this book are to provide the practitioner with (i) an overview of the important ML/DL techniques, (ii) an exposition of the technical challenges associated with developing ML methods for RF applications, and (iii) implementation of ML techniques on synthetic aperture radar (SAR) imagery and communication signals classification.

It is important to become familiar with the terms AI, ML, and DL as these words are sometimes used interchangeably in today's literature. Broadly speaking, DL is the subset of ML, and ML is the subset of AI (a more in-depth definition can be found in Chapter 1). The methodologies of ML have experienced an explosion on the scene of multiple domains solving a variety of problems. Many of these technologies apply to physical reasoning to recognize particular objects in the sensing domain, discerning voice patterns in the audio domain, or that of distinguishing language phrases in the text domain. ML has also had a significant impact in various aspects involving semantic reasoning of intent based upon human kinematic, audio, or written behavior. The enabling technologies for these advances have resulted from the development of faster computers and more storage for training data. Many ML architectures are inspired from biological neural networks to emulate the human cerebral cortex. Here, recurrent training involving sensing operations, feature recognition, and cognitive reasoning enables a given human individual to function well within the environment. Likewise, ML architectures based upon neural networks in silicon-based computer chip sets are trained in a similar fashion through extensive processing involving many data sets.

This book is intended to provide a broad overview of ML concepts relevant to modern radar and communications systems which typically generate and apply complex-valued in-phase and quadrature data channels. A taxonomy of the various methodologies is provided according to a hierarchy of pattern recognition techniques. The discussion proceeds according to *supervised learning* methods, for which labeled training datasets are available, through the use of *semisupervised*, and *unsupervised* methods that apply datasets for which labeled truth information is not available. The taxonomy is decomposed further according to the specific details of the processing architectures. For some elements presented in this book, further details are given to provide the reader with insight pertaining to the strengths of the various techniques. As such, the reader will be able to glean intuition about the selection of specific ML approaches that are likely to yield favorable results for a particular radar or communications application. We sought to provide a comparative analysis of the techniques as a focus of the book, since currently there do not appear to be any books on the market that provide the reader with an understanding of the various AI/ML/DL approaches for radar and communications applications. Simplified examples and diagrams give the reader further insight into the similarities and differences among the many ML strategies that can be developed and utilized. Furthermore, the book includes a comprehensive reference list per chapter for each ML approach, so that the reader is easily steered toward additional auxiliary information to accomplish needed goals on radar and communications projects.

There have been some surprises in the use of ML for the processing of complex-valued radar imagery and in communications signals exploitation. In particular, it could be argued that machine-based neural networks are actually more flexible than the corresponding biological versions of the human prefrontal cortex, at least in some regard. Specifically, ML algorithms can use complex-valued in-phase and quadrature data—or equivalently, magnitude and phase data—which are generated within many radar and communications receivers. It is possible to use the raw phase history data chips as input for both training and testing ML algorithms (as presented in Chapter 6). The actual phase history images appear as random noise to a human, regardless of whether one examines the data as in-phase and/or quadrature images. The statistical uncertainty from noise follows from the general holographic property of complex-valued synthetic aperture radar imagery that is transduced in the collection of the radar signal returns. It is remarkable that ML architectures are able to provide a nontrivial level of classification capability, whereas any human examining the same input phase history image data chips is likely to have extremely low classification capability.

Clearly, ML has arrived on the technological scene, and it is here to stay. This book provides the reader insight into the many approaches for ML architectures applied to complicated problems in radar and communications applications for emerging students, academic researchers, and engineering practitioners. We hope that you will benefit greatly from the composition, organization, and presentation of ML methodologies provided herein applied to RF data.

Chapter 1 provides an evolution of ML algorithms and the past approaches for implementing automatic target recognition (ATR) technology to RF one-dimensional signals and two-dimensional imagery. The chapter highlights the shortcomings of the past approaches and provides the motivation and benefits of modern DL techniques for RF signal and imagery classification. In Chapter 2, the mathematical preliminaries provide the basics to understand ML algorithms. The goal is to review the basics of linear algebra, optimization, and probability theory commonly used ML algorithms. Chapters 3 and 4 depict ML/DL taxonomy in detail. Specifically, Chapter 3 discusses basic ML algorithms, whereas Chapter 4 develops the common deep neural networks (DNN) algorithms. The book categorizes ML algorithms based upon the delineations of *supervised*, *semisupervised*, and *unsupervised* learning, while emerging DL algorithms include *reinforcement, adversarial, and transfer learning*. Within these categories, there are subclasses and subcategories. The book highlights the details of these algorithms. In particular, we provide the implementation examples with a common SAR/RF data set as a use case of ML algorithms in RF spectrum for algorithm comparison. We hope that the ML practitioners will find these two chapters useful, not only for understanding and implementing ML algorithms, but for the entire family of ML algorithms.

Chapter 5 illustrates that radio frequency exploitation is a "big data" problem. From RF imagery to RF communication systems involving Internet of Things (IoT) devices, RF systems are generating between petabytes (10^{15}) to exabytes (10^{18}) of data to be exploited for intelligent decision making. ML plays a key role in decision making by reducing the number of image analysts, computational time, and monetary cost. We stress that in addition to the benefits of increased accuracy, credibility and availability of results, intelligent methods are crucial for secure, robust, and reliable ML-based classification algorithms development. For method comparisons and future developments, we present available RF data sources for ML algorithms research.

Chapter 6 presents algorithms implementation for single RF target classification, often known as image chip classification. This chapter is focused on classification accuracy with confusion matrix analysis. Chapter 7 follows with ML algorithms involving multiple targets classification. Multiple targets in RF imagery can be laid out in a complicated manner (i.e., masking parts of each other or hiding in clutter), thus making the classification tasks extremely challenging. Since multiple target recognition requires a two-step process of detection and classification, the chapter presents various detection algorithms applied to imagery and the methodologies required in order to incorporate these techniques in the RF domain.

Chapter 8 entails ML algorithms implementation for RF signal classification. This chapter details the features of RF communication signals such as Wi-Fi and Universal Software Radio Peripheral (USRP) radios and the methods by which ML algorithms can be applied to identify and fingerprint a given RF device.

Chapter 9 emphasizes the importance of ML algorithms performance evaluation. Often researchers implement DL models on one dataset and receive the desired accuracy. However, a single data approach is only guaranteed to work well on a slightly different dataset (caused by variations in the lighting condition, RF sensors' frequency, bandwidth, etc.) or for a dataset that is perturbed by noise uncertainty. ATR performance modeling allows an understanding of classification degradation as a function of data corruption and adversarial attacks (from pristine to noisy/adversarial data). We illustrate various approaches for ML algorithms' performance evaluation.

Chapter 10 introduces the reader to emerging topics affecting ML applications such as adversarial ML (AML) techniques and methods of mitigation from data, modeling and algorithm attacks. Other important emerging topics include transfer learning (TL) for limited measured data availability, energy-efficient computational architectures for ML systems, and algorithmic approaches for near real-time training.

This book could not have been written without the contributions from several key individuals. We are indebted to Edmund Zelnio (considered as the father of the

AFRL ATR technology) whose guidance and thoughts propelled us to write this book. We acknowledge DARPA program manager Dr. John Gorman for collaborative research with the multi-faceted ATR deep learning (MADRLEARNING) and Target Recognition and Adaption in Contested Environmnets (TRACE) programs that were the first two major programs that embraced the DL technology for SAR imagery classification. We also appreciate DARPA PM Mr. Paul Tilghman and his SETA Dr. Esko Jaska for collaborative research on the RFMLS program. Over the years, we were privileged to work with many great students of which their work is referenced in the text. Specifically, we greatly appreciate the contributions of our recent AFRL interns Nathan "Nate" Inkawhich and Matthew "Matt" Inkawhich and their professors Dr. Yiran Chen and Dr. Hai Li. Nate and Matt are phenomenal doctoral students at Duke University that perfected the DL approaches. They worked with us extensively on RF ML research. We considered them as "Innovative Superstars" for the advancement of next generation AI/ML technology. We also acknowledge the contributions of Chris Capraro, Eric Davis, and Darrek Isereau of SRC, Inc. for contributions in the areas of SAR imaging and adversarial machine learning, Dr. Mark Peot, Dr. John Spear of Teledyne Scientific and Imaging for transfer learning research, and Prof. Dhireesha Kudithipudi of University of Texas and Syed Humza of Rochester Institute of Technology for contributions in near real-time training algorithms development. We are grateful to AFRL Information Directorate leaderships, especially, Dr. Mike Hayduk, Mr. Joe Caroli, and Dr. Don Telesca for their encouragement and support.

We appreciate the mentorship of Dr. Vince Velten, Dr. Mike Bryant, Dr. Mike Minardi, Patti Ryan Westerkamp, and especially Dr. Tim Ross from the AFRL Sensors Directorate–COMprehensive Performance Assessment of Sensor Exploitation (COMPASE) center.

Current AFRL Sensors Directorate researchers contributing to SAR ATR research include: Dr. Eric Branch, Dr. Theresa Scarnati, Benjamin Lewis, Dr. Christopher Paulson, Daniel Uppenkamp, Elizabeth Sudkamp, Dr. Paul Sotirelis, Dr. Linda Moore, Dr. LeRoy Gorham, Steven Scarborough, and Dr. John Nehrbass.

As the book captures many developments, lessons learned, and practical results not least of which is the data, we acknowledge those who created the MSTAR vision. Parts of the MSTAR vision were shared by a number of individuals throughout DoD, but the important fact was that Dr. Richard Wishner of DARPA was primarily responsible for the belief, construction, and implementation of the vision. He convinced Mr. Larry Lynn, the Director of DARPA, to fund the MSTAR program and then constructed the program true to the tenets of his vision. The preference for a model-based approach was motivated by the excellent work performed by Professor Tom Binford at Stanford University. Both Richard Wishner and Ed Zelnio were introduced to the model-based approach by Professor Binford.

On the government side, several individuals played key roles. Major Tom Burns started as program manager for AFRL and then became the DARPA program manager later in the MSTAR program. Mr. Martin Justice took over for Major Burns and continued as program manager until the program ended. Mr. Mark Minardi served as the program technical director, Dr. Bill Pierson served as the ATR performance evaluation lead, and Dr. Tim Ross spearheaded the system engineering function key to the program success. AFRL provided continuity throughout the program while MSTAR was guided by four consecutive DARPA program managers from beginning to end. Dr. Bob Douglass, Mr. John Gilmore, Major Tom Burns, and Dr. Robert Hummel—each contributed in different and special ways to MSTAR directions.

On the contractor side, there were numerous key contributors. The key leaders for the teams from each organization were Dr. Gil Ettinger, Dr. John Wissinger, and Dr. Bob Tenney, Alphatech; Mr. Dennis Andersh, Demaco; Mr. Eric Keydel, ERIM; Mr. Dave Morganthaler, Lockheed Martin; Dr. Tom Ryan, SAIC; Dr. Ron Dilsavor, Sverdrup; and Dr. Bob Hummel, New York University, who later became the MSTAR program manager at DARPA.

Hence, the MSTAR program that contributed to the foundational material of the book was a product of numerous people, those listed above as well as many who played support roles throughout the MSTAR program. We encourage readers to rediscover over 200 available papers from the program, mostly available in the 25+ years of Ed Zelnio's and professor Fred Garber's SPIE Algorithms for Synthetic Aperture Radar Conference.

Finally, as with the 100s of researchers that contributed to the *Deep Learning for Radar and Communications Automatic Target Recognition* techniques, we sought to reference key research publications that reveal insights that follow from tremendous presentations.

In closing, gratitude goes again to Mr. Edmund Zelnio for his sustained leadership for the *SPIE Algorithms for Synthetic Aperture Radar Imagery* conference that provided the forum for researchers, students, and practitioners to continual debate the technology development.

Uttam K. Majumder
Erik P. Blasch
David A. Garren

CHAPTER 1

Machine Learning and Radio Frequency: Past, Present, and Future

1.1 Introduction

American science fiction writer Robert A. Heinlein [1] stated that "when railroading time comes, you can railroad—but not before." From Heinlin's quote we can say that it's the time for applying machine learning (ML) to develop intelligent radio frequency applications. With astounding speed, ML algorithms have been evolving to analyze big data for real-time decision making. The explosion of ML algorithms pervades the public consciousness from multimedia needs for image classification, speech recognition, and cybernetics. However, applying ML to radio frequency (RF) big data problems is a recent development, but is evident in the emerging radar technology in autonomous self-driving cars, communication networks, and imagery data. This chapter focuses on introducing RF signatures, contemporary RF methods, and current RF exploitive challenges. By providing a background on RF solutions without ML algorithms, the reader can appreciate the problems where ML (and more broadly artificial intelligence (AI)) can be applied to efficiently solve RF-signatures classification problems.

1.1.1 Radio Frequency Signals

Radio frequency applications work from the extremely low frequency (ELF) range 3–30 hertz (Hz) to the extremely high frequency (EHF) range 30–300 gigahertz (GHz) as shown in Figure 1.1 [2]. Digital communications signals (e.g., radio, television) are from 300 kilohertz (kHz) to 300 megahertz (MHz). Cellular networks, global positioning systems (GPSs), local area networks, and some radar systems work around 300 MHz to 3 GHz. Most radar, satellites, and some data transmission systems operate between 3 GHz to 30 GHz. Very high-resolution frequency (VHF) radar, automotive, and data communication systems function at 30–300 GHz.

Synthetic aperture radar (SAR) exists from 0.3–100 GHz and is derived from the formation of a two-dimensional (2D) image from radar pulses. SAR aggregates the motion of the radar antenna over a target region to provide a finer resolution

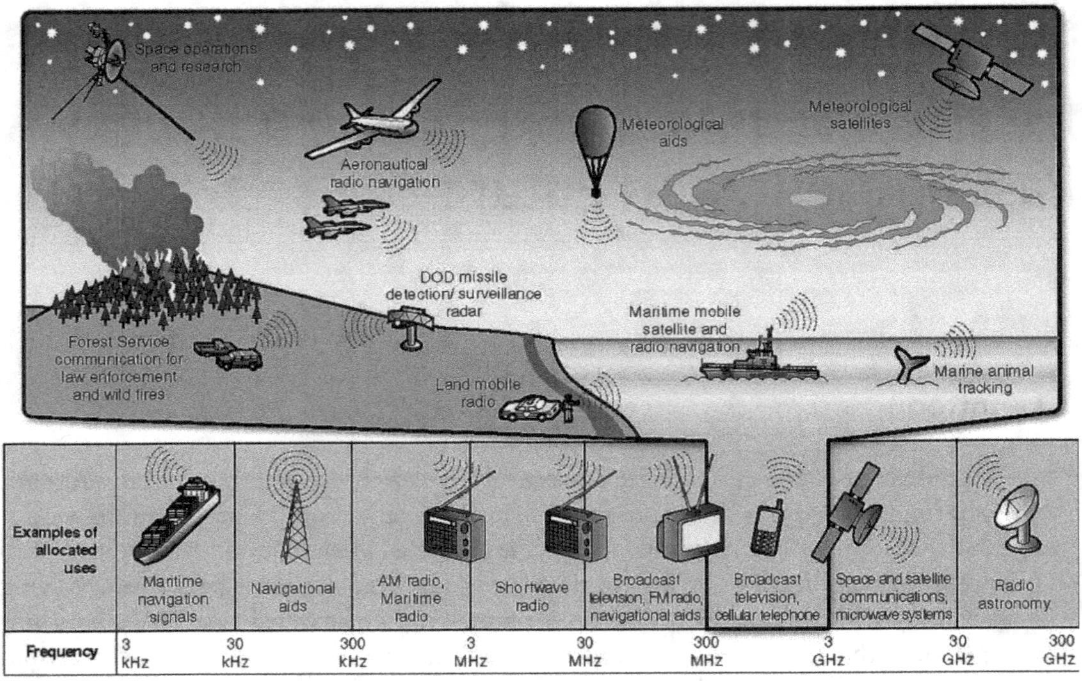

Source: GAO analysis of NTIA, federal agencies, and industry information.

Figure 1.1 Radio frequency data from the electromagnetic spectrum.

of the target illumination. Over 20 books have been published toward SAR image collection, formation, and analysis. The echoed radar waveforms with knowledge of the antenna positions at the pulse transmission times are combined to create images of a region of interest on the surface of the earth. Typical SAR applications include sensing of environmental and man-made objects [3], as shown in Table 1.1. This book will focus on applications of machine learning at X-band with examples to showcase advances of deep learning.

Table 1.1 Radio Frequency Sensing Applications

Frequency Band	*Frequency Range (GHz)*	*Wavelength (cm)*	*Application*
P	0.3–1	60–120	Foliage penetration Soil moisture
L	1–2	15–30	Soil moisture Agriculture
C	4–8	4.0–8.0	Agriculture Ocean
X	8–12	2.4–4.0	Ocean High-range resolution radar
Ku	14–18	1.7–2.5	Snow cover
Ka	27–47	0.75–1.2	High-frequency radar
W	56–100		Remote sensing Communications

The Defense Advanced Research Projects Agency (DARPA) *Moving and Stationary Target Acquisition Recognition (MSTAR)* dataset was collected with 9.6-GHz X-band SAR with a 1 foot range resolution × 1 foot cross-rang resolution over 1° spacing of 360° articulations [4]. The data was collected by Sandia National Laboratory in 1995 and consists of 13 target types including situations of articulation, obscuration, and camouflage, which resulted in a standard dataset of 10 targets (Figure 1.2) for depression angles of 15° and 17°. Another collection in 1996 included 15 target types from 27 actual targets [5]. Table 1.2 provides the numbers and types of targets available in the public MSTAR dataset. Throughout the text, the MSTAR dataset will be used to compare and contrast the many methods of ML, deep learning (DL), and performance evaluation.

Another RF imagery application for which ML developments are prominent includes polarimetric synthetic aperture radar (POLSAR). POLSAR data typically includes L-band and C-band radar for land cover, sea-ice mapping, and terrain analysis. As with X-band radar, there has been a resurgence of POLSAR analysis using deep learning methods. In 2019, a special *IEEE Access Journal* Vol. 7 included more than 10 papers with different developments toward POLSAR developments (e.g., [6]) along with MSTAR analysis (e.g., [7]).

Electro-optic images of objects

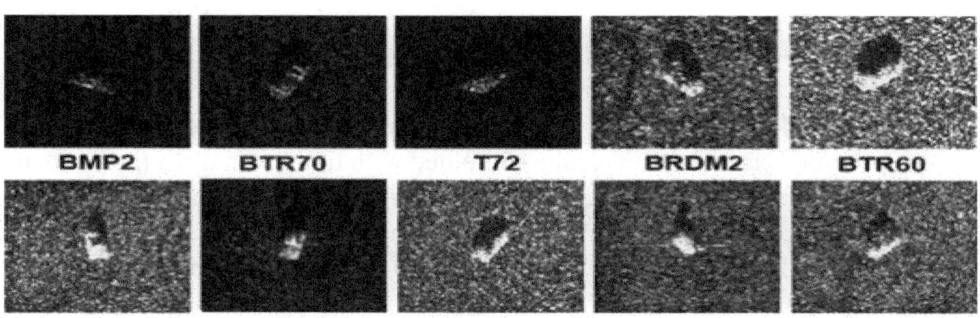

SAR images of objects

Figure 1.2 MSTAR target cCollection.

Table 1.2 Public Release MSTAR Data Set

Target	*Depression Angle (15°)*		*Depression Angle (17°)*	
	#Images	*Vehicles*	*#Images*	*Vehicles*
BMP-2	587	9563, 9566, c21	299	9563, 9566, c21
BTR-70	196	c71	697	c71
T-72	582	132, 812, s7	298	132, 812, s7
BTR-60	195	k10yt7532	256	k10yt7532
2S1	274	b01	233	b01
BRDM-2	263	E-71	299	E-71
D7	274	92v13015	299	92v13015
T62	582	A51	691	A51
ZIL131	274	E12	299	E12
ZSU234	274	d08	299	d08

1.1.2 Radio Frequency Applications

ML can be applied to RF systems for solving three big data problems: (1) object detection and classification from radar imagery, (2) signal classification from digital systems, and (3) developing a cognitive radar or communication system where transmitted energy can be optimized (for better results) on a specified area instead of radiating throughout the environment wasting scarce and expensive resources. Let's explore these three problems further.

First, RF object detection and classification have significant impact on three types of applications: autonomous cars, surveillance security, and biomedical diagnostics. Self-driving cars, or autonomous vehicle technology use RF signatures to integrate scans for detecting scene variations. Unlike other sensor modalities (e.g., video/electro-optic, infrared), RF sensors operate in all-weather (snow, rain, sandstorms, cloud) and day-night conditions. Hence, for self-driving vehicle technology, the RF sensor integration improves object detection and classification to enhance overall safety and performance robustness of systems. Analyzing RF imagery to find objects of interest is important in many security applications such as detecting moving objects within the vehicle vicinity. Finally, for biomedical applications, detection and classification of tumors can be greatly improved by applying ML technology to the X-ray or magnetic resonance imaging (MRI) imagery. All of these applications involve analyzing terabytes (10^{12}) to petabytes (10^{15}) of image data. ML algorithms, and more recently DL, can solve these problems in a fraction of time and more accurately than human analysts. One thing to mention is that ML-provided classification results combined with human analysts (i.e., human-machine interaction) yields optimal (reduced error) decision making.

Autonomous cars are expected to have cameras, light detection and ranging (LIDAR), and radar (radio detection and ranging) as shown in Figure 1.3. The multisensor controller manages inputs from the camera, LIDAR, and radar to analyze data in motion. With mapping and navigation techniques using the multisensor data, the controller can confirm decisions of object detection. For example, the camera takes images of the road that are interpreted by a computer. However, the camera is limited by what the camera can observe. The LIDAR sends out light pulses that are reflected off objects which can define lines on the road and works

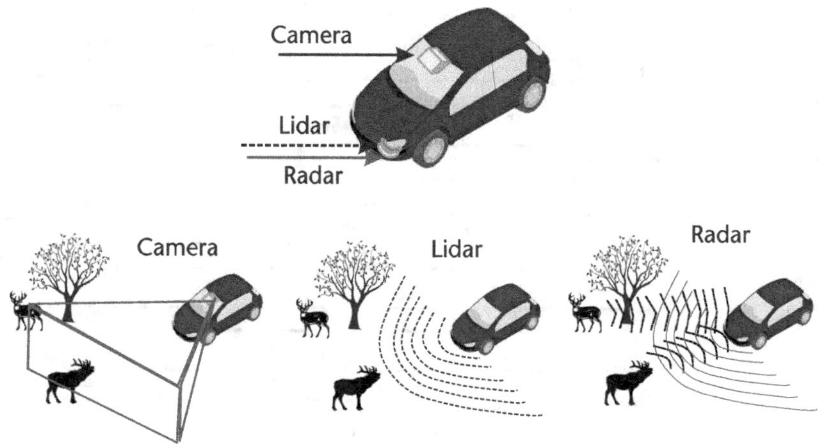

Figure 1.3 Autonomous cars see the road from multisensor fusion.

in the dark. The radar sends out radio waves that bounce off objects in all weather, but the limitation is not fine enough to differentiate objects. Combing the sensors yields object detection and classification in all-weather, day-night conditions.

The second applications of ML to RF signals involve *digital signal classification* for radio frequency identification (RFID) of devices and micro-Doppler of vibration signatures. To take advantage of these emerging systems, DARPA initiated a program called radio frequency machine learning systems (RFMLS) [8]. The goal of the RFMLS program was to develop the foundations for applying modern data-driven ML to the RF spectrum domain. Traditional wireless security relies on a software identity for each wireless device, as shown in Figure 1.4. One of the current challenges with the proliferation of RFID devices is that they can often be hacked or otherwise cloned. ML can be applied to learn features of a RF device, but care must be applied to ensure to avoid memorizing the hardware address of the device. Hence, when conducting detection, recognition, classification, and identification of RFID systems, advanced deep learning methods can detect variations in the data, align of signals to devices, and detect anomalies in expected results.

Figure 1.5 shows an RF device that is at the network edge as related to the current trends in Internet of Things (IoT), radar analysis, and automatic target

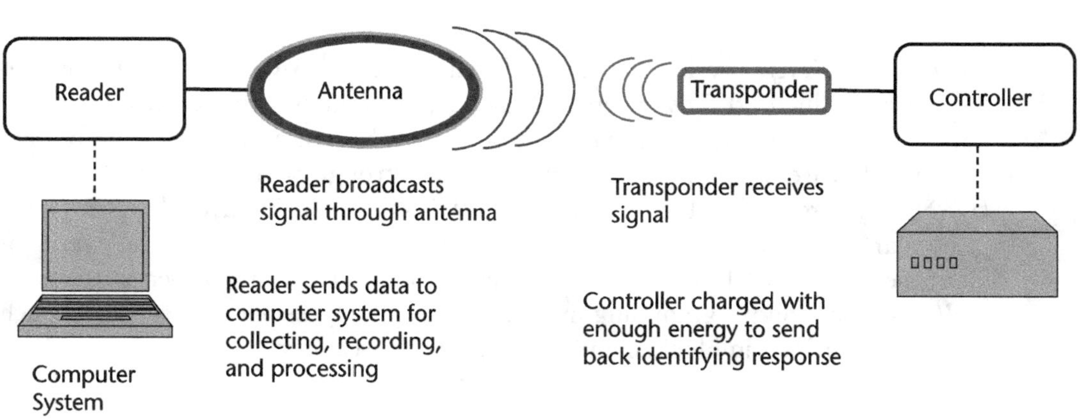

Figure 1.4 Radio frequency identification.

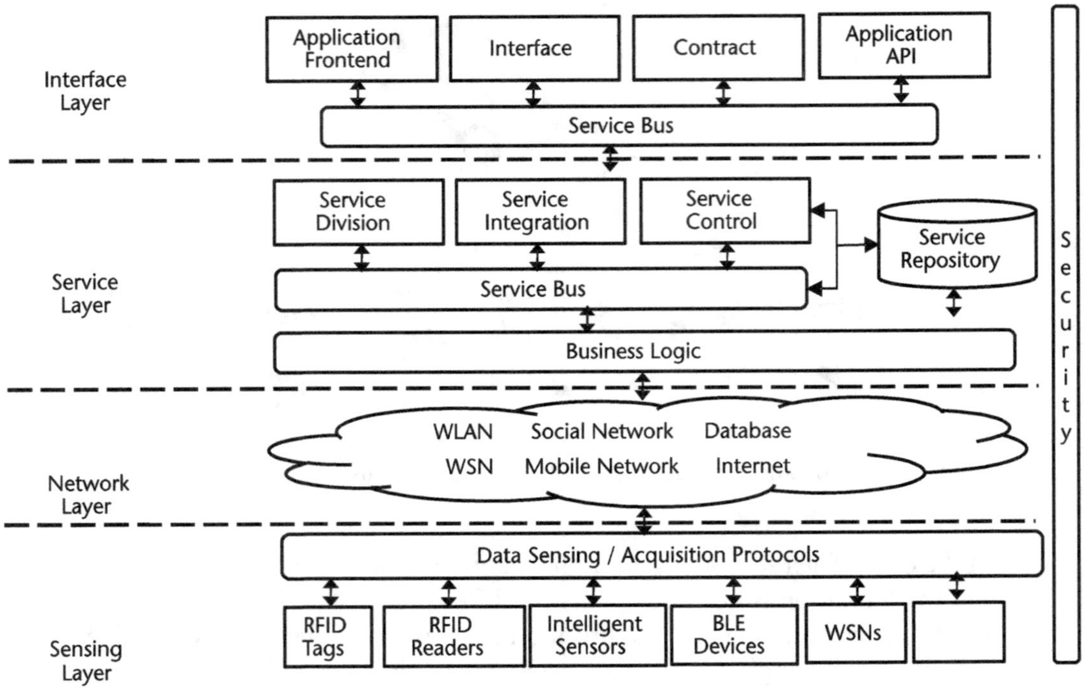

(a)

(b)

Figure 1.5 RFID radar system.

recognition (ATR). The RF devices work with other measurement systems as sensors such as Bluetooth Low Energy (BLE) devices. Together, the sensors send their data to the network layer that has the storage of information in a database of data at rest. Likewise, the data transits the internet through wireless sensor networks (WSNs) or wireless local area networks (WLANs). A service layer determines the particular data routed through logic to an interface layer. The interface layer provides the highest level of arbitration to relate to the platform front-end, interface, and/or application programming interface (API). An agreement is negotiated such as a smart contract in blockchain [9] to ensue data security.

Related to these investigations is the use of micro-Doppler to detect the signature of the target, and the vibration energy of the system. Figure 1.5 demonstrates an example for ATR for various human, animal, and clutter ATR [10].

The third application of the ML to RF system involves developing a *cognitive RF communication system* [11]. Traditionally, the electro-magnetic spectrum is transmitted omnidirectionally and then received signals are processed to find an object of interest. The broadcast approach is a very inefficient use of RF spectrum and expensive. An ML-based cognitive system can be developed with a feedback loop where the signal will be transmitted based on the functionality (e.g., utility and policy value function). For example, if the object of interest is small and already located, then high-frequency special waveforms are transmitted only at the direction of the object. In this manner, object detection and classification can be greatly enhanced. The introduction of software-defined radios (SDRs), software-defined networks (SDNs), and communications has led to cognitive radio and cognitive radar [12], as shown in Figure 1.6.

1.1.3 Radar Data Collection and Imaging

As mentioned above, radar stands for radio detection and anging. Radar imaging radar is commonly known as synthetic aperture radar (SAR). A radar system transmits a sequence of pulses on targets of interest that are reflected back into the receiver antenna. The receiver system records the echoed signal. The time delay between the transmit signal and the receive signal is used to calculate range information based upon the constancy of the speed of the propagation of radar pulses.

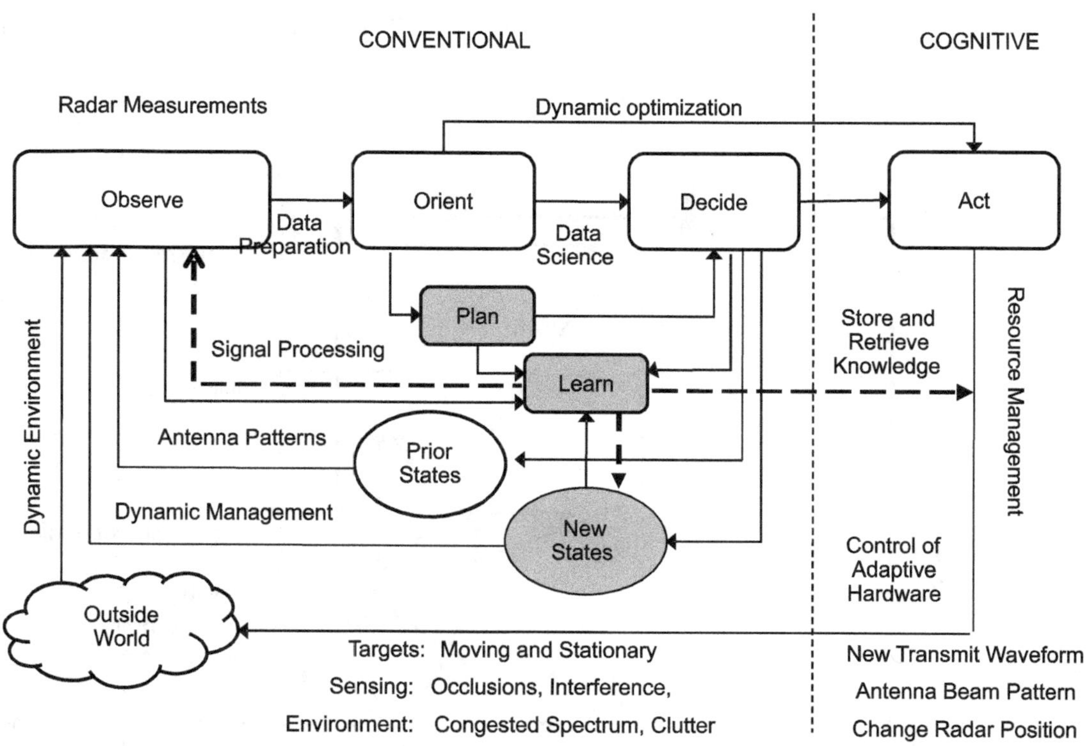

Figure 1.6 Cognitive radar sensing.

The fundamental SAR imaging geometry is shown in Figure 1.7, where an aircraft carrying the radar flies a straight flight path or stripmap mode with a velocity at a certain altitude [13]. The slant range is the distance from the radar to the imaging scene center. The *azimuth angle* φ is defined to be the angle around the scene center which encompasses 0 to 360 degrees. Azimuth angle is also known as aspect angle. The *elevation angle* θ is defined to be the angle measured from the ground plane toward the positive z-axis which encompasses 0 to 90 degrees. The angle between the *slant range* Δ and the ground plane is typically a measure of the elevation angle. Elevation angle is equivalent to the depression angle. Ground squint angle is the angle between the aircraft's flight path to the projection of the slant range on the ground. After radar data have been collected (raw phase history data), some preprocessing and image formation algorithms are applied to the data to form an image. The image formation algorithms could be as simple as range-Doppler processing (to the more accurate polar format algorithm) or to the extremely accurate back-projection algorithms [14]. These methods provide various imagery quality depending on the cost of computation.

A side-looking airborne radar (SLAR) processing approach develops a 2-D reflectivity map of the imaged area. The reflectivity is the illumination of targets where the high signal-to-noise ratio (SNR) backscattered signals are the bright spots in the radar images and flat smooth surfaces are shown as dark areas. The flight direction is denoted as azimuth, and the line-of-sight as the slant range direction. The azimuth resolution decreases as the range increases. For example, a 9.6-GHz X-band SLAR system with a wavelength λ of 0.31m and a 1.77m antenna d has an azimuth antenna beamwidth of

$$\theta_a = \frac{\lambda}{d_a}[= 0.031 \text{ m}/1.77 \text{ m}] = 0.0175 \text{ rad} \approx 1° \tag{1.1}$$

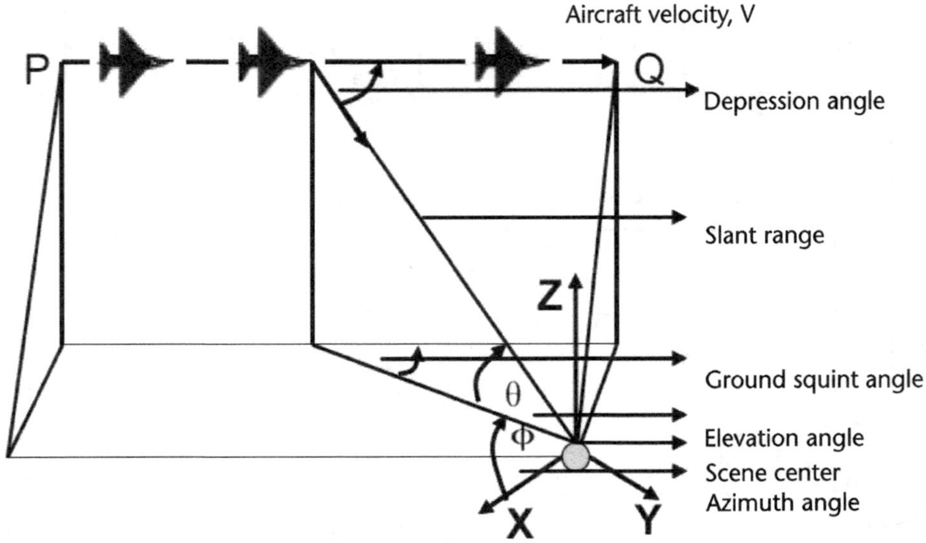

Figure 1.7 SAR imaging geometry and terminology. Azimuth angle is the angle around the scene center. Elevation angle is defined to be the angle measured from the ground plane to the z-axis.

The azimuth resolution δ_a is the smallest distance between two targets that can be detected by the radar which is (for a range of $r_0 = 17.1$m)

$$\delta_a = \frac{\lambda}{d_a} r_0 = \theta_a r_0 [= 0.0175 \text{ rad} \cdot 17.1 \text{ m}] = 0.30 \text{ m} \approx 1 \text{ ft} \tag{1.2}$$

The distance between the radar moving at constant velocity v and a point on the ground, described by its coordinates $(x, y, z) = (x_0, 0, \Delta h)$, with Δh equal to the local terrain height, is obtained by

$$r(t) = \sqrt{r_0^2 + (vt)^2} \approx r + \frac{(vt)^2}{2r_0} \text{ for } vt/r_0 \ll 1 \tag{1.3}$$

where, assuming $t = t_0 = 0$ is the time of closest approach, when the distance is minimum and

$$r(t_0) = r_0 = \sqrt{x_0^2 + (H - \Delta h)^2} \tag{1.4}$$

in terms of the platform height H. In general from (1.3), the distance r_0 is much larger than vt during the illumination time T for which a scattering center point on the ground is interrogated with radar pulses.

Radar imaging seeks 2D imagery. The slant-range resolution δ is inversely proportional to the system bandwidth according to $\delta_r = c_0/2B_r$,where c_0 is the speed of light, B_r is the bandwidth (GHz), and the factor of 2 is due to the two-way path from transmission to reception.

$$\delta_r = \frac{c}{2B_r} = \frac{3 \times 10^8 \, m/s}{2(9.6 \text{ GHz})} = 0.016\text{m} \approx 0.63 \text{ in} \tag{1.5}$$

The bandwidth is related to the difference between the maximum and minimum instantaneous frequency for a given chirp signal transmission waveform. Assuming the transmitted waveform amplitude is constant during the pulse time τ and the instantaneous frequency is linear over a chirp rate κ_r; then the bandwidth is $B_r = \kappa_r \tau$. The radar transmits and receives the echo signal and stores the received signal. The azimuth resolution d_a is provided by the construction of the synthetic aperture, which is the path length over which the radar receives echo signals from a given point target scattering center. With an antenna beamwidth θ_a, the corresponding synthetic aperture length is given by $L_{SA} = {}_a r_0 = (\lambda/d_a) r_0$. A narrow virtual beamwidth results from a long synthetic aperture $\theta_{SA} = \lambda/2L_{SA}$, leading to a high azimuth resolution:

$$\delta_a = r_0 \theta_{SA} = r_0 \left(\frac{\lambda}{2L_{SA}} \right) = \frac{d_a}{2} \tag{1.6}$$

Notice that a shorter antenna sees any point on the ground for a longer time, which is equivalent to a longer virtual antenna length and thus a higher azimuth resolution.

The received echo signal data form a 2D data matrix of complex samples, where each complex sample is given by its real and imaginary part, thus representing an amplitude and phase value. One matrix dimension corresponds to the range direction (or fast time); a range line consists of the complex echo signal samples after being amplified, downconverted to baseband, digitized and stored in memory. One other matrix dimension represents the azimuth (or slow time) as the radar acquires a range line whenever the signal travels a distance $v \bullet PRI$ (pulse repetition interval). Thus, the return *echoes* from the illuminated scene are sampled in both fast time (range) and slow time (azimuth).

A series of echoed radar pulses form a coherent processing interval (CPI), which is also commonly called a *radar dwell.* Each pulse within a CPI is sampled adequately to retrieve meaningful target information. Each sampled point of a pulse is known as *range bin* or range gate, which is often reselected to correspond to a resolution cell that is intrinsic to the details of the sensor collection system. Since a radar system samples a pulse as it arrives, the range bin is also known as fast time sample and range axis is also known as fast time axis (see Figure 1.8).

After sampling the first pulse, the next pulse is sampled and the process continues until the last pulse of the CPI is received. Hence, the pulse axis is known as slow-time axis. Two-dimensional matched filtering in slow time and fast time is called SAR imaging (Figure 1.9).

Besides the collection of a straight path with SLAR, there is also circular SAR (Figure 1.10), which enhances spot-mode radar by focusing on a single point. The single point could the target of interest.

The radar-carrying aircraft moves along a circular path with radius R on the plane $z = z_0$ with respect to the ground plane. Thus, the coordinates of the radar in the spatial domain as a function of the slow-time are $(x, y, z) = (R \cos \theta, R \sin \theta, z_0)$, where $\theta \in [-\pi, \pi)$ represents the slow-time domain. As the radar moves along

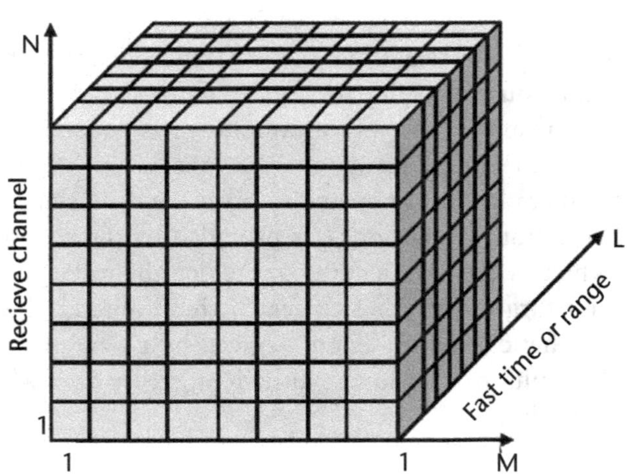

Figure 1.8 Radar data cube. The horizontal axis is the pulse number or slow-time axis. It contains M pulses for a CPI. The axis into the page is the fast time or range axis. It contains L fast time samples. The vertical axis is the number of the receive channel. It contains N receive channels.

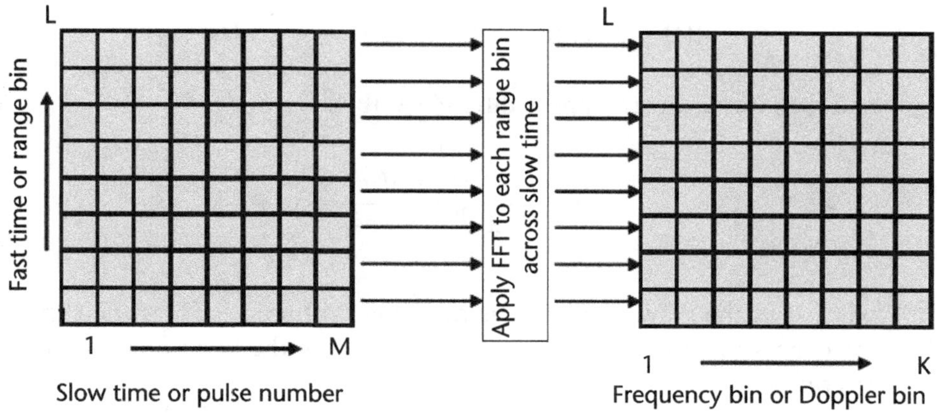

Figure 1.9 Simple radar image formation by applying range-Doppler processing concept. A range-Doppler image is created by applying a fast Fourier transform (FFT) along pulse/slow-time axis for each range bin.

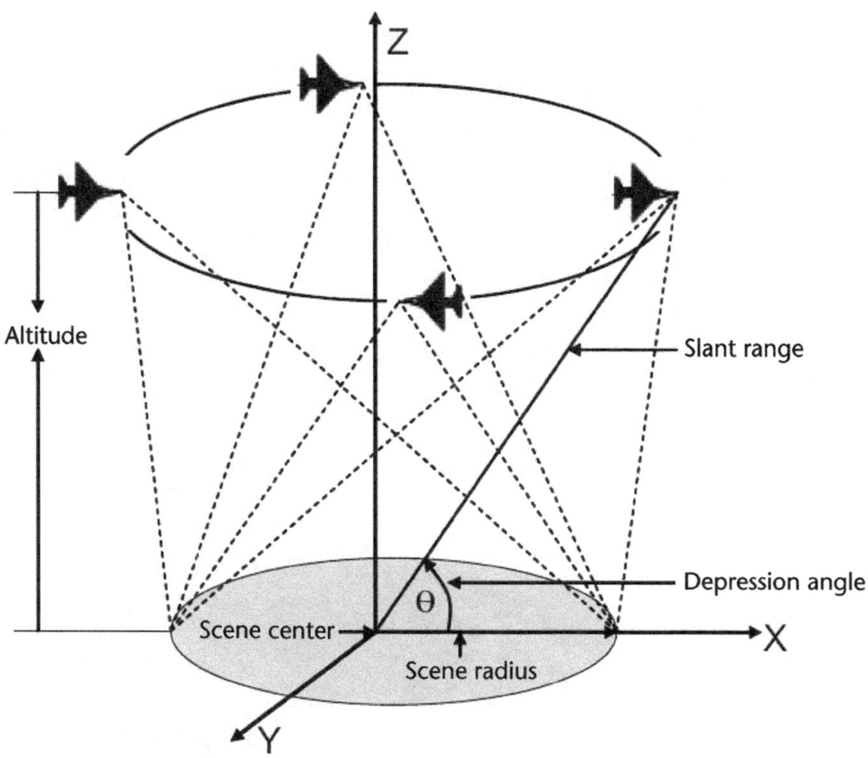

Figure 1.10 Circular SAR (CSAR) data collection geometry. The aircraft carrying the radar flies around a scene of interest. The radar radiation pattern remains constant or varies on the scene.

the circular synthetic aperture, its beam is spotlighted on the disk of radius (note: it is radius r and diameter is $2r$; however D is not used) centered at the origin of the spatial (x,y) domain on the ground plane (the target region's support). With a side-looking SAR, starting at (0, R, H), then the instantaneous distance for slow-time t such that $v = \omega t$, is

$$R(t) = (R\cos\omega t - x_0)^2 + (R\sin\omega t - y_0)^2 + (H - z_0)^2 \tag{1.7}$$

and the interferometric phase difference (IPD) is

$$\phi \approx -\left(\frac{4\pi}{\lambda}\right)\left(\frac{B\sin(\alpha+\beta)}{R_s\cos\alpha}\right)\Delta h \tag{1.8}$$

where λ is the wavelength, Δh is the height difference, R_S is the slant range, α is the depression angle, and β is the angle between the flight horizontal plan and the aircraft altitude.

On example of CSAR is the Gotcha data set, shown in Figure 1.11. The Gotcha data set is fully polarimetric and consists of eight complete circular (360°) passes, with each pass being at a different elevation angle, θ; where the radar used in the Gotcha data collection has a center frequency of 9.6 GHz, giving a center wavelength of λ_c= 0.031, and a bandwidth of 640 MHz [15]. Each pass has a planned (ideal) separation of $\Delta\theta$= 0.18° in elevation with a planned elevation $\theta \in$ [43.7°, 45.0°]. Actual flight paths differ from the planned paths, with elevation varying as a function of azimuth angle.

Given the parameters of a SAR data collection, then two common examples are aerial SAR and space-based SAR. The different altitude determines the range from which different resolutions result. For aerial systems, targets can be discerned whereas for satellite imagery it is typically used for foliage and vegetation analysis (see Table 1.1).

Finally, multisensor fusion acknowledges that there are many types of data that support analysis including radar and electro-optical imagery such as laser, thermal,

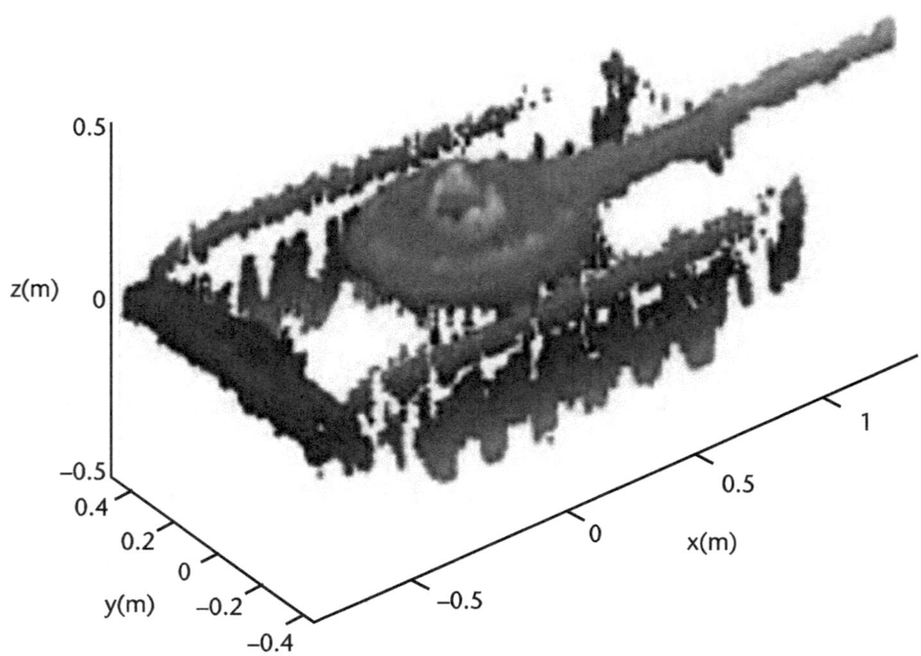

Figure 1.11 3D reconstructed image from Gotcha data [16].

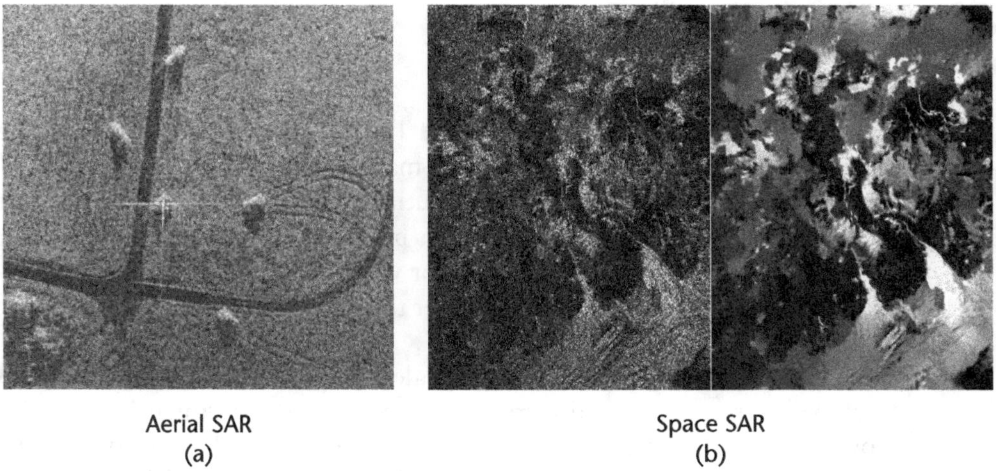

Figure 1.12 (a) Aerial SAR, and (b) space-based SAR. (From Space SAR, https://www.jpl.nasa.gov/spaceimages/details.php?id=pia19054.) (Courtesy NASA/JPL-Caltech.)

and visual data (Figure 1.13). The case for each sensor is based on the available sensors, the situation, and the mission of the sensor collection. Low-level information fusion (LLIF) includes the sensor data for object assessment including classification and tracking; whereas high-level information fusion (HLIF) incorporates the ability to understand the object assessment results over the situational context and sensor management constraints [17]. Hence, information fusion includes many emerging concepts, and more specifically opportunities for RF and EO image fusion [18].

Imagery collection results in target exploitation. Target exploitation has a rich history designated as automatic target recognition (ATR).

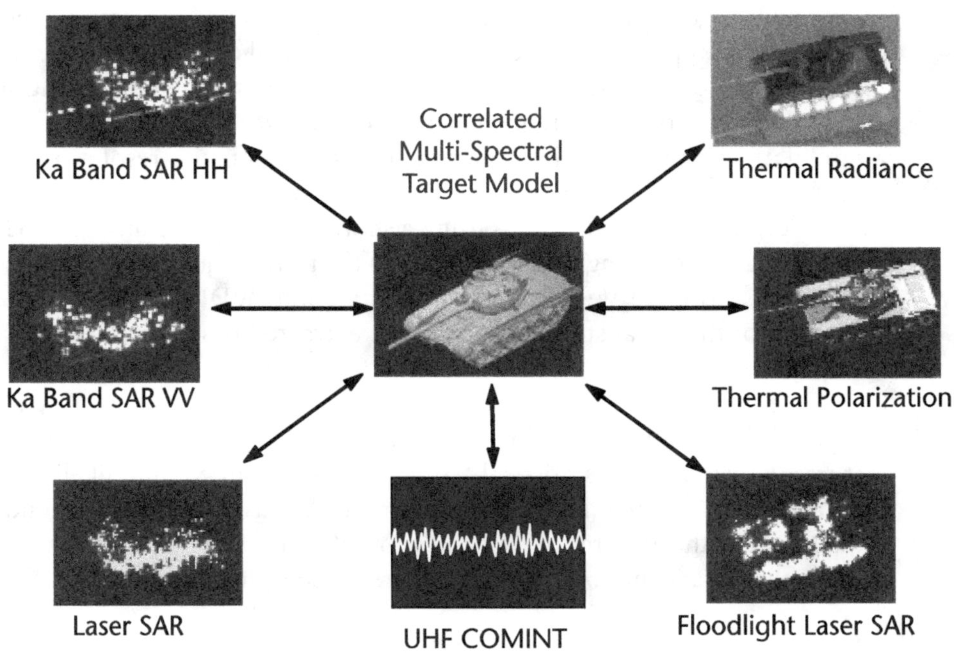

Figure 1.13 Multispectral RF target signatures. (From [19.]

1.2 ATR Analysis

1.2.1 ATR History

Radar methods took prominence in the 1940s to support detection, recognition, classification, and identification of man-made entities such as people, vehicles, and planes. As operators sought methods to utilize the radar signatures for target analysis; the sensors, methods, and approaches became known as ATR. Early approaches to ATR consisted of alerting the operator with an audible representation of the radar signal. Operators learned to decipher the sounds to label the illuminated target on the radar scope.

Enhancing ATR from a 1D signal alert from the RF device, a 2D SAR was developed in the 1950s and increased in resolution and dimensionality. Using the knowledge and strategies of humans, algorithms were developed to increase the speed, accuracy, and confidence in recognition. ATR distinguishes man-made, biological, and natural entities. Man-made objects such as ground and air vehicles produce a strong signal from the radar return reflection from metal objects. Biological targets such as animals, humans, and vegetative clutter can be discerned by filtering out interference caused by large flocks of birds on Doppler weather radar. Other environment affects include particulates in the air for weather analysis.

Three fundamental concepts support ATR including the Doppler effect, Fourier transform, and Bayesian analysis. Radar transmits a signal and the time for the signal to return from an illuminated target determines the distance. If the target is moving, a shift in the returned signal frequency is known as the *Doppler effect*. Different movements include object translational motion from kinematics such as a tank velocity, and vibration motion from dynamics such as a rotating turret and centrifugal spinning from kinetics of the engine. Multiple movements of the Doppler effect cause modulation of the signal, which is known as the micro-Doppler effect. ATR algorithms seek to exploit the modulation by assessing the pattern, signature, or template. The micro-Doppler effect will change over time, depending on the motion of the target, thus causing a time- and frequency-varying signal.

To determine the time-frequency trade-off, a *Fourier transform* (FT) analysis of a modulated signal determines the frequency/time domain, resolution, and intensity. If a target is moving, then the FT of the signal is not constant, and the signal can be decomposed into intervals using a short-time Fourier transform (STFT). Robust approaches building on the FT toward the time-frequency analysis include the Gabor transform, the Wigner distribution function (WDF), and wavelet functions. With each of these transforms, features are extracted based on the signal location, intensity, and time.

While there are many prominent methods for feature analysis, *Bayes' theorem* is a popular method for decision making, termed Bayesian analysis. With multiple features, Bayes employs statistical estimation such as maximum likelihood, majority voting (MV), or maximum a posteriori (MAP) using the a priori and received signal to make the a posterior decision about determining the particular target template in the library that best matches the received signal model.

1.2.2 ATR from SAR

The 1980s leveraged technology in statistical artificial intelligence models and computer-aided design to develop *model-based ATR*. Typically, as with early radar, operators were employed to label the targets, changes, and/or shadows. In the 1990s, the DARPA Moving and Stationary Target Recognition (MSTAR) program sought to expand the ATR methods. To develop the techniques for model-based ATR, three issues were considered:

- *Building methods for model-based, template-based, and human-based recognition*: As operators had been exploiting data since the inception of SAR, then those learned relational elements would be incorporated into the algorithms. Likewise, the model-based approach builds on first-principle physics model of the target. The template approach uses the statistical analysis of the received signal.
- *Devising systems-level ATR approaches*: As ATR is part of a larger architecture that includes contextual information, operator workflow, and signals analysis, the goal was to design an ATR pipeline. The eventual architecture was known as the predict, extract, model, and search (PEMS) loop.
- *Conducting rigorous test and evaluation*: As with deployment of systems, concerns are placed on determining deployment robustness of confidence, accuracy, and timelines of decisions. To provide comparative analysis, the MSTAR target dataset was collected in 1995 and released in 1996.

As the MSTAR program initiated many ideas that have led to deep learning in SAR target recognition, the program forged the initial analysis with model-based approaches [20] using machine learning methods. The MSTAR community included government, industry, and academic interest with the initial analysis; which has subsequently exploded in 2012 with the advent of deep learning. In the early published discussions, methods of known radar exploitation [21] were documented in 1997 using the 1996 data.

1.3 Radar Object Classification: Past Approach

Traditionally, radio frequency objects (signals and images) were classified using *template-based* approaches that developed signature profiles from representative data. Template-based is contrasted with model-based approaches in Figure 1.14.

1.3.1 Template-Based ATR

In the template-based approach, the goal is to find the occurrence(s) of the template of a single target or image in a large image (i.e., find the matches of the template by correlation) [53]. The correlation can be defined as

$$c = \Sigma_{x,y} I(x,y)t(x,y) \tag{1.9}$$

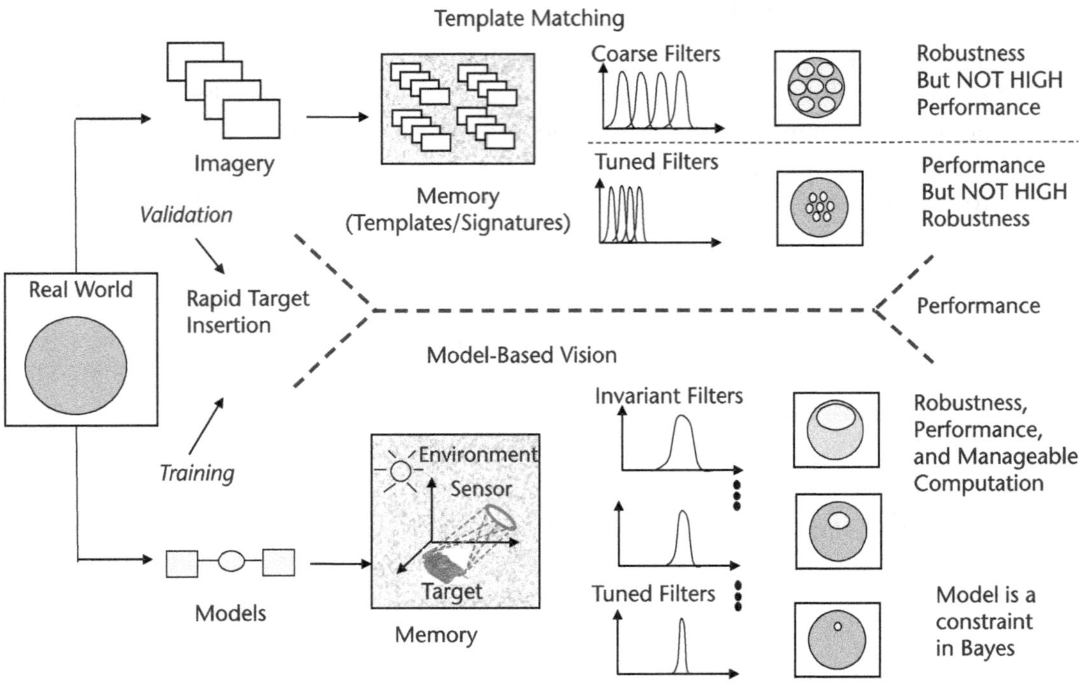

Figure 1.14 Template matching versus model-based ATR.

where I is the large input image and t is the template image.

For a template image that is characterized by width (W) and height (H) parameters as shown in Figure 1.15, then the size $[2W, 2H]$ of the rectangular window in (1.9) for which the summation can be defined precisely as

$$c(x,y) = \Sigma_{k=-W}^{W} \Sigma_{l=-H}^{H} I(x+k, y+l)\, t(k,l) \tag{1.10}$$

Essentially, $c(x,y)$ returns value of search operation on the large image I with the template t. Thus, the maximum value of $c(x,y)$ can declare the target presence in the large image, if this maximum lies above the relevant threshold.

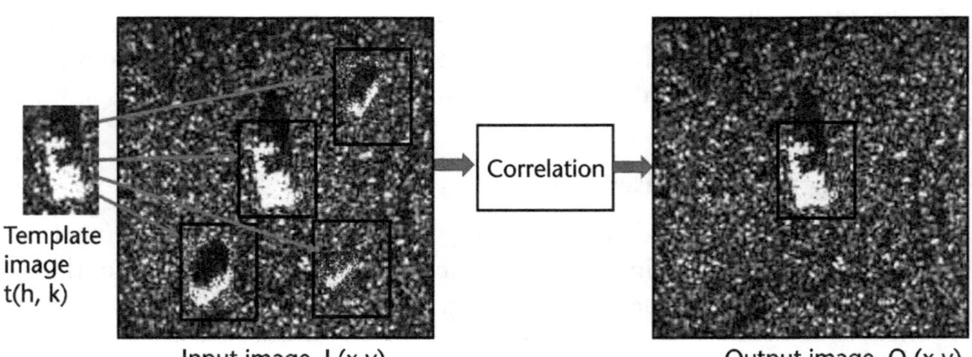

Figure 1.15 Template-based image classification illustration. Input image contains four objects that need to be classified. The template image correlates the entire input image space to find maximum response.

There are many issues with the template-based approach. First, if the image intensity varies with position, the correlation operation may not yield accurate matching. For example, the correlation value between the template and an exactly matched region can be less than correlation between the template and a bright spot. Second, the correlation value depends on the size of the object's feature. Third, correlation is not invariant to changes in image intensity such as lighting conditions. To deal with these issues, normalized correlation and nonmaximum suppression algorithms can be applied at the cost of the large computation. Other than normalized correlation, the sum of the squared difference (SSD) and the sum of the absolute difference (SAD) can be used to minimize the computational burden.

The major limitations of the template-based image classification method demonstrate brittleness in the template approach. First, templates are not scale- or rotation-invariant. That is, a small change in the template size or orientation of the object will yield very low classification accuracy. Hence, templates should represent all possible orientations (azimuth and elevation) of the target, possibly requiring the collection of thousands to millions of templates for a target class in order to expect high classification accuracy. Additionally, there is a need to maintain a large database of template objects. Second, the template-based approach is non-real-time and is computationally very slow, requiring very fast computing resources (data storage, input/output, and fast memory). Finally, the template-based classification approach is essentially a memorization of an object signature. Therefore, the template approach is not well adapted to varying conditions which requires adaptive learning.

1.3.2 Model-Based ATR

The DARPA MSTAR program thoroughly investigated and implemented a model-based approach using the predict, evaluate, match, search (PEMS) loop between 1996 and 1999. In the model-based approach [52], Figure 1.16 highlights both the images being evaluated as well as models for synthetic generation. The models are physics-based computer-aided design (CAD) models used to predict features [22]. The PEMS cycle begins with predicted models that provide a search of related features to match [23]. From the search function, both features from the observed image and the predicted signature are hypothesized and tested to determine the match for target recognition.

The MSTAR analysis enhanced these functions by establishing criteria for performance modeling [24]. Ross et al. [25] established two criteria of performance (accuracy, extensibility, robustness, and utility) and cost (efficiency, scalability, and synthetic trainability). Figure 1.17 presents the evaluation strategy of the performance metrics. These evaluation criteria determined the standard operating conditions (SOCs) and the extended operation conditions (EOCs). The SOCs are the training and test conditions from the available data, while the EOCs are from the modeled conditions. Essentially, the measures of performance (MOPs) determine the object recognition performance, while the measures of effectiveness (MOEs) determine the operational utility.

Using the SAR data of Table 1.3, the MSTAR architecture was developed [26]. Different statistical approaches were presented in 1998 as classifiers including mean-square error (MSE) [27], information theory [28], neural networks [29], and

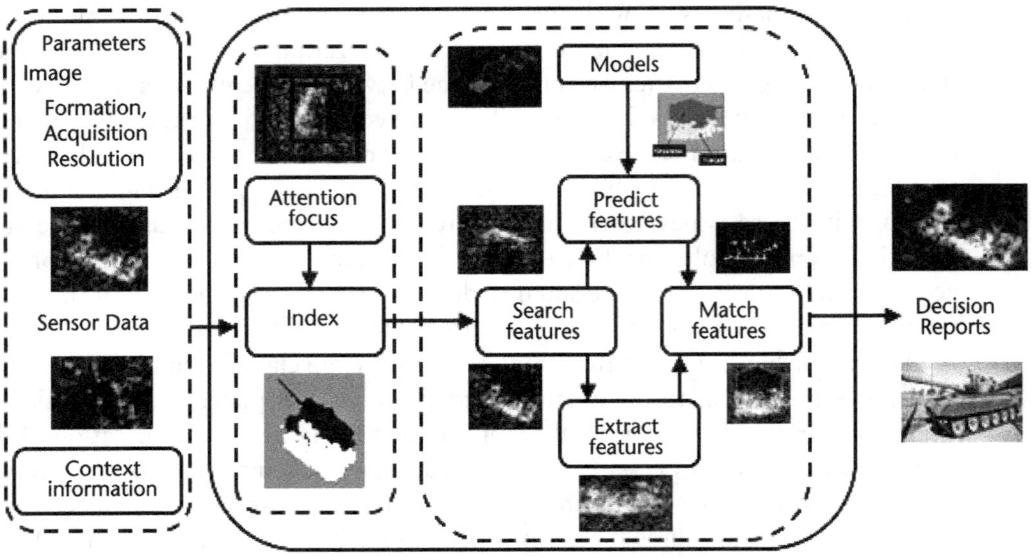

Figure 1.16 Model-based predict, extract, match, search (PEMS) loop.

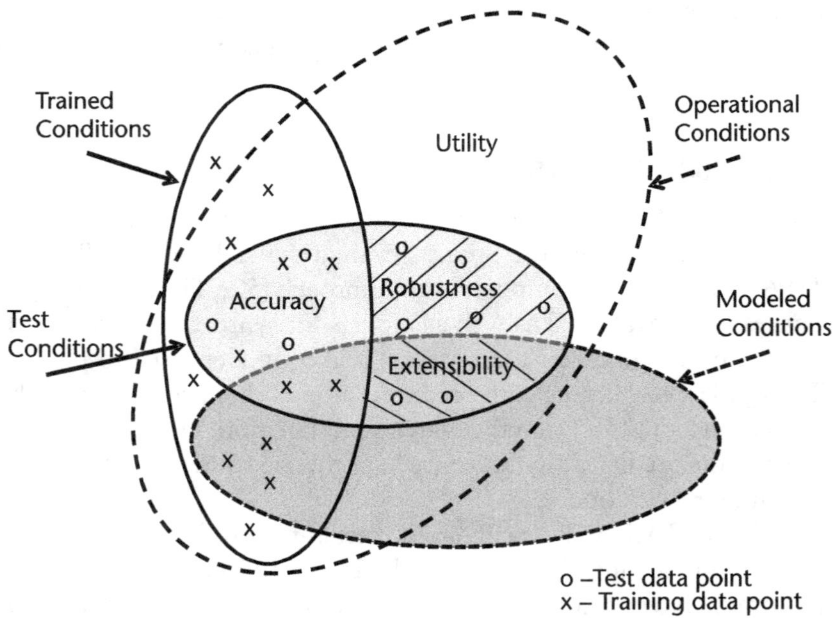

Figure 1.17 Evaluation methods.

evidential reasoning [30]. Additionally, results were presented for the evaluation of the data sets with standard [31] and extended [32] operating conditions. The standard operating conditions were of three categories: sensor (depression 15° to 45°, aspect 0° to 360°, and squint -45° to135°), environment (background, revetment) and target (vehicle classes, target articulation). Hence, these variations were the "SET" of operating conditions.

Table 1.3 MSTAR Data Set

Vehicle Class	*Public Data Set*	*Sequestered Data Set*
Tank	T72	M1, M60
Armored personnel carrier	BMP2, BTR70	M2
Artillery	M110	2S1
Air defense	ZSU23	
Truck	ZIL131	M35, M548

The standard operating conditions test and evaluation (T&E) determine the probability of detection (PD) and the probability of false alarm (PFA) to present the combined results in a receiver operating characteristic (ROC) curve for accuracy analysis. Good results were demonstrated by using a modeling approach, chip registration, cluttered targets, and pose and target type analysis. Included were target variants over a scene to form the extended operating conditions where multiple targets were collected in the same scene for PD, probability of correct classification (PCC), and PFA. Results concluded that PD was not affected by the EOCs, while the PCC was sensitive to the extended operating conditions.

As the MSTAR program proceeded, numerous examples demonstrated baseline techniques using a variety of machine learning methods. Evaluating ATR performance is difficult as there are numerous exemplars needed to train and test over all conditions [33, 34]. Determining the evaluation methods provide challenges such as the model classes [35], sample size variations of different SOCs [36], as well as methods to determine the performance manifold [37]. Another discussion resolved around the type of features from the signal characteristics such as fractal features [38] and scattering features [39]. The baseline approach includes the support vector machine (SVM) [40] that establishes a comparison for machine learning to the deep learning methods presented in the book.

The feature analysis from the MSTAR data set also included the assessment of the high-range resolution (HRR) radar features. Using the MSTAR data, the 1D HRR profiles (110 range bins) were extracted and compared to 2D SAR analysis (128 × 128 chips) that showed less accuracy using the HRR features [41]. A key contribution of HRR is developing the pose estimate of the target [42] to support tracking and classification/identification [43] as well as target selection and rejection using vector quantification [44] and belief filtering [45]. A concluding discussion develops comparisons of 1D HRR, 2D SAR, and 3D SAR (for height analysis) [46].

1.4 Radar Object Classification: Current Approach

With the successful demonstrations of DL algorithms to classify video imagery from the ImageNet dataset, it became an ardent task to test DL on classifying radar imagery. Through video imagery, pristine features demonstrate excellent results and targets can be identified visually (radar images don't visually resemble an object). In other words, object features in RF imagery are not easily discernable to the untrained human. Hence, using the scientific method, the null hypothesis, H_0, corresponds to the hypothesis that DL algorithms can find distinct features in RF images while humans can't see the phenomenon. Alternatively, H_1 is the hypothesis that the radar image does not reveal targets that a human cannot discern. Hence, the potential for DL to augment traditional radar analysis to enhance classification robustness led researchers to experiment with DL on a small subset of SAR imagery. The promising results within the last few years establish the current techniques that rejected the null hypothesis. With the online availability of the MSTAR data in 2015 [47] combined with the interest in deep learning in 2012, Figure 1.18 shows the resurgence of the analysis over the common MSTAR dataset for comparative analysis of machine learning (~1999) to that of deep learning (~2019).

A simple deep neural networks (DNN) that classifies Modified National Institute of Standards and Technology (MNIST) objects was applied to the MSTAR dataset. Using the deep learning approach, objects are classified by developing a labeled data set and a trained model. The trained weights are saved and used to classify test images. Figure 1.19 shows the algorithmic flow diagram to classify RF imagery using the Caffe software [48]. It was demonstrated that DL could be applied to the RF data to classify RF objects and the classification results were promising, as highlighted in the book. Subsequently, contemporary DL methods were initiated to further advance the state-of-the-art RF object classifications using the DL approach.

1.5 Radar Object Classification: Future Approach

With intelligent systems being employed for big data analytics, elements of AI include ML and DL (or DNNs). AI is the superset of intelligent algorithms trying to

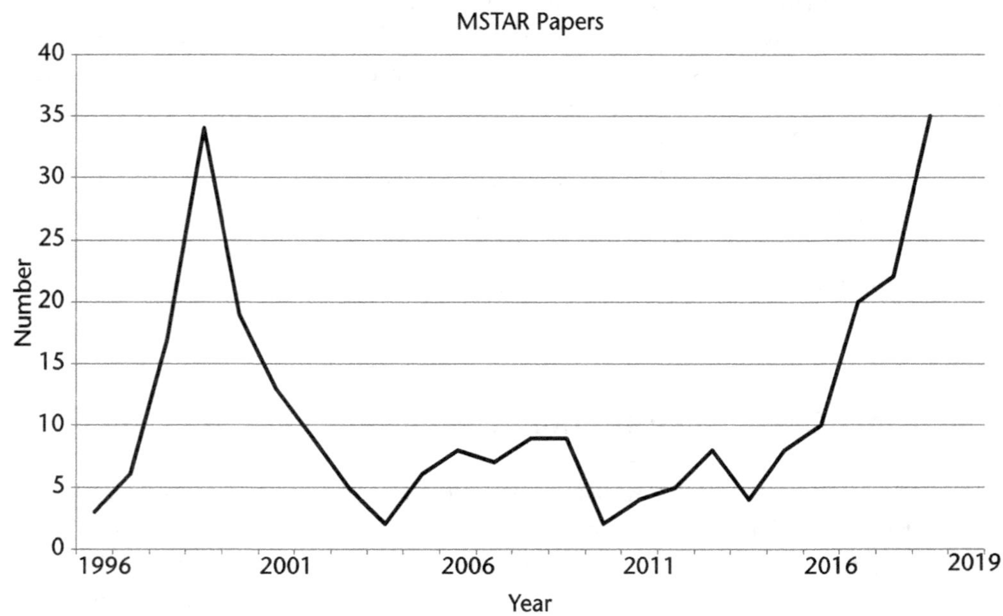

Figure 1.18 MSTAR publications.

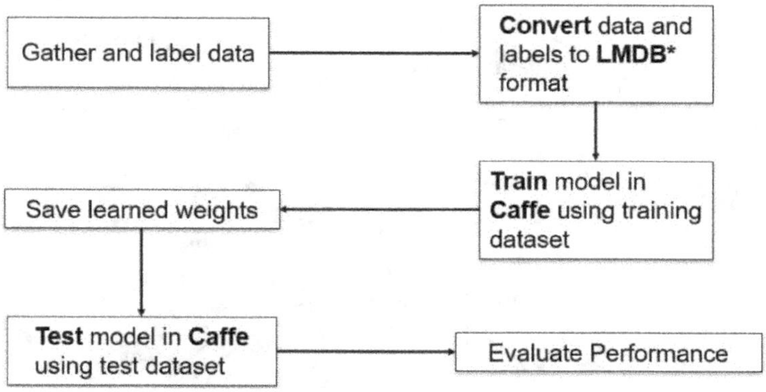

Figure 1.19 Algorithmic flow to classify RF imagery using the deep learning approach. LMDB = Lightning Memory-Mapped Database Manager.

capture human reasoning capabilities. ML algorithms are the subfield of AI algorithms and DL is a special kind of ML algorithms. Machine learning as a subset of AI provides machines with the data they need to learn how to do a task. AI includes computer programs able to "think," reason, behave, and act like humans. Hence, future ATR approaches would leverage contextual information along with DL techniques.

1.5.1 Data Science

AI, ML, and DL each have very distinct technical definitions. Figure 1.20 illustrates the relationships among the techniques of learning, data science, and big data problems. Big data surfaced with the advent of high performance computers around 2010 following advances in data mining, deep learning in 2012, and current AI approaches in 2018. Together, many centers for higher-level education are establishing programs in data science and data analytics. A summary of the key terminology includes

Data:

- *Data mining*: Seeking relevant information from a database of information;
- *Big data*: Corpus of data with velocity, variation, and volume along with performance metrics of veracity and value;
- *Data science*: Systems approach toward management of preparing, processing, and visualizing information.

Learning:

- *Machine learning*: Includes statistical techniques that enable machines to improve tasks with experience;
- *Deep learning*: Composes algorithms that permit software to train itself to performance tasks by exposure to multilayered neural networks over vast amounts of data;

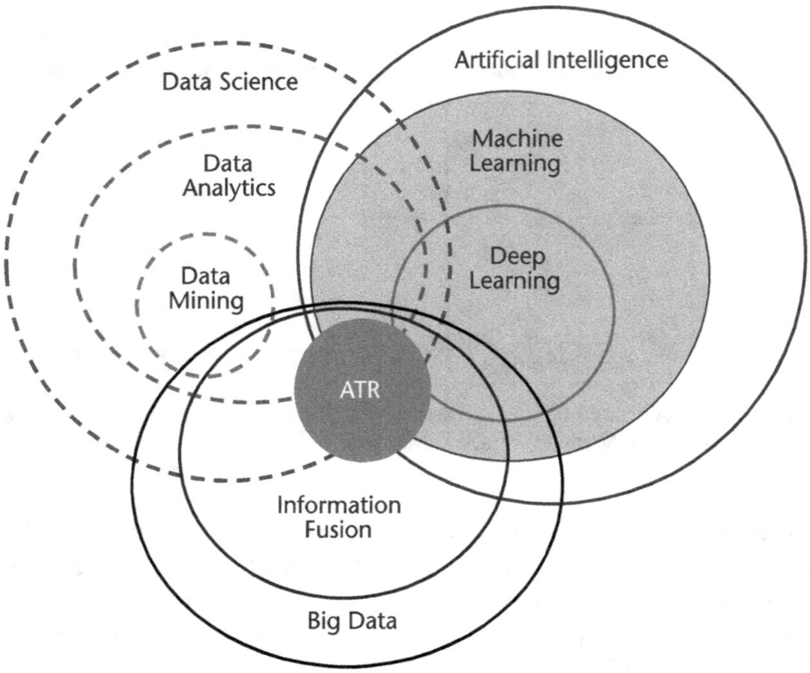

Figure 1.20 AI, big data, and data science.

- *Artificial intelligence*: Describes all aspects (algorithms, hardware) of a reasoning system that enables computers to emulate human intelligence, using logic, if-then rules, decision trees, and learning.

1.5.2 Artificial Intelligence

DARPA published advancement of AI as three waves in terms of its capabilities and limitations [49]. These three waves are: (1) handcrafted knowledge (first wave), (2) statistical learning (second wave), and (3) contextual adaptation (third wave) (Figure 1.21). These concepts along with other information fusion approaches lead to the future of ATR [50].

The first-wave AI systems are developed based on very clear rules given to a machine for decision making. The programmer develops a system with explicit logic to accomplish a task. The logic or rules are identified in advance by human experts. As a result, first-wave systems find it difficult to tackle new kinds of situations. They also have a hard time abstracting—taking knowledge and insights derived from certain situations and applying them to new problems. In a nutshell, the first-wave AI systems are capable of implementing simple logical rules for well-defined problems, but are incapable of learning, and have a hard time dealing with uncertainty. Hence, template-based image classification falls into the first-wave category of AI.

The second-wave AI systems do not to provide precise and exact rules to the expert system. Instead, the programmer develops statistical models of specific problems using big data. The systems learn and adapt themselves to different situations, if they're properly trained. However, unlike first-wave systems, these systems are limited in their logical capacity: they don't rely on precise rules, but instead go for

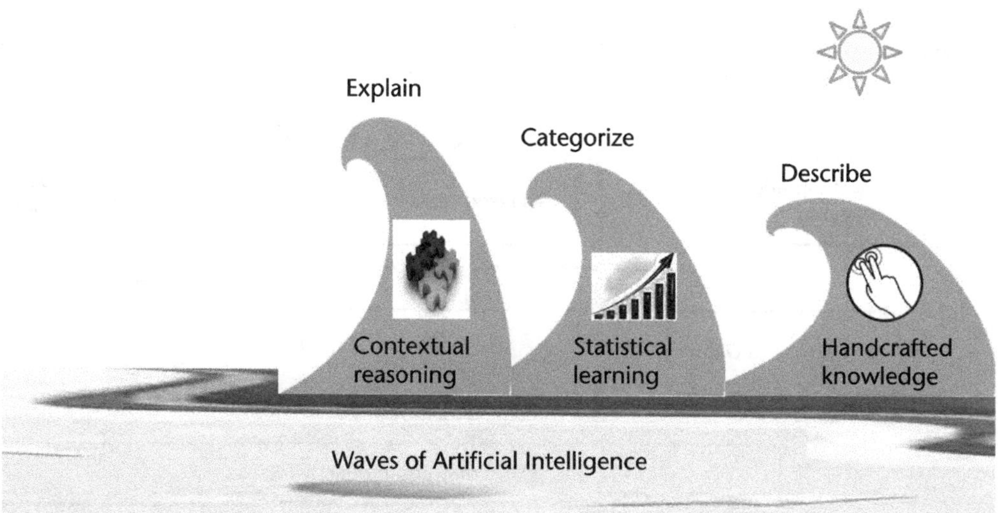

Figure 1.21 The three waves of AI.

the solutions that work well enough, usually. The second-wave systems have been successful in facial recognition, at speech transcription, and at identifying animals and objects in pictures. The biggest issue with the second wave of AI is that it can't explain the reasons it comes to the final decision. Hence, modern ML algorithms (DL, DNN) for object recognition are based on second-wave category of AI.

The third wave of AI involves human-computer interaction/collaboration. In the third wave, the AI systems themselves will construct models that will explain how the world works. In other words, they'll discover by themselves the logical rules which shape their decision-making process. As with the model-based approach and techniques provided in the text, the modern ML algorithms (DL, DNN) for object recognition are utilized in systems approach third-wave systems. Additionally, with extensive evaluation for robustness, the new methods have potential. Hence, emerging ML algorithms (e.g., Generative Adversarial Networks, GAN) that utilize humans for data collection and results confirmation along with model-based and machines for data generation for object recognition emerge as a third-wave category of AI.

1.6 Book Organization

The rest of this book is divided into nine chapters, with each providing specific details on ML algorithms and RF data presented in Figure 1.22. Chapter 2 describes a mathematical foundation for ML providing an overview of linear algebra, probability theory, and optimization algorithms needed to understand various ML algorithms. Chapter 3 provides a review on ML algorithms and their applications to various problems. Chapter 4 presents DL algorithms. Chapter 5 initiates a discussion on RF big data analytics presenting technical challenges associated with applying ML to RF data. Chapter 6 provides implementation details on deep learning- based RF for single object image classification and results. Chapter 7 follows with multiple object classification. In Chapter 8, deep learning on signal

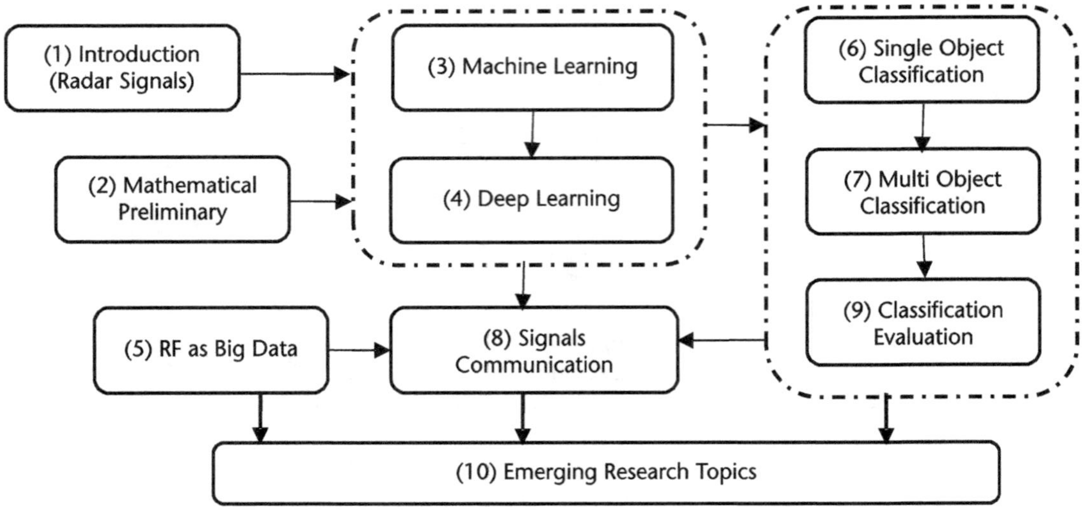

Figure 1.22 Book overview.

classification is briefly discussed. Chapter 9 highlights elements associated with robust evaluation of ML/DL algorithms. Chapter 10 provides a discussion on emerging trends for ML/DL algorithms.

1.7 Summary

This introduction provided the motivation from past results of ML methods, current DL approaches, and future concepts of AI [51]. RF one-dimensional data examples include sensors mounted on autonomous cars, IoT devices, as well as communications. Additionally, 2-D RF examples include SAR and polarimetric SAR from airborne and spaceborne platforms. The resurgence of analysis for RF development using DL methods provides a comparative example using the MSTAR data set throughout the book for the reader to appreciate how deep learning achieves state-of-the-art performance for ATR. The rest of the book is organized around learning methods, classification evaluation, and emerging techniques, so that the reader can get a comprehensive understanding of the issues of ML applied to RF data.

References

[1] Heinlein, R. A., "The Door Into Summer," https://www.goodreads.com/quotes/557199-when-railroading-time-comes-you-can-railroad-but-not-before.

[2] Kraus, J. D., and D. A. Fleisch, *Electromagnetics with Applications,* Fifth Edition, McGraw-Hill, 1999.

[3] Moreira, A., P. Prats-Iraola, M. Younis, G. Krieger, I. Hajnsek, and K. P. Papathanassiou, "A Tutorial on Synthetic Aperture Radar," *IEEE Geoscience and Remote Sensing Magazine*, Vol. 6, No. 43, pp. 6–43, March 2013.

[4] MSTAR Overview, https://www.sdms.afrl.af.mil/index.php?collection=mstar.

[5] Keydel, E. R., S. W. Lee, and J. T. Moore, "MSTAR Extended Operating Conditions, A Tutorial," *SPIE*, Vol. 2757, March 1996.

[6] Liang, X., Z. Zhen, Y. Song, L. Jian, and D. Song, "Pol-SAR Based Oil Spillage Classification with Various Scenarios of Prior Knowledge," *IEEE Access*, Vol. 7, pp. 66895–66909, 2019.

[7] Gao, F., Q. Liu, J. Su, A. Hussain, and H. Zhou, "Integrated GANs: Semi-Supervised SAR Target Recognition," *IEEE Access*, Vol. 7, 2019, pp. 113999–114013.

[8] DARPA Public Release: RF Machine Learning Systems (RFMLS) Industry Day, https://www.darpa.mil/attachments/RFMLSIndustryDaypublicreleaseapproved.pdf.

[9] Xu, R., Y. Chen, E. Blasch, and G. Chen, "BlendCAC: A BLockchain-ENabled Decentralized Capability-based Access Control for IoTs," *IEEE Blockchain*, 2018.

[10] Van Eeden,W. D., J.P. de Villiers, R.J. Berndt, W. A. J. Nel, and E. Blasch, "Micro-Doppler Radar Classification of Humans and Animals in an Operational Environment," *Expert Systems with Applications*, Vol. 102, 15, pp. 1–11, July 2018.

[11] Blasch, E., T. Busch, S. Kumar, and K. Pham, "Trends in Survivable/Secure Cognitive Networks," *IEEE Int'l Conf. on Computing, Networking, and Communications*, January 2013.

[12] Gurbuz, S. Z., H. D. Griffiths, A. Charlish, M. Rangaswamy, M. S. Greco, and K. Bell, "An Overview of Cognitive Radar: Past, Present, and Future," *IEEE Aerospace and Electronic Systems Magazine*, Vol. 34, No. 12, pp. 6–18, December 2019.

[13] Layne, J. R., and E. P. Blasch, "Integrated Synthetic Aperture Radar and Navigation Systems for Targeting Applications," USAF-WPAFB, Tech Report, WL-TR-97-1185, 1997.

[14] Soumekh, M., *Synthetic Aperture Radar Signal Processing with MATLAB Algorithms*, Wiley, 1999.

[15] Ertin, E., C. Austin, S. Sharma, R. Moses, and L. Potter, "GOTCHA Experience Report: Three-Dimensional SAR Imaging with Complete Circular Apertures," *Proc. SPIE*, Vol. 6568, Algorithms for Synthetic Aperture Radar Imagery XIV, May 2007.

[16] Lin, Y., Q. Bao, L. Hou, L. Yu, and W. Hong, "Full-Aspect 3D Target Reconstruction of interferometric circular SAR," *Proc. SPIE 10004*, Image and Signal Processing for Remote Sensing XXII, 2016.

[17] Blasch, E. P., E. Bosse, and D. A. Lambert, *High-Level Information Fusion Management and Systems Design*, Norwood, MA: Artech House, 2012.

[18] Zheng, Y., E. Blasch, and Z. Liu, *Multispectral Image Fusion and Colorization*, SPIE Press, 2018.

[19] Blasch, E., "Information Fusion Performance Evaluation, Tutorial," *International Conference on Information Fusion*, 2004.

[20] Ross, T. D., L. A. Westerkamp, E. G. Zelino, and T. J. Burns, "Extensibility and Other Model-Based ATR Evaluation Concepts," SPIE'97, Algorithms for Synthetic Aperture Radar Imagery IV, 1997.

[21] Novak, L. M., G. R. Benita, G. J. Owirka, and J. D. Popietarz, "Classifier Performance Using Enhanced Resolution SAR Data," Radar Systems (RADAR97) 1997.

[22] Keydel, E. R., and S. W. Lee, "Signature Prediction for Model-Based Automatic Target Recognition," *Proc. SPIE 2757*, Algorithms for Synthetic Aperture Radar Imagery III, 1996.

[23] Wissinger, J., R. B. Washburn, N. S. Friedland, et al., "Search Algorithms for Model-Based SAR ATR," *Proc. SPIE,* Vol. 2757, Algorithms for Synthetic Aperture Radar Imagery III, 1996.

[24] Catlin, A., L, R. Myers, K, W. Bauer, S, K. Rogers, and R, P. Broussard, "Performance Modeling for Automatic Target Recognition Systems," *Proc. SPIE,* Vol. 3070, Algorithms for Synthetic Aperture Radar Imagery IV, 1997.

[25] Ross, T. D., L. A. Westerkamp, E. G. Zelnio, and T. J. Burns, "Extensibility and Other Model-Based ATR Evaluation Concepts," *Proc. SPIE*, Vol. 3070, Algorithms for Synthetic Aperture Radar Imagery IV, 1997.

[26] Diemunsch, J. R., and J. Wissinger, "Moving and Stationary Target Acquisition and Recognition (MSTAR) Model-Based Automatic Target Recognition: Search Technology for a Robust ATR," *Proc. SPIE,* Vol. 3370, Algorithms for Synthetic Aperture Radar Imagery V, 1998.

[27] Novak, L. M., G. J. Owirka, and W. S. Brower, "An Efficient Multi-Target SAR ATR Algorithm,"' in *Proc. 30th IEEE Asilomar Conf. Signals, Syst. Computer*, 1998.

[28] Blasch, E., and M. Byrant, "Information Assessment of SAR Data for ATR," *IEEE National Aerospace and Electronics Conference*, 1998.

[29] Principe, J. C ., M. Kim, and M. Fisher, "Target Discrimination in Synthetic Aperture Radar using Artificial Neural Networks," *IEEE Transactions on Image Processing,* Vol. 7, No. 8, pp. 1136–1149, 1998.

[30] Blasch E. P., and J. Gainey, Jr., "Physio-Associative Temporal Sensor Integration," *Proc. SPIE,* Vol. 3390, April 1998.

[31] Ross, T. D., S. W. Worrell, V. J. Velten, J. C. Mossing, and M. L. Bryant, "Standard SAR ATR Evaluation Experiments Using the MSTAR Public Release Data Set," *Proc. SPIE,* Vol. 3370, Algorithms for Synthetic Aperture Radar Imagery V, 1998.

[32] Mossing, J. C., and T. D. Ross, "Evaluation of SAR ATR Algorithm Performance Sensitivity to MSTAR Extended Operating Conditions," *Proc. SPIE,* Vol. 3370, *Algorithms for Synthetic Aperture Radar Imagery V*, 1998.

[33] Ross, T. D., and J. C. Mossing, "MSTAR Evaluation Methodology," *Proc. SPIE,* Vol. 3721, Algorithms for Synthetic Aperture Radar Imagery VI, 1999.

[34] Ross, T. D., J. J. Bradley, L. J. Hudson, and M. P. O'Connor, "SAR ATR: So What's the Problem? An MSTAR Perspective," *Proc. SPIE,* Vol. 3721, Algorithms for Synthetic Aperture Radar Imagery VI, 1999.

[35] Williams, W. D., J. Wojdacki, E. R. Keydel, et al., "Development of Class Models for Model-Based Automatic Target Recognition," *Proc. SPIE,* Vol. 3721, Algorithms for Synthetic Aperture Radar Imagery VI, 1999.

[36] Blasch, E. P., S. Alsing, and R. Bauer, "Comparison of Bootstrap and Prior Probability Synthetic Data Balancing Method for SAR Target Recognition," *SPIE Int. Sym. On Aerospace/Defense Sim. & Control,* Vol, 3721, 1999.

[37] Alsing, S., E. P. Blasch, and R. Bauer, "Three-Dimensional Receiver Operating Characteristic (ROC) Trajectory Concepts for the Evaluation of Target Recognition Algorithms Faced with the Unknown Target Detection Problem," *Proc. SPIE,* Vol. 3718, 1999.

[38] Kaplan, L. M.,. R. Murenzi, and K. R. Namuduri, "Extended Fractal Feature for First-Stage SAR Target Detection,'" *Proc. SPIE*, Vol. 3721, 1999.

[39] Chiang, H. C., and R. L. Moses, "ATR Performance Prediction Using Attributed Scattering Features,'" *Proc. SPIE*, Vol. 3721, 1999.

[40] Bryant, M., and L., F. Garber. "SVM Classifier Applied to the MSTAR Public Data Set," *Proc. SPIE*, Vol. 3721, 1999.

[41] Williams, R. L., J. J. Westerkamp, D. C. Gross, A. P. Palomino, T. Kaufman, and T. Fister, "Analysis of a 1D HRR Moving Target ATR," *Proc. SPIE,* Vol. 3721, Algorithms for Synthetic Aperture Radar Imagery VI, 1999.

[42] Williams, R., J. Westerkamp, D. Gross, A. Palomino, and T. Fister, "Automatic Target Recognition of Time Critical Moving Targets Using ID High Range Resolution (HRR) Radar," *IEEE Radar Conference*, 1999.

[43] Blasch, E., and L. Hong, "Simultaneous Feature-Based Identification and Track Fusion," *Proceedings of the 37th IEEE Conference on Decision and Control*, 1999.

[44] Ulug, B., and S. C. Ahalt, "HRR ATR Using VQ Classification with a Reject Option," *Proc. SPIE,* Vol. 3718, Automatic Target Recognition IX, 1999.

[45] Blasch, E., *Derivation of a Belief Filter for Simultaneous High Range Resolution Radar Tracking and Identification,* Ph.D. thesis, Wright State University, 1999.

[46] Horowitz, L. L., and G. F. Brendel, "Fundamental SAR ATR Performance Predictions for Design Trade-Offs: 1D HRR versus 2D SAR versus 3D SAR," *Proc. SPIE,* Vol. 3721, Algorithms for Synthetic Aperture Radar Imagery VI, 1999.

[47] "MSTAR Public Dataset." *The Sensor Data Management System.* U.S. Air Force, September 20, 2015.

[48] Caffe, http://caffe.berkeleyvision.org/.

[49] DARPA Perspective on AI, https://www.darpa.mil/about-us/darpa-perspective-on-ai.

[50] Blasch, E., I. Kadar, L. L. Grewe, G. Stevenson, U. K. Majumder, and C.-Y. Chong, "Deep Learning in AI and Information Fusion Panel Discussion," *Proc. SPIE,* Vol. 11018, 2019.

[51] Zelnio, Edmund, and Anne Pavy. "Open set SAR target classification," *Algorithms for Synthetic Aperture Radar Imagery XXVI,* Vol. 10987, International Society for Optics and Photonics, 2019.

[52] Friedlander, Robert D., et al. "Deep learning model-based algorithm for SAR ATR," *Algorithms for Synthetic Aperture Radar Imagery XXV.* Vol. 10647, International Society for Optics and Photonics, 2018.

[53] Zachmann, Isaac, and Theresa Scarnati, "A comparison of template matching and deep learning for classification of occluded targets in LiDAR data," *Automatic Target Recognition XXX,* Vol. 11394, International Society for Optics and Photonics, 2020.

CHAPTER 2

Mathematical Foundations for Machine Learning

In this chapter, we present the essential mathematics needed to understand ML theory. Quoting from Shakuntala Devi, mathematics prodigy, "mathematics is only a systematic way of solving puzzles posed by nature"[1]. Hence, it is important to know the fundamental mathematics behind the ML algorithms that solves the classification puzzles. Mathematics enables the analysis of ML theories, their limitations, and future research (algorithms, computing hardware). Although multiple mathematics books can be written to describe ML algorithms, our intention here is to cover the core topics necessary for implementing ML algorithms. Hence, in this chapter, we provide the basic linear algebra concepts, multivariate calculus and probability theory.

2.1 Linear Algebra

This section describes the essential linear algebra concepts required to perform the computations of machine learning. Many of these calculations are performed in terms of vectors, matrices, and tensors [2].

2.1.1 Vector Addition, Multiplication, and Transpose

First, define a general column vector $\mathbf{x}$ that contains N scalar elements:

$$\mathbf{x} = \{x_n\} = \begin{bmatrix} x_1 \\ x_2 \\ \vdots \\ x_N \end{bmatrix} \tag{2.1}$$

Lowercase English letters are used to describe vectors, with bold letters for the vectors and nonbold letters for the elements of the vectors. It is often convenient to describe a given vector in terms of the general form of one of its elements by using appropriate subscripts. Here, the notation $\{x_n\}$ denotes the nth element of the vector $\mathbf{x}$. The transpose of the column vector $\mathbf{x}$ is a row vector:

$$\mathbf{x}^T = \begin{bmatrix} x_1 & x_2 & \cdots & x_N \end{bmatrix} \tag{2.2}$$

The addition of two vectors $\mathbf{x}$ and $\mathbf{y}$ is performed by adding the corresponding elements; that is,

$$\mathbf{x} + \mathbf{y} = \begin{bmatrix} x_1 \\ x_2 \\ \vdots \\ x_N \end{bmatrix} + \begin{bmatrix} y_1 \\ y_2 \\ \vdots \\ y_N \end{bmatrix} \tag{2.3}$$

Similarly, the multiplication of a vector $\mathbf{x}$ by a scalar λ is performed by multiplying each of the elements of $\mathbf{x}$ by λ; that is,

$$\lambda\mathbf{x} = \begin{bmatrix} \lambda x_1 \\ \lambda x_2 \\ \vdots \\ \lambda x_N \end{bmatrix} \tag{2.4}$$

2.1.2 Matrix Multiplication

The following notation describes a matrix $\mathbf{A}$ of scalars in terms of a general number of rows M and columns N:

$$\mathbf{A} = \{A_{m,n}\} = \begin{bmatrix} A_{1,1} & A_{1,2} & \cdots & A_{1,N} \\ A_{2,1} & A_{2,2} & \cdots & A_{2,N} \\ \vdots & \vdots & \ddots & \vdots \\ A_{M,1} & A_{M,2} & \cdots & A_{M,N} \end{bmatrix} \tag{2.5}$$

Here, $A_{m,n}$ can be used to describe the element of the matrix $\mathbf{A}$ corresponding to the mth row and the nth column. Matrix multiplication is accomplished by summing the products of the elements of a given row of a matrix $\mathbf{A}$ with the corresponding elements of the vector $\mathbf{x}$; that is,

$$\mathbf{y} = \mathbf{A}\mathbf{x} = \sum_{n=1}^{N} A_{m,n}x_n = \begin{bmatrix} A_{1,1}x_1 + A_{1,2}x_2 + \cdots + A_{1,N}x_N \\ A_{2,1}x_1 + A_{2,2}x_2 + \cdots + A_{2,N}x_N \\ \vdots \\ A_{M,1}x_1 + A_{M,2}x_2 + \cdots + A_{M,N}x_N \end{bmatrix} = \begin{bmatrix} y_1 \\ y_2 \\ \vdots \\ y_N \end{bmatrix} \tag{2.6}$$

It is common to omit the explicit sigma symbol, (denoted the sum) and instead use the convention that summation is implied over repeated indicies, which is also called Einstein notation. Thus, the operation of matrix multiplication in (2.6) can be expressed in the form:

$$\mathbf{y} = \mathbf{A}\mathbf{x} = \sum_{n=1}^{N} A_{m,n}x_n = A_{m,n}x_n \tag{2.7}$$

The use of implied summations facilitates the notation of products involving vectors, matrices, and tensors. The concept of a tensor is merely an extenstion to that of one-dimensional (1D) vectors and two-dimensional (2D) matrices to that of ordered sets involving three or more dimensions. Thus, each dimension of tensor is allocated a separate subscript index. For example, a three-dimensional (3D) tensor $B_{p,m,n}$ can be multiplied by a 2D matrix **A** to yield a vector **x**; that is,

$$\mathbf{B}\mathbf{A} = \sum_{m=1}^{M}\sum_{n=1}^{N} B_{p,m,n}A_{m,n} = B_{p,m,n}A_{m,n} = x_p \tag{2.8}$$

2.1.3 Matrix Inversion

The inverse matrix $\mathbf{A}^{-1}$ corresponding to particular matrix **A** is the matrix that satisfies:

$$\mathbf{A}^{-1}\mathbf{A} = \mathbf{A}^{-1}\mathbf{A} = \mathbf{I} \tag{2.9}$$

Here, the identity matrix, I, has ones along the diagonal and zeros for all other elements; that is,

$$\mathbf{I} = \begin{bmatrix} 1 & 0 & \cdots & 0 \\ 0 & 1 & \cdots & 0 \\ \vdots & \vdots & \ddots & \vdots \\ 0 & 0 & \cdots & 1 \end{bmatrix} \tag{2.10}$$

There are numerous methods for computing the inverse of a matrix, so they will not be repeated here.

2.1.4 Principal Components Analysis

A common methodology which is applied in ML is that of principal component analysis (PCA). These techniques determine a basis for the representation of a given matrix **A** and facilitate various operations on it, including the inverse. The PCA for a given matrix **A** involves finding eigenvalues λ and corresponding eigenvectors **x**, which solve the following equation:

$$[\mathbf{A} - \lambda\mathbf{I}]\mathbf{x} = 0 \tag{2.11}$$

Consider a specific example in which the matrix **A** is given by:

$$\mathbf{A} = \begin{bmatrix} 2 & 1 \\ 1 & 2 \end{bmatrix} \tag{2.12}$$

Solutions to the eigen-equation (2.11) exist if the determinate of $\mathbf{A} - \lambda\,\mathbf{I}$ vanishes:

$$\det(\mathbf{A} - \lambda \mathbf{I}) = 0 \tag{2.13}$$

Thus, solutions exist if:

$$\det\left(\begin{bmatrix} 2 & 1 \\ 1 & 2 \end{bmatrix} - \lambda \begin{bmatrix} 1 & 0 \\ 0 & 1 \end{bmatrix}\right) = \det\left(\begin{bmatrix} 2-\lambda & 1 \\ 1 & 2-\lambda \end{bmatrix}\right) = \{2-\lambda\}^2 - 1 = 0 \tag{2.14}$$

Equation (2.14) yields the following solutions for the two eigenvalues:

$$\lambda_1 = 3, \qquad \lambda_2 = 1 \tag{2.15}$$

Next, compute the eigenvector $\mathbf{x}_1$ for the eigenvalue λ_1 by solving the system of equations of (2.11):

$$[\mathbf{A} - \lambda_1 \mathbf{I}]\mathbf{x}_1 = 0 \tag{2.16}$$

Use the form of the eigenvector $\mathbf{x}_1 \equiv [y_1\ y_2]^T$, so that (2.11) with $\lambda_1 = 3$ gives:

$$[\mathbf{A} - \lambda_1 \mathbf{I}]\mathbf{x}_1 = \left\{\begin{bmatrix} 2 & 1 \\ 1 & 2 \end{bmatrix} - 3\begin{bmatrix} 1 & 0 \\ 0 & 1 \end{bmatrix}\right\}\begin{bmatrix} y_1 \\ y_2 \end{bmatrix} = \begin{bmatrix} -1 & 1 \\ 1 & -1 \end{bmatrix}\begin{bmatrix} y_1 \\ y_2 \end{bmatrix} = 0 \tag{2.17}$$

This system of equations yields $y_1 = y_2$, so that the eigenvector of the form $\mathbf{x}_1 = [y_1\ y_2]^T$ solves (2.16). The corresponding normalized eigenvector satisfies $\mathbf{x}_1^T \mathbf{x}_1 = 0$ and thus has the form:

$$\mathbf{x}_1 = \frac{1}{\sqrt{2}}\begin{bmatrix} 1 \\ 1 \end{bmatrix} \tag{2.18}$$

Computation of the other eigenvector of the form $\mathbf{x}_2 \equiv [y_3\ y_4]^T$ using (2.11) with $\lambda_2 = 1$ yields:

$$[\mathbf{A} - \lambda_2 \mathbf{I}]\mathbf{x}_2 = \left\{\begin{bmatrix} 2 & 1 \\ 1 & 2 \end{bmatrix} - 1\begin{bmatrix} 1 & 0 \\ 0 & 1 \end{bmatrix}\right\}\begin{bmatrix} y_3 \\ y_4 \end{bmatrix} = \begin{bmatrix} 1 & 1 \\ 1 & 1 \end{bmatrix}\begin{bmatrix} y_3 \\ y_4 \end{bmatrix} = 0 \tag{2.19}$$

The result is $y_3 = -y_4$, giving a general corresponding eigenvector of the form $\mathbf{x}_2 \equiv [y_3\ y_3]^T$. Use of the normalization equation $\mathbf{x}_2^T\mathbf{x}_2 = 0$ implies that the final eigenvector corresponding to the eigenvalue $\lambda_2 = 1$ is:

$$\mathbf{x}_2 = \frac{1}{\sqrt{2}}\begin{bmatrix} 1 \\ -1 \end{bmatrix} \tag{2.20}$$

Once the eigenvalues and corresponding eigenvectors have been found, then it is possible to compose various forms of interest in matrix manipulations. For example, the spectral decomposition of a matrix **A** can be expressed in terms of the eigenvalues λ_i and the corresponding eigenvectors $\mathbf{x}_i$ via:

$$\mathbf{A} = \sum_i \lambda_i \mathbf{x}_i \mathbf{x}_i^T \tag{2.21}$$

Thus, for the current 2 × 2 matrix example, **A** is expressed via:

$$\mathbf{A} = \lambda_1 \mathbf{x}_1 \mathbf{x}_1^T + \lambda_2 \mathbf{x}_2 \mathbf{x}_2^T = 3\frac{1}{2}\begin{bmatrix} 1 \\ 1 \end{bmatrix}[1 \quad 1] + 1\frac{1}{2}\begin{bmatrix} 1 \\ -1 \end{bmatrix}[1 \quad -1] \tag{2.22}$$

Thus, the matrix reduces to:

$$\mathbf{A} = \frac{3}{2}\begin{bmatrix} 1 & 1 \\ 1 & 1 \end{bmatrix} + \frac{1}{2}\begin{bmatrix} 1 & -1 \\ -1 & 1 \end{bmatrix} = \begin{bmatrix} 2 & 1 \\ 1 & 2 \end{bmatrix} \tag{2.23}$$

which agrees with the original definition of (2.12). In a similar fashion, the matrix inverse can be computed in terms of the eigenvalues and eigenvectors via:

$$\mathbf{A}^{-1} = \sum_i \lambda_i^{-1} \mathbf{x}_i \mathbf{x}_i^T \tag{2.24}$$

The matrix inverse for the current example gives:

$$\mathbf{A}^{-1} = \frac{1}{\lambda_1}\mathbf{x}_1 \mathbf{x}_1^T + \frac{1}{\lambda_2}\mathbf{x}_2 \mathbf{x}_2^T = \frac{1}{3}\frac{1}{2}\begin{bmatrix} 1 \\ 1 \end{bmatrix}[1 \quad 1] + 1\frac{1}{2}\begin{bmatrix} 1 \\ -1 \end{bmatrix}[1 \quad -1] \tag{2.25}$$

Therefore, the matrix inverse becomes:

$$\mathbf{A}^{-1} = \frac{1}{6}\begin{bmatrix} 1 & 1 \\ 1 & 1 \end{bmatrix} + \frac{1}{2}\begin{bmatrix} 1 & -1 \\ -1 & 1 \end{bmatrix} = \frac{1}{3}\begin{bmatrix} 2 & -1 \\ -1 & 2 \end{bmatrix} \tag{2.26}$$

The results for the matrix **A** and the matrix inverse $\mathbf{A}^{-1}$ satisfy (2.9).

2.1.5 Convolution

An operation frequently encountered in signal processing is that of convolution. This operation measures the response of a system based upon a specific input. For example, suppose that a signal x is transmitted in a multipath channel characterized by an impulse response h. Perhaps, this impulse response results in a particular delay profile. The convolution model enables the determination of the output signal g based on the transmission of x through the channel h. This convolution operation uses the symbol * and has the form:

$$\mathbf{g} = \mathbf{h} * \mathbf{x} = \sum_{m=0}^{M-1} h_m x_{n-m} = g_n \tag{2.27}$$

The convolution operation can be performed via matrix multiplication in terms of the convolution matrix, which is constructed by using the elements of the impulse response vector **h** along the matrix diagonals:

$$\mathrm{H} = \begin{bmatrix} h_1 & 0 & 0 & \cdots & 0 \\ h_2 & h_1 & 0 & \cdots & 0 \\ h_3 & h_2 & h_1 & \cdots & 0 \\ \vdots & \vdots & \vdots & \ddots & \vdots \\ h_N & h_{N-1} & h_{N-2} & \cdots & h_1 \\ 0 & h_N & h_{N-1} & \cdots & h_2 \\ 0 & 0 & h_N & \cdots & h_3 \\ \vdots & \vdots & \vdots & \ddots & \vdots \\ 0 & 0 & 0 & \cdots & h_N \end{bmatrix} \tag{2.28}$$

Then, the convolution operation can be expressed in the following matrix form:

$$\mathbf{g} = \mathbf{h} * \mathbf{x} = \mathbf{H}\mathbf{x} \tag{2.29}$$

2.2 Multivariate Calculus for Optimization

Multivariate calculus (MC) is the heart of machine learning algorithm development. It is needed for optimizing the weight parameters by minimizing the error function via differentiation during the learning process. On the other hand, integration is used for calculating probabilities. Scientists and engineers initially learn single-variable calculus (i.e., differentiation and integration involving functions of a single variable). Subsequent courses involve multivariate calculus (i.e., differentiation and integration involving functions of many variables). For machine learning algorithms, multivariate calculus is important. Consider an image of size 8 × 8 pixels. This image contains 64 input parameters (i.e., a vector of size 64). For a typical image classification problem, thousands of input images are used for learning salient features. Hence, a review of vector calculus involving multiple variables is crucial. We will provide fundamental rules and nomenclature involving MC. We will then illustrate gradient descent algorithm used for optimizing a function [3].

2.2.1 Vector Calculus

If a function $f:\mathbb{R}^n \to \mathbb{R}$ is differentiable, then the function ∇f or gradient of f is defined by:

$$\nabla f(x) = \begin{bmatrix} \dfrac{\partial f}{\partial x_1}(x) \\ \vdots \\ \dfrac{\partial f}{\partial x_n}(x) \end{bmatrix} = Df(x)^T \tag{2.30}$$

The gradient of a function of two variables is defined by:

$$\nabla f(x,y) = \left[\frac{\partial f(x,y)}{\partial x}, \frac{\partial f(x,y)}{\partial y}\right] \tag{2.31}$$

Consider a vector function of the form:

$$f(x) = \begin{bmatrix} f_1(x) \\ \vdots \\ f_m(x) \end{bmatrix} \tag{2.32}$$

Then, the gradient is given by:

$$\frac{\partial f}{\partial x_k}(x_0) = \begin{bmatrix} \dfrac{\partial f_1(x_0)}{\partial x_j} \\ \vdots \\ \dfrac{\partial f_m(x_0)}{\partial x_k} \end{bmatrix} \tag{2.33}$$

The derivative matrix or *Jacobian* matrix of $f(x)$ can be defined as:

$$\left[\frac{\partial f}{\partial x_1}(x_0) \;\; \cdots \;\; \frac{\partial f}{\partial x_n}(x_0)\right] = \begin{bmatrix} \dfrac{\partial f_1}{\partial x_1}(x_0) & \cdots & \dfrac{\partial f_1}{\partial x_n}(x_0) \\ \vdots & \vdots & \vdots \\ \dfrac{\partial f_m}{\partial x_1}(x_0) & \cdots & \dfrac{\partial f_m}{\partial x_n}(x_0) \end{bmatrix} \tag{2.34}$$

Given $f:\mathbb{R}^n \rightarrow \mathbb{R}$, if ∇f is differentiable, then f is twice differentiable. The derivative of ∇f is defined by:

$$D^2 f(x) = \begin{bmatrix} \frac{\partial^2 f}{\partial x_1^2} & \cdots & \frac{\partial^2 f}{\partial x_n \partial x_1} \\ \vdots & \ddots & \vdots \\ \frac{\partial^2 f}{\partial x_1 \partial x_n} & \cdots & \frac{\partial^2 f}{\partial x_n^2} \end{bmatrix} \tag{2.35}$$

The matrix $D^2f(x)$ is called the *Hessian* matrix of f at x. This is also denoted as $\boldsymbol{F}(\boldsymbol{x})$.

A matrix, M is symmetric if $M = M^T$. Consider the quadratic form $f:\mathbb{R}^n \rightarrow \mathbb{R}$ that is the function:

$$f(x) = x^T A x \tag{2.36}$$

Let $M \in \mathbb{R}^{mxn}$ and $y \in \mathbb{R}^m$ then we can define derivative D with respect to x as:

$$D\left(y^T M x\right) = y^T M \tag{2.37}$$

$$D\left(x^T M x\right) = x^T M + Mx = x^T M + x^T M^T = x^T\left(M + M^T\right) \tag{2.38}$$

Since M is symmetric matrix, it follows that:

$$D\left(x^T M x\right) = 2x^T M \tag{2.39}$$

$$D\left(x^T x\right) = 2x^T \tag{2.40}$$

2.2.2 Gradient Descent Algorithm

The gradient descent algorithm (GDA) minimizes an objective function iteratively. Consider an objective function $f(x)$. Suppose we are given a point x^k at iteration k. To move to the next point (minimum or optimum) x^{k+1}, we begin at $\underline{x^k}$ and then add an amount $-\eta_k \nabla f\left(x^k\right)$, where η_k is a positive scaler known as *step size* or *learning rate*, and $-\nabla f\left(x^k\right)$ is the direction of maximum rate of decrease. Hence, an equation for GDA can be written as:

$$x^{(k+1)} = x^k - \eta_k \nabla f\left(x^k\right) \tag{2.41}$$

The learning rate η can be selected as a small or large value. A small value of η takes longer compute time to reach the minimum point. On the other hand, a larger value of η results in faster (compute time) convergent to the minimum point, as it requires only few gradient evaluations. There are variations of gradient-based algorithms. Among these, the steepest descent is most commonly used. We can derive a closed form GDA solution for a quadratic function.

Consider a quadratic function of the form:

$$f(x) = \frac{1}{2}x^T Mx - b^T x \tag{2.42}$$

Here, $M \in \mathbb{R}^{nxn}$ is a symmetric positive definitive matrix, $b \in \mathbb{R}^n$, $x \in \mathbb{R}^n$. First, compute the gradient:

$$\nabla f(x) = Mx - b \tag{2.43}$$

Now, the Hessian of $f(x)$ is $F(x) = M = M^T > 0$. For notational simplicity, consider:

$$g^k = \nabla f\left(x^k\right) = Mx^k - b \tag{2.44}$$

Then we can write the steepest descent algorithm for the quadratic function as:

$$x^{k+1} = x^k - \eta_k g^k \tag{2.45}$$

Now for the steepest descent, the step size or learning rate can be computed as:

$$\eta_k = \arg\min_{\eta_k \geq 0} f(x^k - \eta_k g^k) = \arg\min_{\eta_k \geq 0}\left\{\frac{1}{2}\left(x^k - \eta_k g^k\right)^T M\left(x^k - \eta_k g^k\right) - \left(x^k - \eta_k g^k\right)^T b\right\}$$

By taking the derivative (with respect to η_k) of the above minimizer and setting to zero, we find:

$$\left(x^k - \eta_k g^k\right)^T M\left(-g^k\right) + b^T g^k = 0$$

or, equivalently,

$$\eta_k (g^k)^T M g^k = \left((x^k)^T M - b^T\right) g^k$$

However, we also can use:

$$\left((x^k)^T M - b^T\right) = (g^k)^T$$

Hence, the learning rate or step size η_k for the quadratic function can be written as:

$$\eta_k = \frac{(g^k)^T g^k}{(g^k)^T M g^k} \tag{2.46}$$

Finally, we derive the closed form equation for the steepest descent algorithm for the quadratic function as:

$$x^{\{k+1\}} = x^{\{k\}} - \eta_k g^{\{k\}} = x^{\{k\}} - \frac{(g^{\{k\}})^T g^{\{k\}}}{(g^{\{k\}})^T M g^{\{k\}}} g^{\{k\}} \tag{2.47}$$

in terms of the following:

$$g^k = \nabla f\left(x^k\right) = Mx^k - b \tag{2.48}$$

We now consider a specific example of this gradient descent methodology. Assume the following:

$$\mathbf{M} = \begin{bmatrix} 2 & 1 \\ 1 & 2 \end{bmatrix}, \qquad \mathbf{b} = \begin{bmatrix} -2 \\ -1 \end{bmatrix} \tag{2.49}$$

We will select the following initial guess $\mathbf{x}^{\{0\}}$ for seeding the iterative approach:

$$\mathbf{x}^{\{0\}} = \begin{bmatrix} 1 \\ -2 \end{bmatrix} \tag{2.50}$$

Then, (2.33) gives the following:

$$\mathbf{g}^{\{0\}} = \mathbf{M}\mathbf{x}^{\{0\}} - \mathbf{b} = \begin{bmatrix} 2 & 1 \\ 1 & 2 \end{bmatrix}\begin{bmatrix} 1 \\ -2 \end{bmatrix} - \begin{bmatrix} -2 \\ -1 \end{bmatrix} = \begin{bmatrix} 0 \\ -3 \end{bmatrix} - \begin{bmatrix} -2 \\ -1 \end{bmatrix} = \begin{bmatrix} 2 \\ -2 \end{bmatrix} \tag{2.51}$$

Next, the initial value for the learning rate or step size is:

$$\eta_0 = \frac{(\mathbf{g}^{\{0\}})^T \mathbf{g}^{\{0\}}}{(\mathbf{g}^{\{0\}})^T \mathbf{M}\mathbf{g}^{\{0\}}} = \frac{8}{(2)(2)+(-2)(-2)} = 1 \tag{2.52}$$

Thus, (2.47) implies the estimate after the first iteration:

$$\mathbf{x}^{\{1\}} = \mathbf{x}^{\{0\}} - \eta_0 \mathbf{g}^{\{0\}} = \begin{bmatrix} 1 \\ -2 \end{bmatrix} - \begin{bmatrix} 2 \\ -2 \end{bmatrix} = \begin{bmatrix} -1 \\ 0 \end{bmatrix} \tag{2.53}$$

This process is continued until the changes are minimal from one iteration to the next.

2.3 Backpropagation

An important component in artificial neural networks (ANNs) is the training of the weights based on labeled training data via backpropagation processing. This theory is based upon optimization theory involving vectors. Here, we provide an understanding of the mathematics in training the values of the weights of an ANN.

The strategy begins with sets of given labeled training data. Assume that the vector $\mathbf{x}$ describes the form of the input data sets. There is much freedom in the form of the selected input data. For example, it may describe the amplitude of a signal as a function of time. Equivalently, it can describe a two-dimensional image that has been flattened to a single dimension. Furthermore, it could be equivalent to a data set collected in more than two dimensions, as with a video sequence of two-dimensional image frames and the third dimension being time.

The output vector $\mathbf{y}$ describes the information desired by human users. One important type of ML problem is attempting to classify an input data set into one of a number of possible classifications. Assume that each element of $\mathbf{y}$ corresponds to one of these classes (e.g., attempting to classify mobile land-based vehicles). Thus, the first element can be chosen to correspond to the class of cars. The second element can be chosen to correspond to the class of trucks. Other elements can correspond to the general types of military vehicles.

The backpropagation technique is an iterative approach based on descent-gradient methods. Fundamentally, an initial guess for the solution is selected, which is often based on random weights. Then, gradient descent techniques are used to compute the updated weights for each iteration. However, at the beginning of each iteration, it is necessary first to perform the forward propagation from $\mathbf{x}$ to $\mathbf{y}$ in order to compute the outputs at each layer. Thus, it is useful to define forward propagation first, followed by the backpropagation algorithm [7].

Consider a general ANN shown in Figure 2.1. Here, there are some number P of initial images, each of which comprises a particular input data vector $\mathbf{x}$. Each of these data vectors is processed by some number L of layers of the ANN. Let the index m be a particular node in layer ℓ. Assume the output $o_{\ell-1,m}$ from a particular node with index m in the previous layer $\ell - 1$ feeds forward into the subject node with index n in the current layer ℓ be given by $w_{\ell,m,n}$. In addition, a non-zero constant bias $b_{\ell,n}$, which does not depend on the outputs of the previous layer, can be applied to the subject node with index n in layer ℓ.

The linear activation $a_{\ell,n}$ is the sum of the product contributions from all $M_{\ell-1}$ connected nodes at the previous layer $\ell - 1$ plus the bias $b_{\ell,n}$ and can be expressed via:

$$a_{\ell,n} = b_{\ell,n} + \sum_{m=1}^{M_{\ell-1}} w_{\ell,m,n} o_{\ell-1,m} \tag{2.54}$$

Next, this linear activation $a_{\ell,n}$ is input into a general nonlinear function $f_\ell(\bullet)$ at layer ℓ in order to yield the output from node n in layer ℓ; that is,

$$o_{\ell,n} = f_\ell\left(a_{\ell,n}\right) \tag{2.55}$$

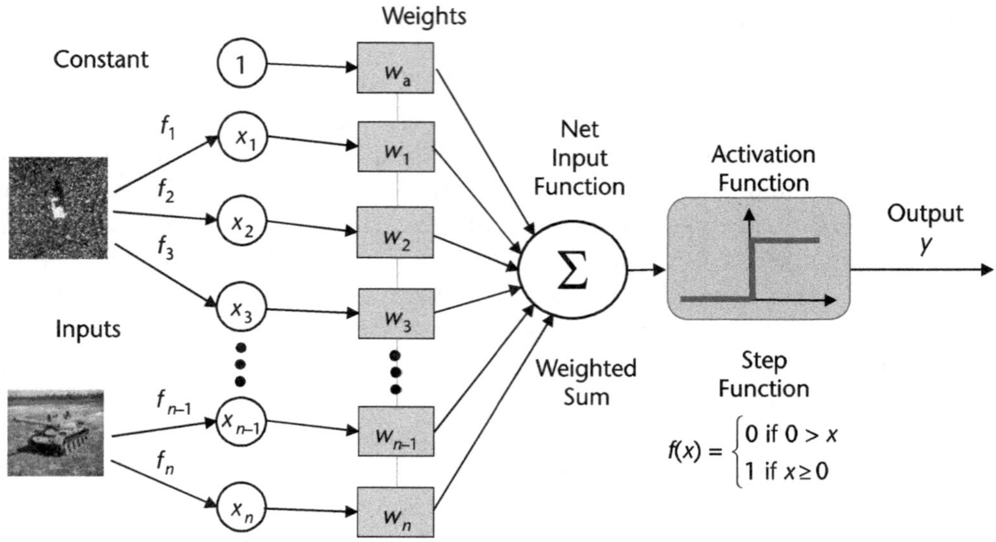

Figure 2.1 Multilayer perceptron.

It is convenient to incorporate the bias $b_{\ell,n}$ into the general weight formulation by defining an output variable corresponding to a nonexistent node $n = 0$ to be:

$$o_{\ell-1,0} = 1 \tag{2.56}$$

Next, the bias weight term is defined by:

$$w_{\ell,0,n} = b_{\ell,n} \tag{2.57}$$

Therefore, the summation in (2.54) can be extended to apply with the lower index equal to zero rather than unity:

$$a_{\ell,n} = \sum_{m=0}^{M_{\ell-1}} w_{\ell,m,n} o_{\ell-1,m} \tag{2.58}$$

The backpropagation weight update is computed often by using the mean squared error of the difference between the measured output data vector $\mathbf{y}$ and the vector predicted at the output of the forward model:

$$E(\mathbf{x},\mathbf{w}) = \frac{1}{2P}\sum_{p=0}^{P}\left\|\hat{\mathbf{y}}_p - \mathbf{y}_p\right\|^2 = \frac{1}{2P}\sum_{p=0}^{P}\left\{\hat{\mathbf{y}}_p - \mathbf{y}_p\right\}\bullet\left\{\hat{\mathbf{y}}_p - \mathbf{y}_p\right\} \tag{2.59}$$

Here, the summation is over the P individual pairs of input-output data vectors; that is, $\{\mathbf{x}_1, \mathbf{y}_1\}$, $\{\mathbf{x}_2, \mathbf{y}_2\}$, ..., $\{\mathbf{x}_P, \mathbf{y}_P\}$. Specifically, the vector $\hat{\mathbf{y}}_p$ denotes the final estimate of $\mathbf{y}_P$ as is obtained from executing the ANN in the forward direction from input to output. Also, the vector $\mathbf{w}$ denotes the set of all of the weight coefficients $\mathbf{w} = \{w_{1,1,1}, w_{1,2,1}, ...\}$ that are to be optimized is this iterative backpropagation processing.

The overall iterative updates to each of the weights $\mathbf{w} = \{w_{1,1,1}, w_{1,2,1}, ...\}$ proceeds by applying the partial derivative of (2.59) with respect to each weight individually. Here, we show a particular iteration using the superscript $\{i\}$, so that the weight $w_{\ell,m,n}^{\{i+1\}}$ at iteration $i+1$ is expressed in terms of that at the previous iteration $w_{\ell,m,n}^{\{i\}}$ and the partial derivative of the optimization function of (2.59) with respect to the weight $w_{\ell,m,n}^{\{i\}}$ via:

$$w_{\ell,m,n}^{\{i+1\}} = w_{\ell,m,n}^{\{i\}} - \eta_0 \frac{\partial E(\mathbf{x},\mathbf{w})}{\partial w_{\ell,m,n}} \tag{2.60}$$

In this equation, η_0 is the learning rate or step size of (2.52). In (2.60), move the partial derivative inside of the summation of (2.59) to give:

$$\frac{\partial E(\mathbf{x},\mathbf{w})}{\partial w_{\ell,m,n}} = \frac{1}{2P}\sum_{p=0}^{P} \frac{\partial\left\{\left\|\hat{\mathbf{y}}_p - \mathbf{y}_p\right\|^2\right\}}{\partial w_{\ell,m,n}} = \frac{1}{2P}\sum_{p=0}^{P} \frac{\partial}{\partial w_{\ell,m,n}}\left\{\left\{\hat{\mathbf{y}}_p - \mathbf{y}_p\right\}\bullet\left\{\hat{\mathbf{y}}_p - \mathbf{y}_p\right\}\right\} \tag{2.61}$$

The analysis proceeds by defining the corresponding mean squared error to be:

$$E_p(\mathbf{x},\mathbf{w}) = \left\|\hat{\mathbf{y}}_p - \mathbf{y}_p\right\|^2 = \frac{1}{2}\left\{\hat{\mathbf{y}}_p - \mathbf{y}_p\right\}\bullet\left\{\hat{\mathbf{y}}_p - \mathbf{y}_p\right\} \tag{2.62}$$

Thus, the optimization of (2.61) can be expressed as:

$$\frac{\partial E(\mathbf{x},\mathbf{w})}{\partial w_{\ell,m,n}} = \frac{1}{P}\sum_{p=0}^{P} \frac{\partial E_p(\mathbf{x},\mathbf{w})}{\partial w_{\ell,m,n}} \tag{2.63}$$

Thus, it is necessary to determine a methodology for computing $\partial E_p(\mathbf{x},\mathbf{w}) / \partial w_{\ell,m,n}$ for each weight $w_{\ell,m,n}$. The mean squared error $E_p(\mathbf{x},\mathbf{w})$ depends on the linear activation $a_{\ell,n}$ of (2.58). In order to prevent the notation from becoming too cumbersome, we drop the explicit subscript p for a given input-output vector pair from (2.62) and elsewhere. Thus, the chain rule of calculus implies that the desired partial derivative is:

$$\frac{\partial E(\mathbf{x},\mathbf{w})}{\partial w_{\ell,m,n}} = \frac{\partial E(\mathbf{x},\mathbf{w})}{\partial a_{\ell,n}} \frac{\partial a_{\ell,n}}{\partial w_{\ell,m,n}} \tag{2.64}$$

The first factor in (2.64) is called the *error* in the lexicon of ANN terminology:

$$\varepsilon_{\ell,n} \equiv \frac{\partial E(\mathbf{x},\mathbf{w})}{\partial a_{\ell,n}} \tag{2.65}$$

Equation (2.58) implies that the second factor in (2.64) can be expressed in terms of the output $o_m^{\ell-1}$ of node m of the previous layer $\ell - 1$ via:

$$\frac{\partial a_{\ell,n}}{\partial w_{\ell,m,n}} = \frac{\partial}{\partial w_{\ell,m,n}} \left\{ \sum_{m=0}^{M_{\ell-1}} w_{\ell,m,n} o_{\ell-1,m} \right\} = o_{\ell-1,m} \tag{2.66}$$

Combining (2.65) and (2.66) implies that (2.64) can be expressed as the product of two contributions:

$$\frac{\partial E(\mathbf{x}, \mathbf{w})}{\partial w_{\ell,m,n}} = \varepsilon_{\ell,n}\, o_{\ell-1,m} \tag{2.67}$$

That is, the desired partial derivative is proportional to the energy $\varepsilon_{\ell,n}$ of the subject node n at the current layer ℓ and the output $o_{\ell-1,m}$ of the node m at the previous layer $\ell - 1$. These calculations actually begin at the last layer $\ell = L$ and then proceed in the reverse or backward direction through the ANN all the way to the beginning input vector **x**. Ergo, this aspect of the analysis earns the name *backpropagation.*

Next, use (2.55) to describe the output due to all the nodes $n = 1, \ldots, N$ in the final layer $\ell = L$ via:

$$o_L = \mathbf{f}_L(\mathbf{a}_L) \tag{2.68}$$

Here, the activation vector is $\mathbf{a}_L = \{a_{L,1}, a_{L,2}, \ldots, a_{L,N}\}$, and the output vector is $\mathbf{o}_L = \{o_{L,1}, o_{L,2}, \ldots, o_{L,N}\}$. Likewise, the vector $\mathbf{f}_L = \{f_{L,1}, f_{L,2}, \ldots, f_{L,N}\}$ results from applying (2.55) to each element of $\mathbf{a}_L$. Therefore, the mean squared error corresponding to a given input-output data vector pair is:

$$E(\mathbf{x}, \mathbf{w}) = \frac{1}{2}\{\mathbf{f}_L(\mathbf{a}_L) - \mathbf{y}\} \bullet \{\mathbf{f}_L(\mathbf{a}_L) - \mathbf{y}\} \tag{2.69}$$

Then, the elements of the error vector $\varepsilon_L = \{\varepsilon_{L,1}, \varepsilon_{L,2}, \ldots, \varepsilon_{L,N}\}$ in the final layer are given by:

$$\varepsilon_{L,n} \equiv \frac{\partial E(\mathbf{x}, \mathbf{w})}{\partial a_{L,n}} = \{\mathbf{f}_L(\mathbf{a}_L) - \mathbf{y}\} \bullet \frac{\partial \mathbf{f}_L(\mathbf{a}_L)}{\partial a_{L,n}} \tag{2.70}$$

In this equation, $\mathbf{f}_L(\mathbf{a}_L)$ is known based on the forward computation. Again, at a given iteration, the forward propagation is executed before the backpropagation process in order to compute the outputs $o_{\ell,m}$ at each of the nodes in each of the layers. Therefore, we arrive at the following:

$$\frac{\partial E(\mathbf{x}, \mathbf{w})}{\partial w_{\ell,m,n}} = \varepsilon_{L,n} o_{L-1,m} = \{\mathbf{f}_L(\mathbf{a}_L) - \mathbf{y}\} \bullet \frac{\partial \mathbf{f}_L(\mathbf{a}_L)}{\partial a_{L,n}} o_{L-1,m} \tag{2.71}$$

The last element is the calculation of the energy for the hidden layers with $\ell < L$.

$$\varepsilon_{\ell,n} \equiv \frac{\partial E(x,w)}{\partial a_{\ell,n}} = \sum_{q=1}^{M_{\ell+1}} \frac{\partial E(x,w)}{\partial a_{\ell+1,q}} \frac{\partial a_{\ell+1,q}}{\partial a_{\ell,n}} \tag{2.72}$$

The first substitution follows from (2.65) applied to the next layer $\ell + 1$.

$$\varepsilon_{\ell,n} = \sum_{q=1}^{M_{\ell+1}} \varepsilon_{\ell+1,q} \frac{\partial a_{\ell+1,q}}{\partial a_{\ell,n}} \tag{2.73}$$

Next, notice that the linear activation equation of (2.58) can be expressed as:

$$a_{\ell+1,q} = \sum_{j=1}^{M_\ell} w_{\ell+1,j,q} o_{\ell,j} = \sum_{j=1}^{M_\ell} w_{\ell+1,j,q} f_\ell\left(a_{\ell,j}\right) \tag{2.74}$$

Next, compute the partial derivative of (2.74) with respect to the linear activation at the previous layer:

$$\frac{\partial a_{\ell+1,q}}{\partial a_{\ell,n}} = \frac{\partial}{\partial a_{\ell,n}} \left\{ \sum_{j=1}^{M_\ell} w_{\ell+1,j,q} f_\ell\left(a_{\ell,j}\right) \right\} = w_{\ell+1,n,q} \frac{df_\ell\left(a_{\ell,n}\right)}{da_{\ell,n}} \tag{2.75}$$

Then, (2.75) implies that the energy of (2.73) has the form:

$$\varepsilon_{\ell,n} = \sum_{q=1}^{M_{\ell+1}} \varepsilon_{\ell+1,q} w_{\ell+1,n,q} \frac{df_\ell\left(a_{\ell,n}\right)}{da_{\ell,n}} = \frac{df_\ell\left(a_{\ell,n}\right)}{da_{\ell,n}} \sum_{q=1}^{M_{\ell+1}} \varepsilon_{\ell+1,q} w_{\ell+1,n,q} \tag{2.76}$$

Finally, the partial derivative of the mean squared error for the hidden layers follows from (2.76) in (2.67):

$$\frac{\partial E(\mathbf{x},\mathbf{w})}{\partial w_{\ell,m,n}} = o_{\ell-1,m} \frac{df_\ell\left(a_{\ell,n}\right)}{da_{\ell,n}} \sum_{q=1}^{M_{\ell+1}} \varepsilon_{\ell+1,q} w_{\ell+1,n,q} \tag{2.77}$$

In summary, the equations required for updating the weights at each iteration in the backpropagation processing are (2.60), (2.63), (2.67), (2.70), and (2.76). These equations first require computation in the forward direction in order to calculate the node outputs $o_{\ell,m}$ that are used in these equations.

2.4 Statistics and Probability Theory

Statistics and probability theory are necessary for machine learning and artificial intelligence. Probability theory models uncertain events and estimates parameters from measurements. As more data are available (i.e., additional observations), the accuracy of classification improves significantly. Statistics and probability theory explain these phenomena. In this section, we present some of the important concepts in statistics and probability [5–9] theory applied to machine learning.

2.4.1 Basic Probability

Classical probability theory is often based upon the examination of outcomes under conditions for which the particular results are uncertain or depend on variables that are unknown to analysts. Under these conditions, insight can be obtained through a statistical analysis of possible outcomes. Specifically, begin with the concepts of classical probabilities, which are described in terms of successes and failures of particular events. For example, suppose that it is desired to obtain a "4" on a 6-sided die, so that this outcome is a success and all others are a failure. Thus, the probability of success is equal to the ratio of the success outcomes (i.e., unity), divided by all possible outcomes (i.e., 6), yielding the probability of obtaining a "4" equal to $p_4 = \frac{1}{6}$.

The computation of probabilities for complicated scenarios involves an understanding of combining multiple independent probabilities. Again, consider the scenario of rolling a single die, but now assume that a success is defined as the outcome of obtaining *either* an odd number *or* a "4." This corresponds to a situation in which multiple mutually exclusive probabilities must be combined in order to yield a single composite probability. First, the probability of obtaining an odd number is $\frac{1}{2}$. Second, the probability of obtaining a "4" is $\frac{1}{6}$. Since these two possible outcomes are mutually exclusive, then the resulting composite probability is obtained via the addition of the corresponding individual probabilities: $\frac{1}{2}+\frac{1}{6}=\frac{2}{3}$.

Another scenario to consider corresponds to that of multiple independent outcomes. For example, consider a scenario of rolling a single die and randomly selecting a single card from within a standard 52-card deck. Assume that a success is defined by obtaining a "4" on the roll of the die *and* a card that is a heart. Again, the independent probability of obtaining a "4" is $\frac{1}{6}$. Since the hearts of one of four possible equally probable suites, the probability of randomly drawing a heart is $\frac{1}{4}$. The probability that multiple independent outcomes occur is given by the product of the individual probabilities. Thus, for the particular scenario of interest of obtaining a "4" on a die and randomly selecting a heart is given by the product of these individual probabilities (i.e., $\frac{1}{6}\times\frac{1}{4}=\frac{1}{24}$).

2.4.2 Probability Density Functions

Another important concept in the theory of probability and statistics is that pertaining to a continuous probability density function (PDF). Such a function is intended to provide a convenient and relatively accurate continuous-valued representation of a scenario involving a large number of discrete events. In this concept, a standard histogram with a large number of small bins can be approximated by the curve of a PDF. For example, assume that a given PDF is described by a function $f(x)$ of a single variable. We will further assume that this PDF has been normalized properly, so that total integral under the curve equals unity; that is,

$$\int_{-\infty}^{\infty} f(x)dx = 1 \tag{2.78}$$

In order to compute nontrivial probabilities from a given PDF, it is necessary to calculate the integral under the PDF corresponding to the relevant upper and lower limits of the random variable of interest. For example, suppose that we want to compute the probability that the variable x can have a value that is anywhere between an upper bound of x_{max} and a lower bound of x_{min}. This probability is computed by using x_{max} and x_{min} as the upper and lower limits, respectively, in the integral calculation:

$$P\{x_{min} \le x \le x_{max}\} = \int_{x_{min}}^{x_{max}} f(x)dx \tag{2.79}$$

The standard normal PDF for a single variable is defined as:

$$f(x \mid \mu, \sigma) = \frac{1}{\sqrt{2\pi}\sigma} \exp\left(-\frac{\{x-\mu\}^2}{2\sigma^2}\right) \tag{2.80}$$

This function is commonly called the Gaussian PDF. In this equation, μ and σ are parameters. The symbol | delineates *conditioned upon* to represent the PDF of the random variable x, conditioned upon μ and σ (see Figure 2.2). Equation (2.80) gives a univariate standard normal PDF (i.e., a function of the single random variable).

It is also possible to define and use multivariate PDFs. For example, consider a bivariate standard normal in terms of the two independent random variables x and y:

$$f(x, y \mid \mu_x, \sigma_x, \mu_y, \sigma_y) = \frac{1}{2\pi\sigma_x\sigma_y} \exp\left(-\frac{\{x-\mu_x\}^2}{2\sigma_x^2} - \frac{\{y-\mu_y\}^2}{2\sigma_y^2}\right) \tag{2.81}$$

This joint PDF is formed as the product of two univariate standard normal density functions.

The mean of the univariate standard normal of (2.80) is:

$$\varepsilon\{x\} = \int_{-\infty}^{\infty} x f(x)dx = \frac{1}{\sqrt{2\pi}\sigma} \int_{-\infty}^{\infty} x \exp\left(-\frac{\{x-\mu\}^2}{2\sigma^2}\right) dx = \mu \tag{2.82}$$

Thus, we see that the result of expectation of the standard normal PDF yields the parameter μ, which is equal to the mean of $f(x)$. Similarly, the expectation of the variance about the mean for (2.80) gives:

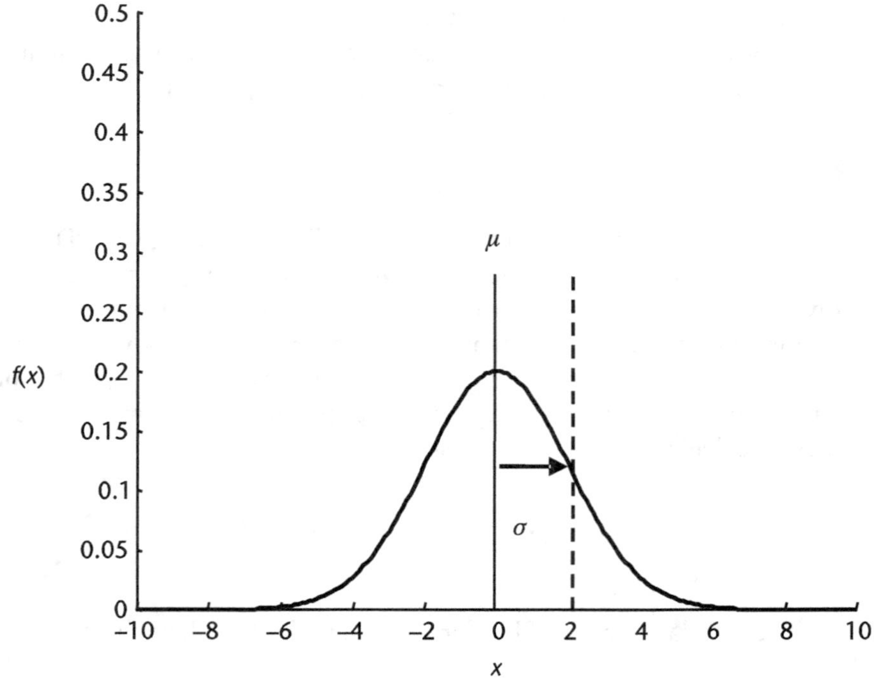

Figure 2.2 Target PDFs with Gaussian normal distribution of $\mu = 0$ and $\sigma = 2$; N(0,2.0).

$$\varepsilon\left\{(x-\mu)^2\right\} = \int_{-\infty}^{\infty}(x-\mu)^2 f(x)dx = \frac{1}{\sqrt{2\pi}\sigma}\int_{-\infty}^{\infty}(x-\mu)^2 \exp\left(-\frac{\{x-\mu\}^2}{2\sigma^2}\right)dx = \sigma^2 \qquad (2.83)$$

Therefore, the expectation of $\varepsilon\{(x-\mu)^2\}$ yields the parameter σ^2, which is called the variance of the PDF. In addition, the square root of the variance (i.e., σ), is called the standard deviation of the PDF.

2.4.3 Maximum Likelihood Estimation

The maximum likelihood corresponding to a PDF is computed by assuming that the probability depends on some number N of parameters $\{\varphi_1, \varphi_2, ..., \varphi_N\}$; that is, $f(x|\varphi_1, \varphi_2, ..., \varphi_N)$. The maximum likelihood value of this PDF, conditioned upon the parameters $\{\varphi_1, \varphi_2, ..., \varphi_N\}$, is found by computing the partial derivative of $f(x|\varphi_1, \varphi_2, ..., \varphi_N)$ with respect to each of the individual parameters φ_n, and setting each resulting equation equal to zero. This process yields a set of N equations in the N unknowns $\{\varphi_1, \varphi_2, ..., \varphi_N\}$. Then, these optimized values of the various parameters are used within the final PDF $f(x|\varphi_1, \varphi_2, ..., \varphi_N)$.

One common example of maximum likelihood theory within the realm of SAR image formation is that of phase gradient autofocus (PGA) [9]. This methodology uses maximum likelihood calculations to generate the optimal estimate of a phase error parameter φ_1 arising from the unknown radar platform motions over the duration of a SAR collection interval. Then, this phase error estimate is used to compensate for the deleterious defocus effects that arise from the unknown radar platform motions. In addition, a recent analysis [10] applies an extension of these concepts in a two-dimensional maximum likelihood analysis in order to estimate two atmospheric defocus parameters $\{\varphi_1, \varphi_2\}$ jointly. These two parameters correspond to the unknown path delay and the unknown bending effects that occur due the refraction of radar waveforms within the earth's atmosphere. Finally, the theory for maximum likelihood estimation for an arbitrary number of parameters is developed in [11].

2.4.4 Bayes' Theorem

Bayes' theorem is used to calculate posterior probability density, which is based on conditional probabilities and prior probabilities. Assume that we have some knowledge of the prior probability of a certain event (this could be very broad). After we observe some data, we can compute the posterior distribution, which could describe the event more accurately. Hence, Bayesian inference updates a probability hypothesis (i.e., calculating posterior distribution) as more data or information become available.

First, it is necessary to understand the concept of conditional probabilities. The notation $P(X|Y)$ represents that conditional probability that X occurs, given that Y actually does occur. Next, $P(Y)$ is that prior probability that outcome Y actually does occur. In many cases for applying Bayes' theory, this prior probability represents information that is brought into the problem, typically with high fidelity based on some truth knowledge, prior to the actual measurements used to generate the final desired posterior probability $P(Y|X)$. Specifically, assume that $P(X)$ is the probability of measurements (features), $P(Y)$ is the probability of features existing, and $P(X|Y)$ is the likelihood. Mathematically, Bayes' rule is written as:

$$P(Y \mid X) = \frac{P(X \mid Y)P(Y)}{P(X)} \tag{2.84}$$

where the denominator $P(X)$ is effectively a normalization factor that encompasses all possible outcomes. Thus, Bayes' rule offers a way to interpret information in a sequential way by accumulating evidence $P(Y|X)$ (likelihood) of events relating to the outcome. Using Bayes' rule, the world determination assesses whether features can exist $\mathrm{P(Y)}$ or do not exist $\mathrm{P(\neg Y)}$:

$$P(Y \mid X) = \frac{P(X \mid Y)P(Y)}{P(X \mid Y)P(Y) + P(X \mid \neg Y)P(\neg Y)} \tag{2.85}$$

Many contextual factors can lead to an understanding of ATR decisions through feature measurement and existence relation. The key is to expedite decision making via the use of knowledge of contextual features (e.g., shadows in SAR images).

For example, a priori information has been collected over time. For this example, the processing begins with the ATR assessing the measurement. Assuming the collection of equivalent data, the distinction between the vehicle classification (1) and the incorrect classification (0), which is p = [0.5, 0.5], is shown in Figure 2.3.

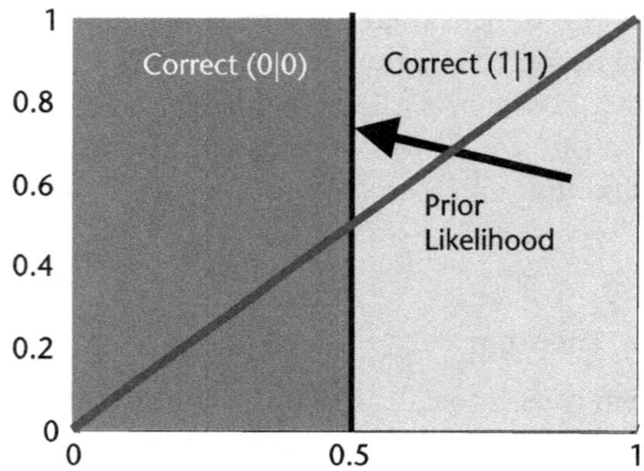

Figure 2.3 A priori target likelihood $p = [0.5, 0.5]$.

After analysis, the likelihood of the analysis determines the regions of the correct analysis, as well as the regions of error. The choice of depiction of the results is twofold, to show the correct regions typical of analysis $P(1|1) + P(0|0)$ and then separate errors from *deception* $P(1|0)$ and *concealment* $P(0|1)$ for situational understanding. For concealment, the user must query other information sources to ascertain the location and classification of the target (T). For deception, the user must address the contextual issues (C). Together the user must address the evaluation (E) issues associated with finding, requiring, and searching for the target. Bayes essentially determines the update. A cycle repeats itself with the evidence supporting the contextual analysis of ATR actions. Since the ATR observes the targets (i.e.,

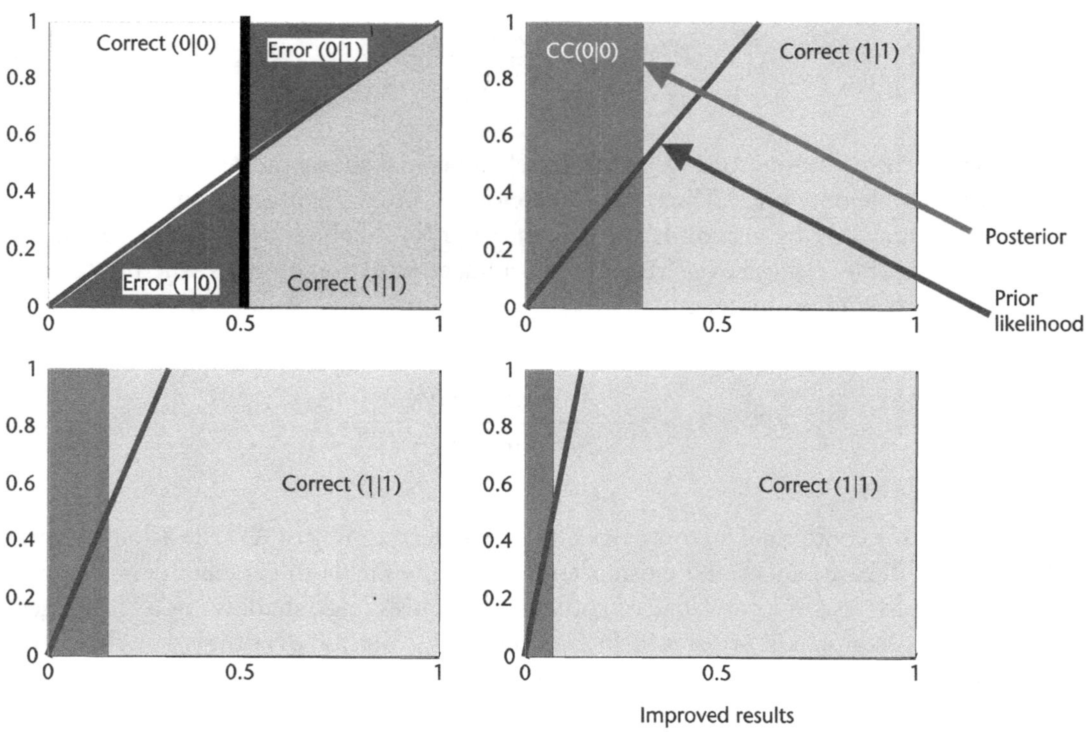

Figure 2.4 Accumulating evidence.

detecting a target), $P(1|1)$, the evidence supports the increased likelihood the target via the large area on the right in Figure 2.4:

$$P(1 \mid 1) = \frac{P(1 \mid 1)P(1)}{P(1 \mid 1)P(1) + P(1 \mid 0)P(0)} \tag{2.86}$$

2.5 Summary

This chapter presented fundamental mathematical concepts applied to machine learning theory. Basic linear algebra notations and some of the important manipulations wer provided to introduce multivariate calculus and optimizations needed for machine learning. An important concept was the theory and implementation of steepest descent algorithm. Finally, this chapter described statistics and probability theory needed for ML algorithms.

References

[1] Shakuntala Devi, https://en.wikipedia.org/wiki/Shakuntala_Devi

[2] Strang, G., *Introduction to Linear Algebra,* Fifth Edition, Wellesley, Massachusetts, 2016.

[3] Chong, K. P., and S. Zak, *An Introduction to Optimization,* 4th Edition, New York: John Wiley and Sons, 2013.

[4] *Backpropagation, Brilliant Math and Science Wiki*, https://brilliant.org/wiki/backpropagation.

[5] Papoulis, A., S. Unnikrishna Pillai, *Probability, Random Variables and Stochastic Processes,* 4th Edition, McGraw-Hill Kogakush, 1981.

[6] MacKay, D., *Information Theory: Inference and Learning Algorithms,* Cambridge UK: Cambridge University Press, 2003.

[7] Freund, J. E., and Ronald E. Walpole, *Mathematical Statistics,* Third Edition, Upper Saddle River, NJ: Prentice-Hall, 1980.

[8] Hastie, T., R. Tibshirani, and J. Friedman, *The Elements of Statistical Learning*, Springer, New York, 2001.

[9] Hayes, M. H., *Statistical Digital Signal Processing and Modeling*, New York: John Wiley, 1996.

[10] Wahl, D. E., P. H. Eichel, D. C. Ghiglia, et al., "Phase Gradient Autofocus—A Robust Tool for High Resolution SAR Phase Correction," *IEEE Trans. Aerospace & Electronic Systems,* Vol. 30, No. 3, 1994, pp. 827–834.

[11] Garren, D. A., "Theory of Data-Driven SAR Autofocus to Compensate for Refraction Effects," *IET Radar, Sonar & Navigation*, Vol. 13, No. 2, 2019, pp. 254–262.

[12] Van Trees, H. L., "Detection, Estimation, and Modulation Theory, Part 1: Detection, Estimation, and Linear Modulation Theory," New York: John Wiley and Sons, 1968.

CHAPTER 3

Review of Machine Learning Algorithms

This chapter reviews common methods in ML that are foundational to ATR and pattern recognition using electro-optical (EO) imagery (e.g., visual and infrared), that are applied to SAR. With an understanding of ML algorithms [1], the goal is to compare methods prominent in video image (e.g., computer vision) classification to those of radar frequency (e.g., SAR) analysis. Many of the ML challenges and opportunities are similar in EO and RF imagery; however, results can differ from the features extracted and their correspondence to the physical phenomena. For example, ML methods typically support detecting, recognizing, classifying, identifying, and fingerprinting objects [2]. If a method of segmentation supports the selection of data, then pattern recognition (i.e., testing) is possible by comparing the observed data to that of a truthed set of data (i.e., training). While RF imagery poses additional challenges to ML methods from EO imagery, it derives from the radar signal collection, preprocessing, and use (e.g., image formation). The benefits of ML and DL for RF object classification have potential, and the resulting improvements in RF ML algorithm hardware and software enable enhancements in RF data analytics.

ML methods support dimensionality reduction, structure prediction, and anomaly detection. Chapter 1 provided a historical overview of ML methods and general inspirations for ML applied to RF imagery. As ML is based on *data modeling plus optimization*, this chapter provides common methods in ML with a brief review of methods applied to radar analysis; though the reader is encouraged to consult other teaching material on signal processing, computer vision, and information fusion. Many of the methods are applied in later chapters for comparative analysis. This chapter discusses traditional ML methods while Chapter 4 presents deep learning methods such as deep neural networks.

3.1 Introduction

Over the years, ML algorithms have evolved from handcrafted knowledge to describe the data, to statistical processing to categorize the data, towards methods of contextual analysis for explainability. The interest in these methods, as for many problems was that solutions proved well for minor variations in the operating conditions from variations in sensors, environment, and targets (SET). For example, the expert systems of the 1970s and 1980s helped physicians'diagnoses based on probabilities from a neural network (NN) classifier, while in the 1990s/2000s, there was a benefit from statistical reasoning combined with models for analysis of large target sets for unique sensors such as SAR. Hence, the ML techniques support (1) analysts who wish to make a decision over a complex space (e.g., buying things on the internet), (2) operators who routinely have to exploit large imagery data (e.g., identifying target behaviors), and (3) users who make decisions in the environment desiring technical help (e.g., driving a car). Recently, deep learning has shown better accuracy in image and text recognition in research challenge problems versus traditional methods. Hence, each of the ML/DL methods provide some benefits that are important including:

- *Accuracy*: Improving performance over analysts conducting similar tasks;
- *Execution*: Solving big data problems in real-time versus manual operation;
- *Solving*: Resolving complex problems over many options that are nearly impossible for everyday users.

Like other ML techniques available for different applications, the ML approaches support RF analysis, including:

- *Image classification*: Automatic target recognition, classification, and identification;
- *Natural language processing*: Target semantic labeling with a dictionary;
- *Autonomous systems*: Pointing the sensor for command and control;
- *Sentiment analysis*: Optimizing the collection parameters for a given scenario;
- *Biomedical applications*: Decision support for a subject matter expert.

The foundation of DL is built from decades of analysis and support, which is evident from most publications in ML. To support the reader, this chapter presents the foremost approaches developed and applied for common SAR image data sets (e.g., MSTAR) that showcase the developments and enhancements. This chapter begins with the neural network motivation, and then reviews the literature associated with linear and nonlinear classifiers. Finally, it highlights current trends in machine learning to support deep learning. With the advancement of methods for SAR, this chapter does not include the methods for software used such as MATLAB (ML toolboxes), TensorFlow (Google), Caffe (UC Berkley), Theano, or Torch (Facebook).

3.1.1 ML Process

The ML techniques highlighted support RF signals for object classification from imagery (e.g., EO/SAR) as well as signal classification (e.g., RF/communications) which is discussed in later chapters. While ML for object classification from video imagery is more known to researchers, these algorithms can and recently have been applied to the RF domain.

One important point to appreciate is that the methods all employ both offline (training) and online (testing) as shown in Figure 3.1. For the offline system, a large set of training data (e.g., 80%) is used to train the methods, while testing includes (20%). The samples for training are typically labeled to resolve the salient features and the resulting statistical classifier. The features represent something in the data which for 1D signals are peaks for 2D images as shapes and lines. As the data becomes more complex, such as in natural language processing (NLP), various sets of features are formed as a bag of words and stored in a dictionary. As a *dictionary* provides a lookup for semantic content, a dictionary of features provides physical attributes of the sensor information based on context. Training thus seeks to build a dictionary offline for the feature set to be used in online analysis.

Using the classifier and the set of features, the online evaluation brings in a new measurement sample and determines (1) if it is in the dictionary to assign a label (e.g., target), (2) if it is not in the dictionary but representative of the content in the dictionary for which a new label is assigned (e.g., confuser), or (3) if it represents a measurement or observation that should not be in the dictionary (e.g., no label). Inherent in the analysis is the label output quality from what traditionally has been probablity of correct classification (PCC), or probability of identification (PID), for a label. Other methods include the F-score, which is presented in Chapter 9 with the deep learning evaluation methods; however, the RF SAR community has traditionally used PCC and PID.

For a SAR example shown in Figure 3.2, objects are analyzed in the image by determining parts of the image (bounding box) for which a trained example with similar sensor, target, and environment is labeled and corresponds to the data provided. Figure 3.2 demonstrates an example, and the bounding box features of a target in the image which is used for analysis.

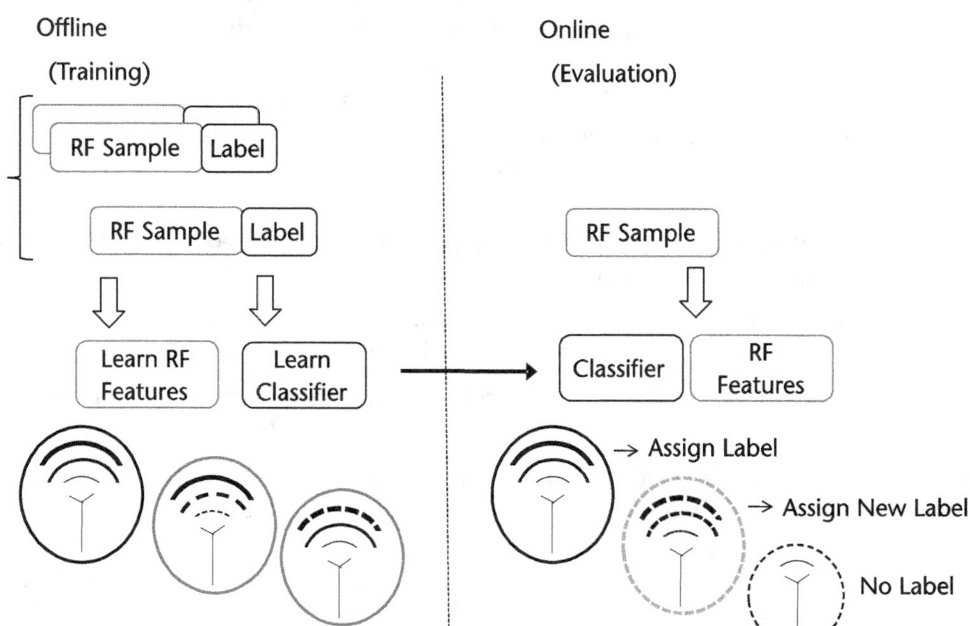

Figure 3.1 Offline training and online evaluation.

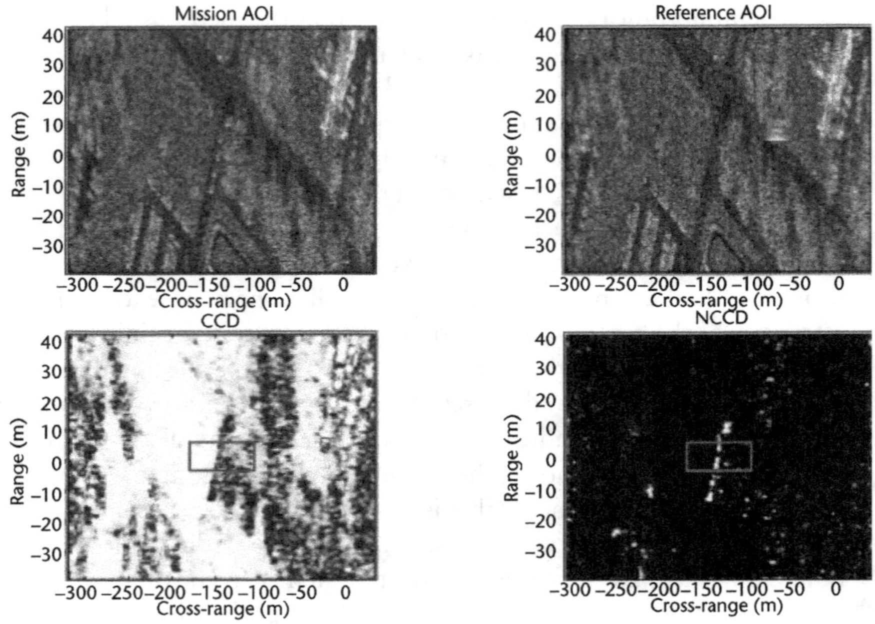

Figure 3.2 Objects in SAR imagery [3].

3.1.2 Machine Learning Methods

The three types of ML techniques build on Figure 3.1 and include supervised, semisupervised, and unsupervised learning. These three traditional ML techniques support the broad categories emerging in DL approaches. For example, supervised learning includes linear and nonlinear classifiers that are the basis of this chapter. As these techniques are common, the chapter provides an overview that allows the reader to appreciate the key concepts as compared to the new DL methods. For the case where there is no label, then unsupervised methods are used. Finally, those that have some labels (e.g., 25% of the data), then semisupervised methods are used as the entire feature and target set for decision making is still unknown at the start of the analysis. The descriptive definitions are:

- *Supervised learning*: Develop predictive models based on learning inputs and output labeled data;
- *Semisupervised learning*: Determine relationships based on input data and a few output labels over unlabeled data;
- *Unsupervised learning*: Group (e.g., cluster) and interpret data geometry based only on input data;
- *Reinforcement learning*: Generate optimal policies from reward functions of outputs from input data.

To organize ML/DL methods, Table 3.1 provides a motivation based on domain support, training, and testing. For example, supervised learning has benefitted reinforcement learning (RL); while unsupervised learning is utilized in new DL approaches. Likewise, transfer learning builds from semisupervised where the systems development starts with a known set of data (e.g., U-band radar) and seeks

Table 3.1 ML Methods Based on Domains, Training, and Testing Types

Domain Support	*Training*	*Testing*
Data	Supervised learning	Reinforcement learning
Scenario specific	Semisupervised learning	Transfer learning
Unknown challenge	Unsupervised learning	Generative adversarial network/learning

to seed the new classifier for a new sensor (e.g., X-band radar). Finally, the power of DL is shown in the unsupervised case for large complex data sets where the decision is unknown. An example is the explosion of generative adversarial networks (GANs) that use a generator network and an evaluator network that compete toward improving classifier robustness. The GAN generator could be from a model, while the GAN discriminator is representative of a human making a decision for unique situations.

Table 3.2 presents the four categories of ML methods along with their purpose. Many ML applications leverage traditional approaches for clustering, classification, and regression, while other attributes include dimension reduction and data association. With the advent of contemporary NN methods, Figure 3.3 organizes these ML techniques from linear to nonlinear classifiers, including many of the popular NN approaches.

Figure 3.3 highlights the techniques in categories developed for RF analysis. Specific to this chapter are methods of supervised learning that consider linear and nonlinear classifiers. The unsupervised methods focus on cluster and mixture models. Chapter 3 focuses supervised methods leading to those presented in Chapter 4 with the recurrent NN (RNN) for temporal data, convolution NN (CNN) for 2D spatial imagery, and long short-term memory (LSTM) for dynamic information.

The development of many ML algorithms derives from the methods used to train the algorithm, explain the results, and/or determine performance. Inherent in the discussion is the data used for training, some of which is labeled (e.g., offline data collection) and some unlabeled (e.g., online streaming analysis). Another factor to consider is the repeatability of the statistical analysis providing verification of the weightings that provide interpretability for explanations. The third factor is performance, which should be conducted not only with standardized conditions (SOCs), but also extended operating conditions (EOCs). The performance analysis is typically provided by metrics such as accuracy (Figure 3.4).

Figure 3.4 presents a representation of the ML/DL methods from which pure statistical methods like linear regression are simple to understand, but yield low accuracies. On the other hand, recent methods in deep learning are essentially *black boxes* from which accuracy has been demonstrated in numerous competitions, but the interpretability is vague. Future methods seek high interpretability and high accuracy, of which the random forest method is one possibility.

Most machine learning methods are derived from the concept of a biological neural network. A biological NN consists of inputs, activation functions, and outputs. The NN is designed to recognize numerical patterns in the input data and ultimately learn the mapping function between the input and output data. The basis of a NN includes that of an artificial NN (ANN) as featured in the text by Priddy [4]. The ANN was devised as a bio-inspired approach to approximate the

Table 3.2 ML Methods-Based on Domains, Training, and Testing

Type	*Subtype*	*Methods*
Supervised	Regression	(Generalized) linear regression Nonlinear regression Gaussian process regression (GPR) Ensemble methods Decision trees Neural networks
	Classification	Support vector machine (SVM) Linear discriminant analysis (LDA) K-nearest neighbor (kNN) Artificial neural networks (ANNs) Naïve Bayes Random forest
Semisupervised		Semisupervised SVM Graph-based label propagation Generative methods Active learning Transfer learning
Unsupervised	Clustering	K-means, K-manifolds Random forest Fuzzy C-means Agglomeration hierarchical Kernel density based Gaussian mixture models (GMM) Hidden Markov models (HMM) Self-organized maps (SOM) Neural networks
	Dimension reduction	Principal component analysis (PCA) Kernel proposal component analysis Multidimensional scaling Maximum variance unfolding Isomap Local linear embedding (LLE) Diffusion maps Local tangent space alignment (LSTA)
	Association	A prior Bayesian data fusion Evidential reasoning Multihypothesis association
Reinforcement learning		Epsilon greedy Prioritize sweeping Temporal difference Q-learning

understanding of how a mammalian brain might process signals derived from auditory, tactile, taste, temperature, visual, vestibular, and/or proprioception sensors. The approximation of a biological neuron resulted in the naming convention of an artificial analysis. Collectively, it is called an ANN as shown in Figure 3.5. Each neuron can receive a signal and transmit a result to another neuron. The collective interaction of the neurons is a network where synapses represent the electrical/chemical messaging between neurons. Since the biological transmission is electrochemical, it includes a spike train of pulses. To represent the data being transmitted, the ANN uses a logical analysis (e.g., $-N$ to N) where N is a real number. The computation is a nonlinear function of the sum of the inputs that output a result. To model the connections used in the computation, the neuron is a node and the

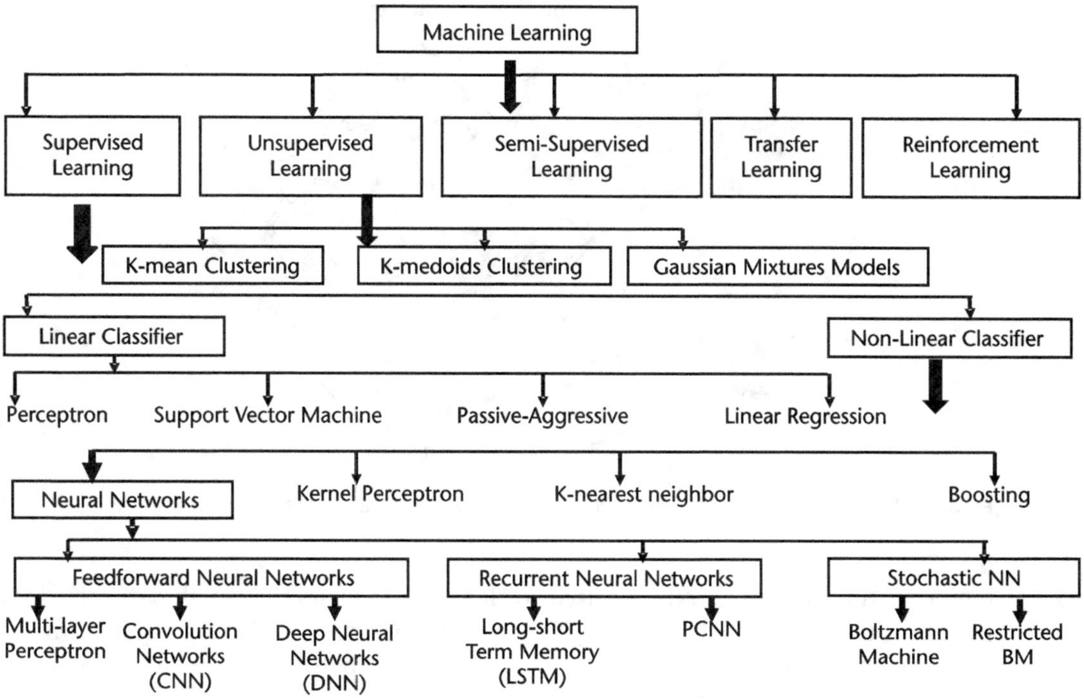

Figure 3.3 ML method categories.

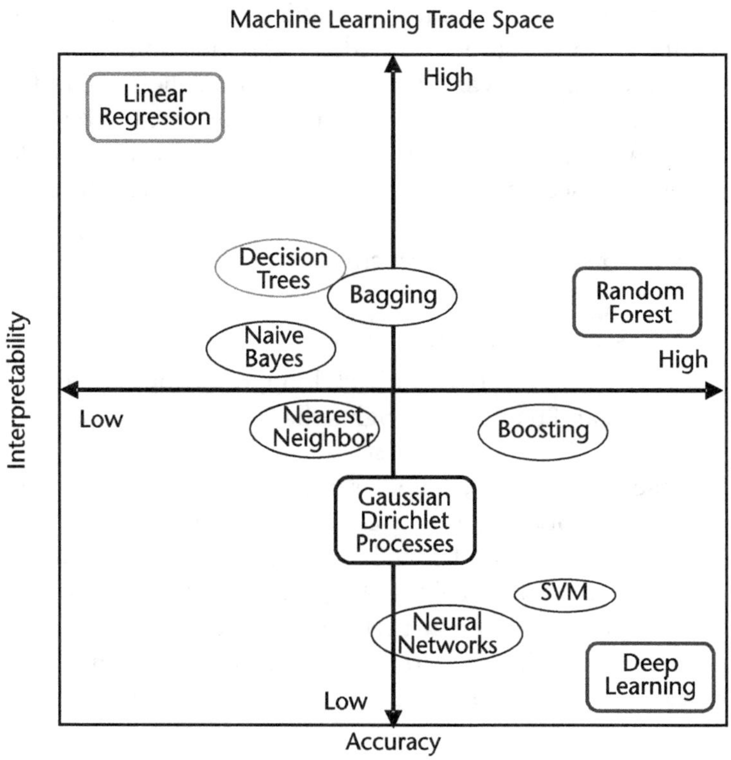

Figure 3.4 ML Trade space.

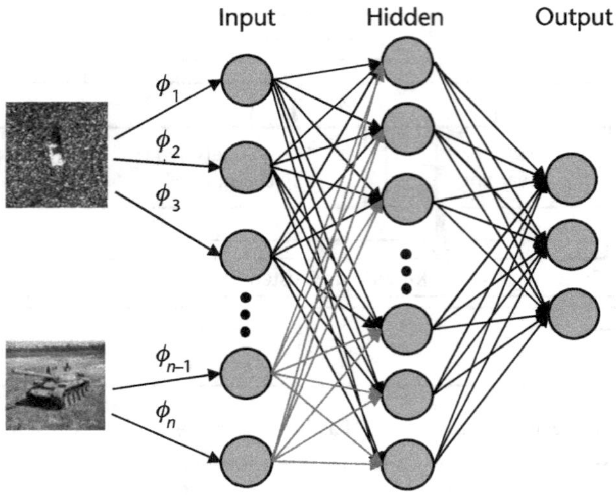

Figure 3.5 Artificial neural network, where $\phi_1 \ldots \phi_n$ are feature extraction functions that input data x.

connection is an edge. ANN, as used for learning, adjusts the weights of the edges based on the signal connect strength/weakness. Another ANN attribute is thresholding where a signal has to achieve a certain value before being considered valid for subsequent processing.

Advances in ML have led to DL where multiple hidden (or latent) layers are included in the network. Different layers provide different functions where a strategy is to have the first layer learn the structure of the signal (e.g., shape) and the subsequent layers fine-tune the additional feature information (e.g., contours). The input-to-output layers constitute the ability of the neurons to regress over the signals analysis as many times as needed to provide an increasing fidelity of the relationship resolved. The variations of NN approaches extend from the ANN; however, what distinguish the many ML approaches are the types of data presented to the input layers, termed *data mining*. Supervised learning uses labeled training data while unsupervised approaches use unlabeled data. Hence, an ANN:

- Attempts to learn a distinctive set of features in each hidden layer based on the previous layer's output;
- Processes data in a feed forward manner without memory;
- Provides a corrective feedback loop, rewarding weights that support correct guesses and punishing weights that lead to error.

In general, the ANN employs hierarchical block architectures, wherein the deeper the network, such as a convolutional neural network (CNN), the more complex features can be learned (feature hierarchy). Likewise, DL methods employ a dynamic set of loops that have memory such as a recurrent NN. Finally, a DL method utilizes different performance metrics beyond error analysis such as throughput, credibility, and timeliness to optimize results. To begin the RF ML approaches, supervised learning assumes you have a labeled data set for training.

3.2 Supervised Learning

Many practical ML methods use supervised learning, which infers a mathematical function from labeled training data. The training data (see Chapter 2) includes the collected examples (e.g., RF images), the labels (e.g., target/no targets), and some metadata context (e.g., sensor pointing angle). For all examples, the resultant training pair includes the input object (e.g., vector) and output value (e.g., weights). Since the process is supervised, the object-signal pair is prelabeled with a tag. Supervised learning generally requires large volumes of labeled training data from which an algorithm develops a mathematical model that explains the signals. The main applications of supervised learning include:

- *Classification:* The output value is a class label or limited set of values;
- *Regression:* The output value is a real or numerical value within a range;
- *Ranking:* The output value is a similarity value for verification.

The ML method seeks to develop a robust mapping function, *f*, such that:

$$\mathrm{Y} = f(\mathrm{X}) \tag{3.1}$$

where **X** is the input data and **Y** is the output data. Each of N exemplars is represented as a vector (or array) x_i, where $i = 1, \ldots, N$, and the data for training is represented as matrix **X**, where $\mathbf{X} = [x_i, \ldots, x_N]$. Using an objective function, iterative optimization determines the classifier or regression function. With the functional mapping, the system predicts the new output based on the new input data. As new data becomes available for training, then the function improves accuracy until it sufficiently learns the task. Additionally, an optimal function (e.g., minimizing the objective function error) allows for determining outputs for data not in the training set.

The ML goal seeks to utilize the new input data **X** (e.g., images), to predict **Y** (e.g., class labels) for that data. Hence, the classifier maps new data into output results. For common examples, if all the training data includes exemplars of the scenario, then the test results should provide near-perfect results. However, a typical use of these ML methods is to recognize objects unseen in the training data. Likewise, when deploying the classifier/regressor to an unknown scenario, the performance results decrease for the unobserved instances. Hence, over the years, methods have been developed to build a generalized classifier/regressor.

The types of linear and nonlinear supervised learning methods approaches are included in Table 3.3.

Figure 3.6 illustrates the different techniques of regression, classification, and clustering. For the linear methods, the first case is *regression*. For a set of data, a line is determined that minimizes the total error (or distance) between all the points and the line. The minimum square error (MSE) represents the optimal boundary. For *classification*, the goal is to determine the line that best distinguishes the data. If the data is highly separable, then multiple lines can be determined to separate the data, from which the line that generates that largest separation (or distance) among

Table 3.3 Linear or Nonlinear Approaches for ML Functions

Function	*Linear*	*Nonlinear*
Classification	Perceptron	Kernel perceptron
Classification	Support vector machine	Neural networks
Ranking	Passive aggressive	Boosting
Regression	Linear regression	k-nearest neighbor

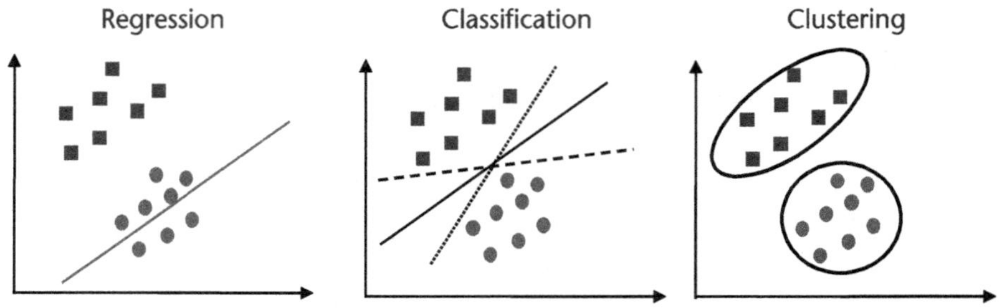

Figure 3.6 Linear regression and classification.

all the points is the optimal. Finally, if the data can be separated, the respective data are organized into *clusters*.

Figure 3.7 demonstrates why linear methods are limited when there is complexity in the data. Feature extraction from the data could reduce outliers, but typically the feature space is multidimensional and complex. The top left of Figure 3.7 highlights that while the data can be separated into classes, multiple decision boundaries might be needed. Likewise, the top right shows where finding a simple linear cluster might not be possible if the features extracted from the same sensor-target pair do not overlap another cluster. The lower left shows an example of a nonlinear function that provides the best separation. Finally, on the lower right, and representing most cases (even with a potential 100% classification), the boundary is highly nonlinear. To get to such a nonlinear representation, it is important to derive the mathematics that extends linear machine learning techniques to nonlinear problems.

3.2.1 Linear Classifier

There are a variety of linear functions/models such as naive Bayes, perceptrons, and regression that seek a relation $y= wx + b$. A linear model applies a linear transformation with matrix w applied to the input x. The models solve problems that are linear in structure where the decision boundary is a hyperplane. Typically, the assumption is that the labels are (nearly) linear-separable. Popular methods for linear classification include (1) linear regression, (2) logistic regression, (3) perceptron, (4) decision tree, (5) random Fforest, and (6) support vector machine (SVM).

3.2.1.1 Perceptron

The perceptron is a binary classifier, invented in 1957 at Cornell Aeronautical Laboratory by Frank Rosenblatt [5]. It is also known as single-layer perceptron. The

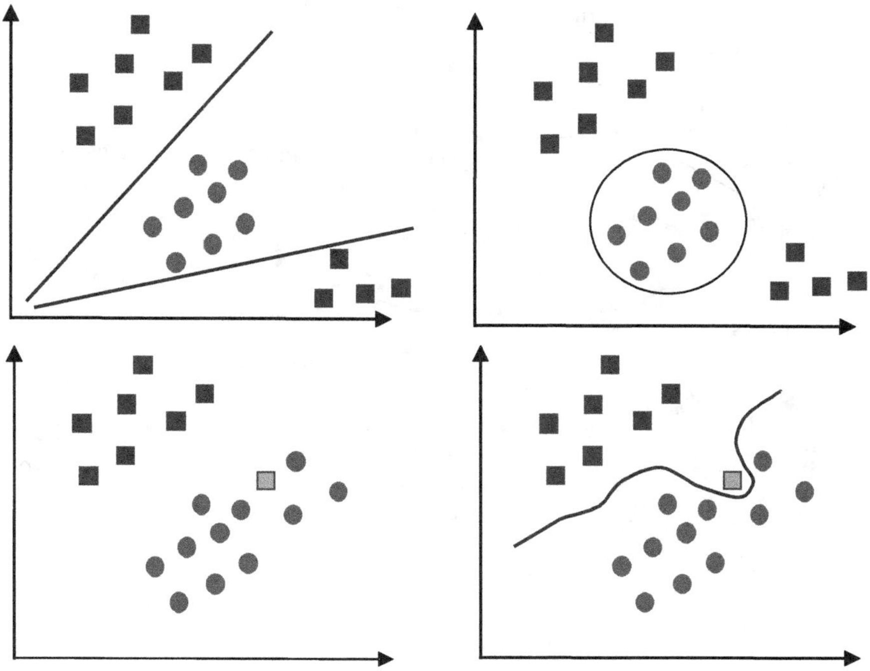

Figure 3.7 Nonlinear classification.

perceptron is the simplest form of a *feedforward neural network* (FNN). The perceptron algorithm attempts to divide the input space into two halves with a single line. However, the perceptron does not consider the entire data set at the same time, but rather looks at one example at a time, processes it, then moves onto the next example (online). The perceptron is like the ANN, with a weighted summation of the FNN and then an activation function (such as step function for a binary decision) shown in Figure 3.8. In Figure 3.8, and inherent in most ML techniques, is the extraction of features using a feature map Φ(x), where *x*s are the features such as intensity values depicted from SAR and EO imagery. The feature extraction functions $\phi_1 \ldots \phi_n$ determine the features to form an independent variable set **x**= $[x_1, \ldots, x_n]$. Each variable is assessed a weight of importance $w_i \in i = 1, \ldots n$, for collective summation. Since the perceptron is a linear classifier, a line seeks to separate the data into clusters (as shown later in Figure 3.9).

SAR Example

Due to the availability and volume of SAR data for training, as soon as the MSTAR imagery was accessible by researchers, the first applications of the ANN (and/or perceptron) sought nonlinear extensions that yielded a multilayer perceptron (MLP). For example, to develop the NL MLP, a linearized version was developed based on a constant false alarm rate (CFAR) to construct a quadratic gamma discriminator—a nonparametric classifier based on the local image intensity [6]. Results demonstrated that the linear classifier had a significant number of false alarms (~25%) at any detection rate, while the ML approach was able remove the false alarms for a probability of detection (PD) of 50% at a probability of false alarm rate (PFA) at 10% and PD = 90% at a PFA of 25%. Hence, a perceptron-based

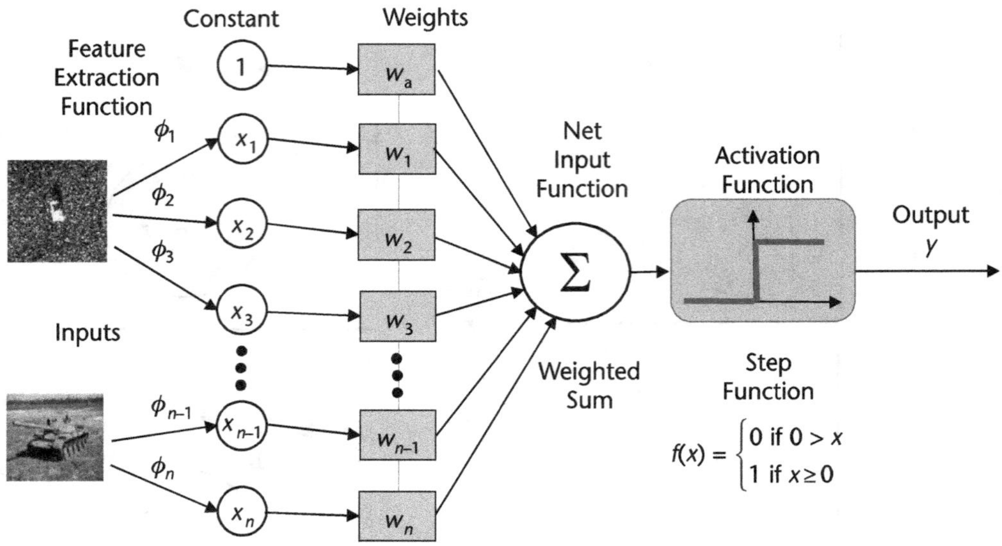

Figure 3.8 Perceptron algorithm.

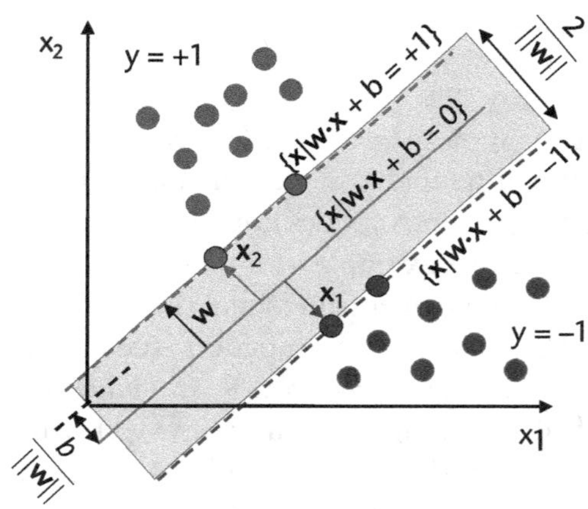

Figure 3.9 Support vector machine boundaries.

ANN has limited performance for SAR classification due to the challenge of finding an accurate distribution of the data that could yield a linear classifier. One of the prominent baseline methods applied to the MSTAR data is the SVM.

3.2.1.2 Support Vector Machine

The SVM method is one of the foundational ML methods used for image classification, character recognition, and many other applications seeking to partition a feature space. Additionally, many variants have been developed and applied to radar applications (e.g., signals, imagery, and communications). SVM uses a global optimization for the loss function and various ML toolboxes exist that include the SVM. The SVM works similarly to the perceptron in that it attempts to divide the input space with a linear separator. The main difference is that the SVM works to maximize the margin between the two classes in the input space and therefore may still update even when the data is nonlinear, yet the perceptron assumes that the data is linear. Hence, the SVM seeks to divide regions of the data with a linear boundary that separates data into categories.

The SVM includes multiple supervised methods for classification/regression. Each method has a different kernel before summation and activation (see Figure 3.14 in the Kernel Perceptron). A *kernel function* computes the inner product (convex optimization) of the pairs of feature data to determine a similarity measure. The SVM is binary and nonprobabilistic, although extensions to the SVM support probabilistic reasoning. The kernel affords nonlinear classification by mapping inputs to a high-dimensional feature space to perform a linear classification on the features.

An infinite number of hyperplanes exist that can separate two classes. The optimal binary hyperplane, is the center line in Figure 3.9, is defined as the hyperplane which has the maximum margin, $p(\mathbf{w},b)$ and is defined by the following set:

$$\{\mathbf{x} | \mathbf{w} \cdot \mathbf{x} + b = 0\} \tag{3.2}$$

The samples closest to the separating hyperplane are the support vectors. The hyperplanes through the support vectors are defined by the following sets:

$$\{\mathbf{x} | \mathbf{w} \cdot \mathbf{x} + b = -1\} \text{ and } \{\mathbf{x} | \mathbf{w} \cdot \mathbf{x} + b = +1\} \tag{3.3}$$

The margin is the minimum distance for class –1 to the separating hyperplane plus the minimum distance for class +1 to the separating hyperplane. The distance of a sample to the separating hyperplane is:

$$d(\mathbf{w};\mathbf{w},b) = \frac{|\mathbf{w} \cdot \mathbf{x} + b|}{\|\mathbf{w}\|} \tag{3.4}$$

Therefore, the margin can be defined by:

$$p(\mathbf{w},b) = \min_{\{\mathbf{x}_i, y_i = -1\}} d(\mathbf{x}_i, \mathbf{w}, b) + \min_{\{\mathbf{x}_j, y_j = +1\}} d(\mathbf{x}_j, \mathbf{w}, b) = \frac{2}{\|\mathbf{w}\|} \tag{3.5}$$

In addition, the objective function is minimized by:

$$\Phi(\mathbf{w}) = \frac{1}{2}\|\mathbf{w}\|^2 \tag{3.6}$$

Some measure of algorithm capacity for the maximum margin classifier is needed in order to use the SVM principle. Many capacity measures have been proposed, such as annealed entropy, fat shattering dimension, and cover numbers.

For the classifier, the vector classifier (VC) dimension will be used as an example. Referring to the Figure 3.10, the VC dimension for a hyperplane is

$$h \le \min\left[R^2, A^2, n\right] + 1 \tag{3.7}$$

where R is the radius of the hypersphere enclosing all the training samples, $1/A$ is the minimum distance the hyperplane can approach a sample point, and n is the dimension of the space. Typically, R is smaller than the dimension of the space, especially when kernels are used to map the input space into a higher dimensional feature space. Maximizing the margin also minimizes the VC dimension:

$$\begin{aligned} \min(h) &\leftarrow \min\left[R^2, A^2\right] \\ &\leftarrow \max(1/A) \\ &\leftarrow \min\left[\frac{1}{2}\|\mathbf{w}\|^2\right] \end{aligned} \tag{3.8}$$

Support vector machines have one more major property. Various kernel functions have been developed to map the input features to a higher dimensional feature space. The kernel function is important because it allows complexity of the decision surface to be more complex in the input space, where the decision surface is still an optimal separating hyperplane in the higher dimensional feature space. For example, since the VC dimension is determined by the min $[R^2, A^2, n]$, the VC dimension and, therefore, the generalization of the algorithm is likely not affected by the *curse of dimensionality*. A hyperplane is used in the higher dimensional feature space to determine the output. In relation to the other techniques, the addition of support vectors through kernels determines various boundaries.

SAR Example

Bryant and Garber [7] compared the SVM versus the standard CFAR method for minimizing the mean square error (MSE). Using the standard MSTAR data set, the SVM distinguishes the target classes (e.g., BMP versus T72). As the SVM is a linear

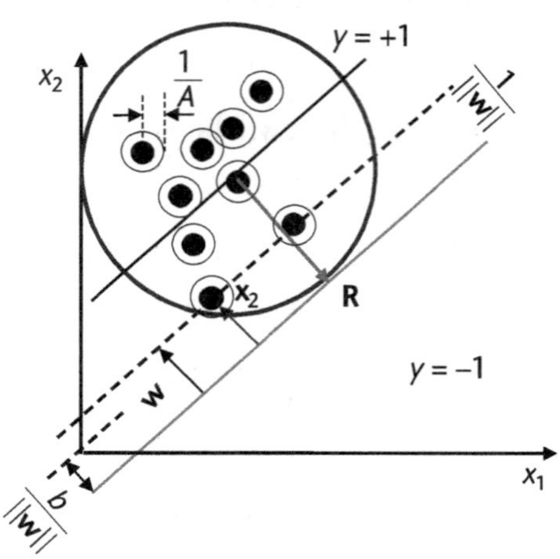

Figure 3.10 Vector classifier.

classifier, the seminal paper by Bryant and Garber for SAR classification departs from classical signal processing detection methods and establishes the new SAR ATR methods using machine learning. The research using the MSTAR data set is a comparative example for the rest of the techniques compared in this book. Additionally, Bryant et al. [7] also established an evaluation method of including both known (e.g., labeled targets) and unknown targets (e.g., confusers that were not part of the training for classification), as shown in Table 3.4.

Since the SVM typically employs a kernel function (see Figure 3.14), then the choice of the kernel determines whether the classifier is linear or nonlinear. For the study, Bryant and Garber [7] use a polynomial kernel:

$$K(\mathbf{x},\mathbf{x}_i) = \left[(\mathbf{x} \cdot \mathbf{x}_i)\right]^d \tag{3.9}$$

Using a polynomial kernel, the results obtained for the MSTAR data for PD and PCC are repeated in Table 3.5. One interesting analysis is that the standard MSE suffers from the image processing pipeline of a higher detection rate that allows more data (useful and confusing) to be presented to the classifier that reduces performance. On the other hand, for the SVM, once the detection of the target has been performed, the classification rate increases. It is also noted in results presented in Chapter 9 that a more comprehensive analysis includes the receiver operating characteristics (ROC) curve that plots the probability of hit rate (PHR) versus the probability of false alarm rate (PFA), where the hit is detection, recognition, or classification. Using the ROC, the area under the curve (AUC) provides a comparative example, as well as a F-score that determines the overall metric.

While the SVM is simple to implement and generally interpretable, Bryant [7] concludes with these observations on deployment limitations:

1. The percentage of the no-decision boundary is difficult to control;
2. The SVM requires that the collected data be roughly aligned from the given registration;
3. Given that the targets are taken from different poses (36 at 10° angles), then the MSE requires binning for decisions, while the SVM benefits from no binning as it is N^*B where N is the number of targets and B is the number of bins, so as to get a decision;
4. The MSE requires a scaling function for the data to shape the distribution to a normal function, while the SVM scaling between –1 and 1 had a slight increase in performance.

Table 3.4 MSTAR Data Categories in the Bryant and Garber Test

Train and Test	*Test*	*Confusers*
BMP2-c21	BMP2-9564	2S1
BTR70-c71	BMP2-9566	D7
T72–132	T72-812	
	T72-s7	

Table 3.5 MSTAR Results from the Bryant and Garber Test

Classification	*PD*	*PCC*
MSE	90%	88.9%
SVM polynomial $d = 1$	85.9%	92.8%
SVM polynomial $d = 2$	85.8%	93.4%

Another example of a baseline approach using the SVM on the MSTAR data was from Zhao [8]. The results in Table 3.6 highlight that the SVM outperformed the standard classifiers at that time, although the implementation only used the three target sets available.

As discussed throughout the text, there are many extensions. The first is *data fusion* (or image fusion), in which data is collected from multiple sources. The inclusion of other sensor sources, when combined for a classifier can help in increasing accuracy. The second variation is gathering *dynamic information.* Using the MSTAR data over the 32-target set and recovering the high range resolution (HRR) radar information, the SVM results using a target tracking method also increased performance [9]. Finally, *contextual information* can be used to better resolve the quality of the features through conditioning, such as using environment information with Moving-SAR data. These ideas will be shown with the appropriate ML method, data set, and comparative results.

As fundamental to most ML methods, regression approaches for data analytics determine performance.

3.2.1.3 Linear Regression

Linear regression models the relationship between a dependent variable and set of independent variables where the input data and the output data are numeric values. The dependent variable Y is a scalar function and the explainable variable if singular, X, is simple linear regression, while from many variables it is multiple linear regression. For multiple independent variables and multiple dependent variables, then *multivariate linear regression* (MLR) supports modeling for prediction. The typical and most popular method for MLR is ordinary least square (OLS) which works to minimize the sum of the squared residuals. Given a regression line through the data, MLR calculates the distance from each data point to the regression line, square that distance, and sum the distances for each point (i.e., the sum of the squared residuals). The important regression techniques include: linear, logistic, polynomial, stepwise, ridge (ℓ_2 regularization), and least absolute shrinkage and selection operator, Lasso (ℓ_1 regularization).

Table 3.6 MSTAR Data Results Based on the Zhao et al. [8] Implementation

Classifier	*PCC (%)*	*Confuser Rejection (%)*
Template matching	90.4%	53.5%
Principle component analysis with multiresolution (PCA-M)	89.7%	60.0%
Quadratic mutual information (QMI)	94.1%	54.5%
Support vector machine (SVM)	94.9%	68.8%

Figure 3.11 depicts the method of regression analysis. Given a data set $\{y_i, x_{i1}, \ldots, x_{ip}\}$ for $i = 1, \ldots n$ units, a linear regression model assumes that the relationship between the dependent variable y and the p-vector of independent variables **x** is linear. This relationship is modeled through *error variables*—an unobserved random variable reflecting noise of the uncertainty (e.g., error) of the linear relationship between the dependent variable and regressors. A linear the model takes the form:

$$y_i = \beta_0 1 + \beta_1 x_{i1} + \ldots + \beta_p x_{ip} + \varepsilon_1 = \mathbf{x}_i^T \beta + \varepsilon_i; \quad i = 1, \ldots n \tag{3.10}$$

where ^{T} denotes the transpose and $\mathbf{x}_i^T\beta$ is the inner product between the vectors $\mathbf{x}_i$ and β. For a simple linear regression, β_0 is the y intercept and is the β_1 slope of the line, as shown in Figure 3.11.

The n equations can be stacked together and written in matrix notation as

$$\mathbf{y} = \mathbf{X}\beta + \varepsilon \tag{3.11}$$

The OLS method minimizes the sum of squared residuals and leads to a closed-form expression for the estimated value of the unknown parameter vector β:

$$\hat{\beta} = \left(\mathbf{X}^T\mathbf{X}\right)^{-1}\mathbf{X}^T\mathbf{y} \tag{3.12}$$

where **y** is a vector whose *i*th element is the *i*th observation of the dependent variable, and *X* is a matrix whose *ij* element is the *i*th observation of the *j*th independent variable. The expression in (3.12) is often called the *pseudoinverse*. The estimator is unbiased and consistent if the errors have finite variance and are uncorrelated with the regressors:

$$\mathrm{E}\left[\mathbf{x}_i\varepsilon_1\right] = 0 \tag{3.13}$$

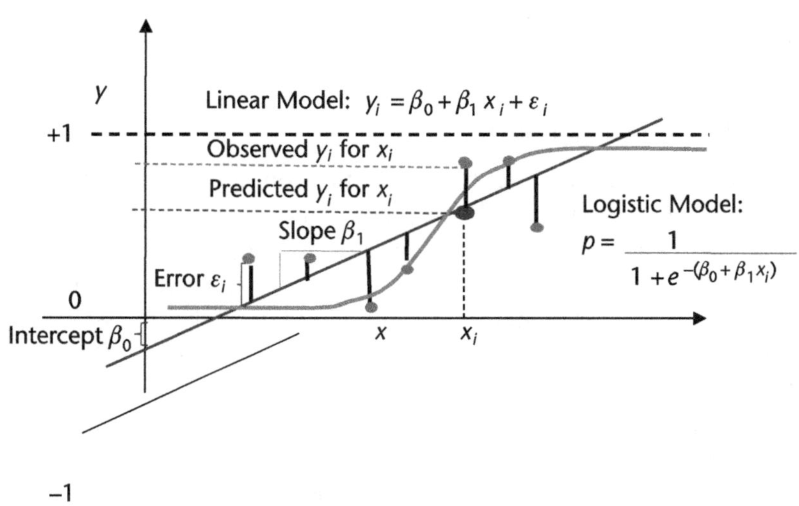

Figure 3.11 Linear regression, the observations (dots) are assumed to be the result of random deviations (vertical line from dots) from an underlying relationship (a linear line minimizing the length of the vertical lines) between a dependent variable (y) and an independent variable (x).

where **X**. OLS is also efficient under the assumption that the errors have finite variance and $E[\varepsilon_i^2|\mathbf{x}_i]$ does not depend on i. That the errors are uncorrelated with the regressors will generally be satisfied in an experiment, but in the case of observational data, it is difficult to exclude the possibility of an omitted covariate that is related to both the observed covariates and the response variable. The existence of such a covariate will generally lead to a correlation between the regressors and the response variable, and hence to an inconsistent estimator of β. The condition of homoscedasticity can fail with either experimental or observational data. If the goal is either inference or predictive modeling, the performance of OLS estimates can be poor if multicollinearity is present, unless the sample size is large.

Logistic regression predicts the probability that an outcome **y** can only have two values (e.g., {0 1}, {−1 +1}). The prediction is based on features, categories, of numerical information. From Figure 3.11, the linear regression prediction does not support the binary value prediction {0, 1} for all values of x_i, does not require equal variance between groups, and the errors ε_i for a binary system will not be normally distributed errors. The logistic regression uses the natural logarithm of the target variable odds to maintain values between 0 and 1, while the linear regression uses the probability.

SAR Examples

Line regression (LR) is used in many cases for statistical analysis of SAR data. One example [10] determines the forest vegetation from SAR imagery using LR to determine the tree stems based on the backscattering amplitudes. Since LR supports other techniques, another example employs a Markov random field (MRF)-based segmentation using a multilevel logistic (MLL) regression model for distribution labeling of sea ice from SAR [11]. Finally, many examples using LR support data fusion. Data fusion consists of pixel-level, feature-level, and decision-level fusion [12]. For feature-level fusion, the SVM is an example as shown in Figure 3.8. An example of decision-level fusion is classifier fusion where multiple classifiers process that data, and the outputs are fused. Hence, different versions on LR classifiers (e.g., principle component analysis) are resolved on the MSTAR data to enhance decision-based classification [13].

3.2.1.4 Passive-Aggressive

One recent technique to extend classification and regression for an online perceptron is a set of methods called *passive-aggressive (PA) learning*. [14] Assume a binary classification in which the decision boundary is yes (+1) or no (−1) as shown in Figure 3.11. An online streaming set of data (e.g., determination of a target from a set of moving SAR images) requires updating the decision process. If the data comes from a similar distribution (e.g., same look angle of a SAR image), then the classifier will learn the slow data variations. However, if there is a delay in the SAR image collection, a new SAR image might have a different distribution (e.g., from Gaussian to heavy-tail) from which the system would forget the previous distribution weights w and learn the new distribution. Using the hinge-loss function (the same from a support vector machine), then the loss function l (bounded between 0 as a match and $C > 0$ as wrong) will instantiate the PA update. If the current weight vector determines the sign and it is correct, then the loss is 0 and the current weight is used from which the correct classification is *passive*. However, if a new point (90° orthogonal) is determined, then the dot product is negative (–1) and the conflict arises besides the label is +1. Hence, the weight update is *aggressive* as it seeks agreement between not losing the previous learned information and seeking $\ell = 0$ to correctly classify the object. A high C is more aggressive (risk of destabilization in presence of noises), while low C is passive that leads to adaptation. For robustness, a balance is needed to avoid rapid changes leading to misclassifications. To support regression, advances use learning methods with a slack variable.

SAR Example

In [15], the authors develop a multiview passive-aggressive method for Polarimetric synthetic aperture radar (PolSAR). The data comes from three real PolSAR data sets, named sanf1, sanf2, and oberp; the subsets of the San Francisco Bay data and the Oberpfaffenhofen Area data, respectively. The images are collected by AIRSAR and E-SAR, respectively, and can be obtained from the website: http://earth.eo.esa.int/polsarpro/datasets.html. Many methods were tested for feature-level fusion over (1) polarimetric features, (2) color-texture features, (3) fused features, (4) AdaPA: two-view PA, and (5) OMVPA: o nline multiview PA.

The results show that the OMVPA method has the lowest average error rate at 1%, compared to the fused-feature PA (2%) and AdaPA (5%). Additionally, the error rate of OMVPA consistently decreases with the data stream size, and the average mistake rates of these methods tend to be stable at 30,000 samples. As most data is complex, then the nonlinear classifiers demonstrate better performance.

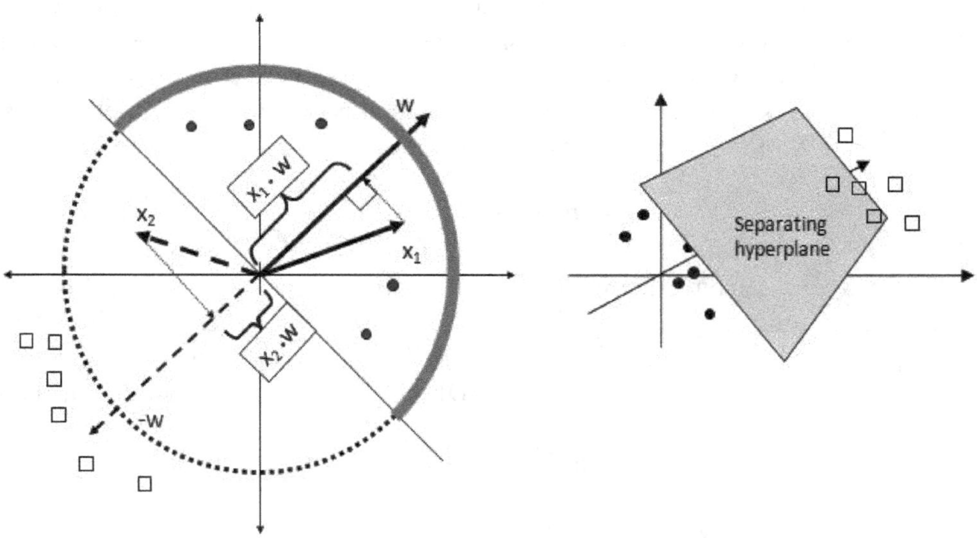

Figure 3.12 Three variants of the passive-aggressive algorithm for binary classification.

3.2.2 Nonlinear Classifier

Nonlinear NNs are designed to further learn the patterns based on extensions to a linear NN. Using a set of linear functions can support a nonlinear analysis such as the ANN and the SVM. Many natural processes are extremely complex, which require an assessment of the various nonlinear effects. For learning tasks such as complex relations, some unknown hidden variables (or ones that simply cannot be observed), need to be resolved. Hence, current ML methods apply multiple layers for uncovering the nonlinear relationships between the variables and the decision output. Some of the popular methods include (1) multilayer (kernel) perceptron (MLP), (2) modular neural network (MNN), (3) radial basis function network (RBF), (4) convolutional neural network (CNN), (5) recurrent neural network (RNN), and (6) Autoencoder (AE).

3.2.2.1 Kernel Perceptron

The kernel perceptron works similarly to the perceptron in that it attempts to divide the input space. The key idea is that the algorithm uses a kernel function to transform the input space. The transformed space does not have to have the same dimensionality as the input space, and in many cases it has a higher dimensionality. The kernel perceptron attempts to construct a n–1 dimensional hyperplane in the n-dimensional transformed space that correlates to a nonlinear separation in the original input space. The kernel perceptron is an effective algorithm and the update rules are the same as the classic perceptron algorithm except for the transform itself.

$$\hat{y} = \text{sgn}(w_1x_1 + w_2x_2 + \ldots + w_nx_n) = \text{sgn}(\mathbf{w} \cdot \mathbf{x}) \tag{3.14}$$

The perceptron method is error-driven learning as it iteratively improves a model with streaming training samples by using a supervised signal to determine if there is an incorrect classification requiring a model update. The standard perceptron method is a linear binary classifier (see Section 3.2.1.1). A vector of weights **w** (and optionally an intercept term b) classifies a sample vector $\mathbf{x} = (x_1, \ldots, x_n)$ as a binary class sign one $y_i = \{+1\}$ or sign class minus one $y_i = \{-1\}$. A linear classified is a vector **X** of the same dimension as **x**, used to make the estimated prediction $\hat{y}$:

$$\hat{y} = \text{sgn}(\mathbf{w} \cdot \mathbf{x}) = \text{sgn}(\mathbf{w}^T\mathbf{x}) \tag{3.15}$$

where a zero is arbitrarily mapped to one or minus one $y_i = \{-1, +1\}$. Visually, $\mathbf{x} \cdot \mathbf{w}$ is the distance where **x** is projected onto **w**, as shown in Figure 3.13(a). The line perpendicular to **w** divides the vectors classified as positive from the vectors classified as negative. In three dimensions (3D) the line is a plane Figure 3.13(b) and in 4D, the line is a hyperplane.

To move origin away from the origin, a modifcation is made:

$$\hat{y} = \text{sgn}(1x_0 + w_1x_1 + w_2x_2 + \ldots + w_nx_n) \tag{3.16}$$

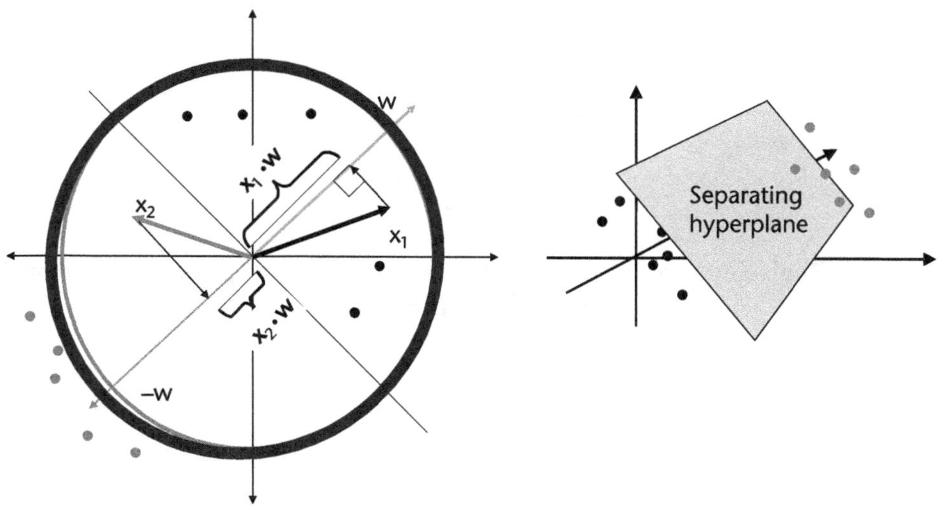

Figure 3.13 Kernel separation: (a) projection of x onto w, and (b) hyperplane. The line perpendicular to w divides the vectors classified as positive from the vectors classified as negative.

ALGORITHM: Perceptron
INPUT: Number of parameter features $p > 0$
Number of parameter iterations $T > 0$
For $i = 1, 2, \cdots, T$
- Receive instance: $\mathbf{x}_i \in \mathbb{R}^n$
- Receive correct label: $y_i \in \{-1, +1\}$
- Precict: $\hat{y}_t = \text{sign}(\mathbf{w}^T \mathbf{x}_i)$,
- If $\hat{y} \neq y_i \mathbf{x}_i$,
- Update:

$\mathbf{w} \leftarrow \mathbf{w} + y_i \mathbf{x}_i$
$\mathbf{w}_{i+1} = \mathbf{w}_i + y_i \cdot \mathbf{x}_i$

By contrast with the linear models learned by the perceptron, a *kernel machine* is a nonlinear classifier that stores a subset of its training examples $\mathbf{x}_i$, associates with each a weight α_i, and makes decisions for new samples $\mathbf{x}'$ by evaluating the data for the decision boundary. Using the dual form from the perceptron, a kernelized version is:

$$\mathbf{w} = \sum_{i=1}^{n} \alpha_i y_i \mathbf{x}_i \tag{3.17}$$

where α_i is the number of times $\mathbf{x}_i$ was misclassified, forcing an update $\mathbf{w} \leftarrow \mathbf{w} + y_i\mathbf{x}_i$. The kernelized perceptron algorithm cycles through samples as the perceptron, making predictions, but instead of storing and updating a weight vector $\mathbf{w}$, it updates a mistake counter vector α. Hence, the prediction of the dual perceptron is of the form:

$$\begin{aligned}\hat{y} &= \operatorname{sgn}\left(\mathbf{w}^{\mathrm{T}}\mathbf{x}\right) \\ &= \operatorname{sgn}\left(\sum_{i=1}^{n} \alpha_i y_i \mathbf{x}_i\right)^{\mathrm{T}} \mathbf{x} \\ &= \operatorname{sgn}\sum_{i=1}^{n} \alpha_i y_i \left(\mathbf{x}_i \cdot \mathbf{x}\right)\end{aligned} \tag{3.18}$$

Replacing the dot product (•) in the dual perceptron by an arbitrary kernel function achieves the effect of a feature map Φ without computing Φ($\mathbf{x}$) explicitly for any samples.

$$\hat{y} = \operatorname{sgn}\sum_{i=1}^{n} \alpha_i y_i K\left(\mathbf{x}_i \cdot \mathbf{x}\right) \tag{3.19}$$

Where K is some kernel function. Formally, a kernel function is a nonnegative semidefinite kernel, representing an inner product between samples in a high-dimensional space, as if the samples had been expanded to include additional features by a function Φ: $K(\mathbf{x}, \mathbf{x}') = \Phi(\mathbf{x}) \bullet \Phi(\mathbf{x}')$. Intuitively, it can be thought of as a similarity function between samples, so the kernel machine establishes the class of a new sample by weighted comparison to the training set. Each function $\mathbf{x}' \mapsto K(\mathbf{x}_i, \mathbf{x}')$ serves as a basis function in the classification.

ALGORITHM: Kernel Perceptron
INPUT: Number of parameter features $p > 0$
    Number of parameter iterations $T > 0$
    Number of training samples $n > 0$
For $i = 1, 2, \cdots, \mathrm{T}$
- Receive instance: $\mathbf{x}_i \in \mathbb{R}^n$

For $j = 1, 2, \cdots, n$
- Receive correct label: $y_{\mathrm{i}} \in \{-1, +1\}$
- Precict: $\hat{y}_t = \operatorname{sgn}\sum_{i=1}^{n} \alpha_i y_i K\left(\mathbf{x}_i, \mathbf{x}_{\mathrm{j}}\right)$
- If $\hat{y} \neq y_i$,
- Update: (Perform by incrementing the mistake counter)
  $a_j \leftarrow a_j + 1$
- Return

The voted perceptron algorithm of Freund and Schapire also extends to the kernelized case [16], giving generalization bounds comparable to the kernel SVM. The sequential minimal optimization (SMO) algorithm used to learn SVMs can also be regarded as a generalization of the kernel perceptron [17]. A kernelized version of the SVM (which is depicted in Figure 3.14):

$$y = f(\mathbf{x}) = \sum_{i=1}^{n} w_{\mathrm{i}} \cdot K\left(\mathbf{x}, \mathbf{x}_i\right) + b \tag{3.20}$$

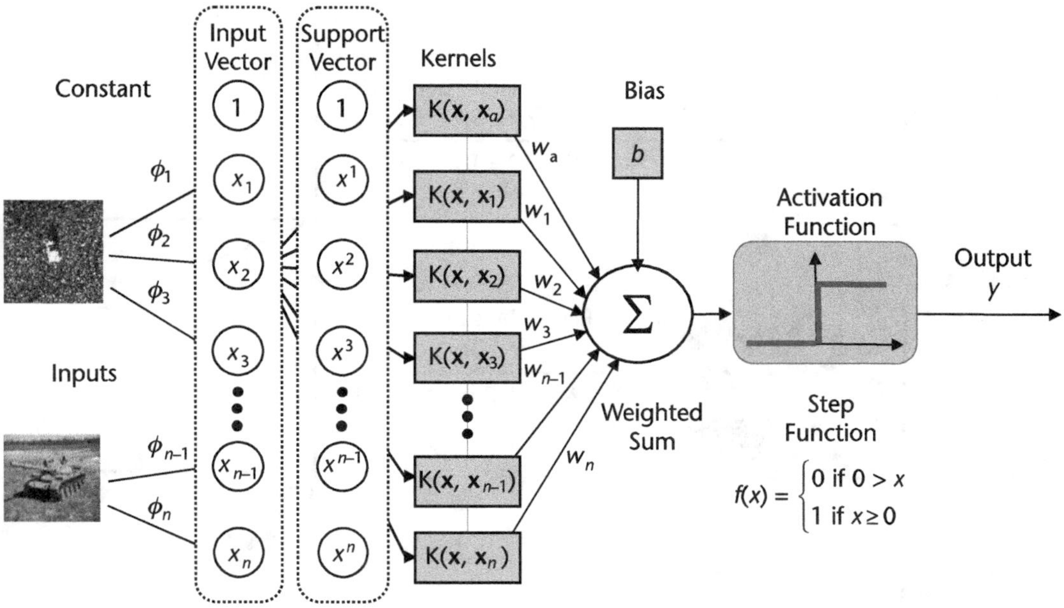

Figure 3.14 Kernelized version of the support vector machine.

And replacing the weights w_i with Lagrange multipliers α_i:

$$y = f(\mathbf{x}) = \sum_{i=1}^{n} \bar{\alpha}_i \cdot K(\mathbf{x}, \mathbf{x}_i) + b \tag{3.21}$$

Example kernel functions include:

- Polynomial (of degree exactly d): $K(\mathbf{x}, \mathbf{x}_i) = [(\mathbf{x} \cdot \mathbf{x}_i)]^d$, where d is user defined
- Polynomial (of degree up to d): $K(\mathbf{x}, \mathbf{x}_i) = [(\mathbf{x} \cdot \mathbf{x}_i) + 1]^d$, where d is user defined
- Gaussian (normalized): $K(\mathbf{x}, \mathbf{x}_i) = \exp[-(\|\mathbf{x} - \mathbf{x}_i\|)/2]$
- Gaussian (σ is variance): $K(\mathbf{x}, \mathbf{x}_i) = \exp[-(\|\mathbf{x} - \mathbf{x}_i\|)/(2\sigma^2)]$, where σ is data defined
- Radial basis function (normal): $K(\mathbf{x}, \mathbf{x}_i) = \exp[-\gamma(\mathbf{x} - \mathbf{x}_i)^2]$, where γ is user defined
- Radial basis Function (σ is variance): $K(\mathbf{x}, \mathbf{x}_i) = \exp[-(\|\mathbf{x} - \mathbf{x}_i\|)/(2\sigma^2)]$, where σ is data defined
- Two-layer neural network: $K(\mathbf{x}, \mathbf{x}_i) = \tanh[b(\mathbf{x} \cdot \mathbf{x}_i) - c]$, where b and c are user defined
- Sigmoid (η is weight, ν is bias): $K(\mathbf{x}, \mathbf{x}_i) = \tanh[\eta(\mathbf{x} \cdot \mathbf{x}_i) + \nu]$, where η and ν are data defined

Figure 3.15 demonstrates how various kernels are determined as nonlinear classifiers. The polynomial kernel provides a boundary separation while the radial basis kernel does a nonlinear transformation. Typically, the RBF kernel is the contemporary method used with the SVM for classification.

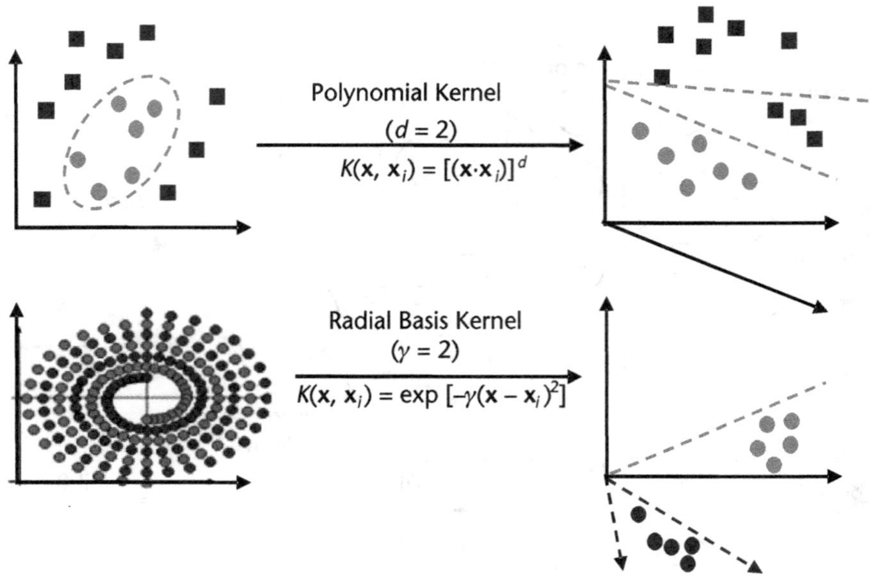

Figure 3.15 Kernel projections.

The kernel perceptron has many benefits such as high accuracy with medium interpretability, discerning complex and high-dimensional classification, and efficient training for static scenarios. However, there are known limitations for this method. For situations with sparse data, kernel machines need modification. The kernel perceptron does not regularize easily, making it vulnerable to overfitting.

SAR Example

An extension to the ANN work looked at variations of the SVM with various kernels [18]. The RBF and SVM improved in a standard perceptron hyperplane, where the authors compared a global versus local discriminant to which they ascribed the notions of empirical and structural risk variations. The local discriminant showed improvement over the global discriminant functions for SAR classification as shown in Table 3.7 for the 3 target MSTAR data set.

As the kernel methods with the SVM continue to be a baseline method, additional papers have revisited the discussion with the MSTAR data [19]. One comparable method is the k-nearest neighbor.

3.2.2.2 K-Nearest Neighbor

The K-nearest neighbor (KNN) is a nonparametric method (i.e., distribution free) of instance learning that works on entire data set at once to discern a pattern in feature space that determines the k closest examples. k is a specified positive integer, which

Table 3.7 Topology of Classifier Versus Training Criterion: Classification Rate

	Empirical Risk Minimization	*Structural Risk Minimization*
Global Discriminant	Perceptron (93%)	Optimal hyperplane (93%)
Local Discriminant	Radial basis function (95%)	Support vector machine (95%)

is usually relatively small. The KNN does not learn any model—rather the model is itself the training set. For each new instance, the algorithm searches through the entire training set, calculating the difference between the new instance and each training model. A classic example is the Voronoi diagram (Figure 3.16). which partitions regions based on a distance to the features. Once the distance regions are determined, then a corresponding boundary set is available for classification.

- For *classification*, the output is the class with the k most similar neighbors;
- For *regression*, the output value is based on the mean or median of the k most similar instances.

Both for classification and regression, the KNN assigns weights to the features, so that the nearer neighbors contribute more to the average than the more distant ones. For example, a common weighting scheme consists of giving each neighbor a weight of $1/d$, where d is the distance to the neighbor. An example is shown in Figure 3.17. The center test sample dot should be classified either into triangles or circles. If $k = 3$ (solid line circle), it is assigned to the triangles, because there are three triangles and no circles inside the inner circle. If $k = 20$ (dashed line circle), it is assigned to the **red** circles, as there are 11 circles vs. 9 triangles inside the outer circle.

The KNN works to classify the training data point x_q (center dot).

- Given a query instance q to be classified;
  - Let $x_1, \ldots, x_k$ be the k training instances in training set $T = (x, f(x_i))$ nearest to q;
  - Return;

$$\hat{f}(q) = \underset{v \in V}{\arg\max} \sum_{i=1}^{n} \delta\left(v, f\left(x_i\right)\right) \tag{3.22}$$

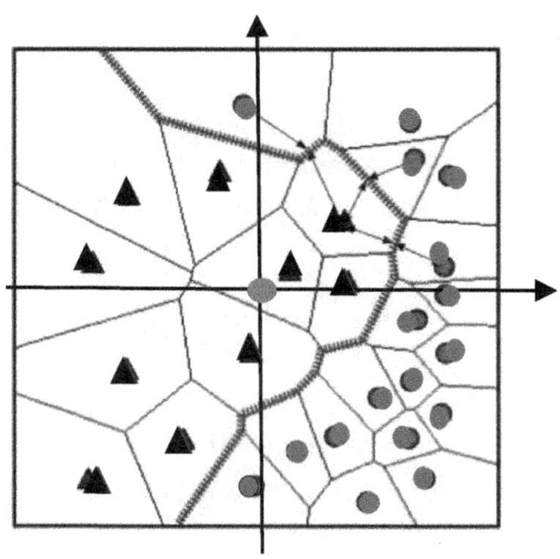

Figure 3.16 K-nearest neighbor from the Voronoi diagram.

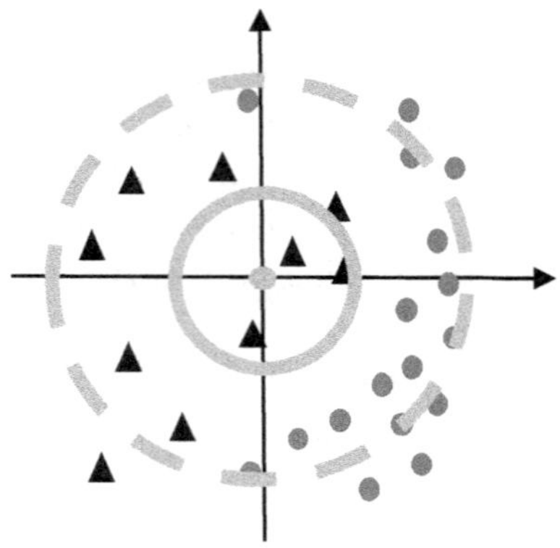

Figure 3.17 Example of KNN classification.

- where V is the finite set of target class values, and $\delta(a,b)=1$ if $a=b$, and 0 otherwise (Kronecker function);
- then, the k-NN algorithm assigns to each new query instance the majority class among its k nearest neighbors.

A distance-weighted KNN uses a difference between the test point x_q and the training data x_i :

$$\hat{f}(q) = \arg\max_{v \in V} \sum_{i=1}^{n} \frac{1}{d\left(x_i, x_q\right)} \delta\left(v, f\left(x_i\right)\right) \tag{3.23}$$

There are many distance functions, such as the city block (Manhattan), Chebyshev, Minkowski, quadratic, correlation, and chi-square. Two popular methods include:

$$\text{Euclidian distance: } d\left(x_i x_q\right) = \sqrt{\left(x_1 - x_q\right)^2 + \left(x_2 - x_q\right)^2 + \ldots + \left(x_2 - x_q\right)^2} \tag{3.24}$$

$$\text{Mahalanobis: } d\left(x_i, x_q\right) = [\det \mathrm{V}]^{(1/m)} \left(x_i - x_q\right)^{\mathrm{T}} V^{-1} \left(x_i - x_q\right) \tag{3.25}$$

where *det* is the determinant and T is the transpose operator.

A practical method is estimating the density at a point x as the reciprocal of the average of the distances to the k nearest neighbors of x:

$$f(x) = \frac{1}{\frac{1}{K} \sum_{n \in N} d(x, n)} \tag{3.26}$$

where N denotes the k nearest neighbors of x, and n is an element of N. Note that increasing the value of k takes more information about a point into consideration. Setting k=1, the estimate is completely local, which will result in producing a high number of clusters. Larger values of k will yield increasingly global estimates, while decreasing the granularity of the result, so that fewer clusters are produced.

KNN makes predictions just-in-time by calculating the similarity between an input sample and each training instance. In general, the KNN is similar to the SVM with a Gaussian kernel. Benefits of KNN include: no assumptions about data distributions, simple (i.e., easy to explain and understand/interpret), accurate (i.e., achieves reasonable results), and versatile (i.e., useful for classification or regression). Limitations of KNN include: computationally expensive, slow run-time, and high memory requirement.

SAR Examples

While the KNN has been applied to SAR target classification, it continues to be a baseline approach along with the SVM. For example, Falcon et al. [20] developed a machine learning fuzzy KNN classifier to determine ship traffic behaviors through information fusion of SAR contact data and automatic identification systems (AIS) reports. In [21], five gray-level co-occurrence matrix (GLCM) texture features were used {contrast, correlation, energy, homogeneity, and entropy} to develop a fusion of texture and shape features for SAR target classification. The geometrical shape features included connected components, area, centroid, axis length, eccentricity, orientation, and convex area. The five stages included preprocessing, salient region detection, feature extraction, feature fusion, and classification. Comparative results include SVM (linear, quadratic, cubic, Gaussian) and the KNN (standard, cubic, weighted). Three results revealed from the experiment over the MSTAR data include that shape features have better classification, SVM was better than KNN, and the fusion was slightly better than the shape features. Another study [22] compared methods using Bayesian decision fusion with the MSTAR data with these results. The various methods were implemented including the principal component analysis (PCA), attributed scattering centers (ASC) and dominant scatter analysis (DSA), as shown in Table 3.8.

3.2.2.3 Boosting

Boosting is a technique that combines several weak classifiers in order to produce a strong classifier. The designation of weak classifier implies that many classifiers are dedicated to identifying certain features versus a strong classifier that includes the fusion of many of these dedicated classifiers. Classifier fusion allows each classifier

Table 3.8 MSTAR Data Results

Method	*PCC(%)*	*Time (ms)*
Morphologic operations	95.74	123.4
Template Matching	88.54	210.2
PCA+NN	89.21	154.5
PCA+SVM	93.74	130.4
DSA+KNN	93.36	164.8
Discriminative Features	92.32	152.1
Model-based	90.33	102.3
ASC matching	94.57	354.6

Adapted from Ding et al. [22].

to vote on the final result, thus improving accuracy, efficiency, and robustness. The analogy is the wisdom and efficiency of multiple experts supporting a common goal each having a weight in the final outcome. Assuming $H(x)$ is the overall classifier model, then $h(\mathrm{x})$ functions are the individual classifiers and weights $\mathbf{w} = \{w_1, ..., w_n\}$'s are assigned to each individual classifier, then

$$H(x) = w_1 * h_1(x) + w_2 * h_2(x) + \ldots + w_n * h_n(x) \tag{3.27}$$

The weights for each of the classifiers are updated based on the error they contribute. The update rule for the weights is remarkably simple, as it is a scaling process. The basic boosting process seeks an optimal solution that minimizes the mean square error (MSE) between the predicted outcomes of the response function $\hat{y}$ and $\mathbf{H}(x)$ such that $\hat{y} \approx \mathbf{H}(x)$. For a training set of size n of actual values of the output variable y, then

$$\mathbf{w} = \frac{1}{n}\sum_{i=1}^{n}(\hat{y}_i - y_i)^2 \tag{3.28}$$

The combination of classifiers is trained over stages m, $1 \leq m \leq M$, starting with an imperfect model $H_m(\mathrm{x})$. Through estimation of $h(x)$, boosting improves the model: $H_{m+1}(x) = H_m(x) + h(x)$. By seeking the perfect outputs y, then $y = H_m(x) + h(x)$ from which the goal is to fit $h(x)$ to the residuals (error):

$$h(x) = y - H_m(x) \tag{3.29}$$

After many iterations, the boosting algorithm combines these weak rules into a single strong prediction rule. There are many intelligent boosting algorithms that seek methods to determine the classifier fusion. Some of the algorithms include gradient decision tree boosting and adaptive boosting (AdaBoost).

Another closely related approach is nonnegative matrix factorization (NMF), where the nonnegativity constraint leads to only additive, not subjective, combinations of the original feature data.

Gradient Boosting

Gradient boosting constructs new base learners that can be maximally correlated with negative gradient of the loss function associated with the whole ensemble. Hence, gradient boosting sequentially improves the model such that $H_{\mathrm{m+1}}(x) = H_{\mathrm{m}}(x) + h(x)$ by choosing $h(x) = y - F_{\mathrm{m}}(x)$ that corrects the errors. Observe that the residuals $y - F_{\mathrm{m}}(x)$ for a given model are the negative gradients (with respect to $F(x)$). Gradient boosting uses a gradient descent algorithm, and generalizing it entails choosing a loss function and its gradient. Using a training set $\{(x_1, y_1), ..., (x_{\mathrm{n}}, y_{\mathrm{n}})\}$ of known values of x and corresponding values of y, the goal is to find an approximation (x) to a function $H(x)$ that minimizes the expected value of the loss function:

$$\hat{H}(x) = \arg\min_{F} \mathrm{E}_{x,y}\left[L(y, H(x))\right] \tag{3.30}$$

The gradient boosting method assumes a real-valued y and seeks an approximation $\hat{H}(x)$ in the form of a weighted sum of functions $h_i(x)$ from weak learners, weighted by γ_i:

$$\hat{H}(x) = \sum_{i=1}^{M} \gamma_i h_i(x) + \text{const} \tag{3.31}$$

Using the *empirical risk minimization* (ERM) principle, gradient boosting starts with a model $H_0(x)$, and incrementally improves in a greedy fashion:

$$\begin{aligned} H_0(x) &= \arg\min_{\gamma} \sum_{i=1}^{n} L(y_i, \gamma) \\ H_m(x) &= H_{m-1}(x) + \arg\min_{h_m \in \mathbb{H}} \left[\sum_{i=1}^{n} L(y_i, H_{m-1}(x_i) + h_m(x_i))\right] \end{aligned} \tag{3.32}$$

where $h_m \in \mathbb{H}$, is a base learner function.

Choosing the best function h at each step for an arbitrary loss function L is a computationally infeasible optimization problem. Therefore, a simplified version of the problem is to apply a steepest descent step to this minimization problem (functional gradient descent) computing the multiplier γ_m by solving a one-dimensional optimization problem.

Gradient boosting is typically used with *decision trees*, especially classification and regression trees (CART), of a fixed size as base learners. Generic gradient boosting at the *m*th step fits a decision tree $h_m(x)$ to pseudoresiduals. Let J_m be the number of its leaves. The tree partitions the input space into J_m disjoint regions $R_{1m}, \ldots, R_{J_m}$ and predicts a constant value in each region. Using the indicator notation, $\mathbf{I}_{R_{jm}}(x)$, the output of h_m for input x can be written as the sum:

$$h_m(x) = \sum_{j=1}^{J_m} b_{jm} \mathbf{I}_{R_{jm}}(x) \tag{3.33}$$

where b_{jm} is the value predicted in the region $\boldsymbol{R}_{jm}$. Then the coefficients b_{jm} are multiplied by some value γ_m. By choosing a separate optimal value of γ_m for each tree regions, instead of a single γ_m for the whole tree, coefficients b_{jm} from the tree-fitting procedure TreeBoost can be then simply discarded and the model update rule becomes:

$$H_m(x) = H_{m-1}(x) + \sum_{j=1}^{J_m} b_{jm} \mathbf{I}_{R_{jm}}(x) \tag{3.34}$$

$$\gamma_{jm} = \arg\min_{\gamma} \sum_{x_i \in R_{jm}}^{n} L\left(y_i, H_{m-1}(x_i) + \gamma\right) \tag{3.35}$$

A common form of the tree procedure is AdaBoost.

AdaBoost

AdaBoost fits a sequence of weak learners with different distributions on different weighted training data. It starts by predicting the original data set and gives equal weight to each observation. If prediction is incorrect using the first learner, then it gives higher weight to observations that have been predicted incorrectly. AdaBoost uses decision steps. The final selection is a combination of weak learners to create a strong learner—which improves prediction. Boosting attends to misclassified (e.g., errors) samples from the weaker rules.

For example, Figure 3.18 shows the iterative selection of weak rules (or decision steps). For the first decision boundary (D1), the weak learning determines that red-gray box for labels of –1 (denoted as circles) are $x > 7$. However, there are still some circles that are not classified correctly. The second decision boundary is $x > 4$, but now there are some triangles that are misclassified. Finally, the third decision boundary (weak classifier) is $y < 4$. Notice that each case, all the classifiers are weak in that they do not exactly label the data.

Combining all the weak classifiers, Figure 3.19 shows a strong classifier.

Figure 3.20 illustrates the decision tree developed from the AdaBoost method.

SAR Example

In [23], Sun et al. classify the MSTAR chips using AdaBoost with the RBF network as the base learner. As the RBF network is a binary classifier, a multiclassifier is derived from the set of binary ones through the error-correcting output codes (ECOC) method, which specify a dictionary of code words for the set of three possible classes. For the three target MSTAR evaluation, they achieved 96% accuracy. Another variation [24] develops network boosting with transfer network learning in which the training and test distributions are different. The final discussion for the supervised methods is the standard neural networks.

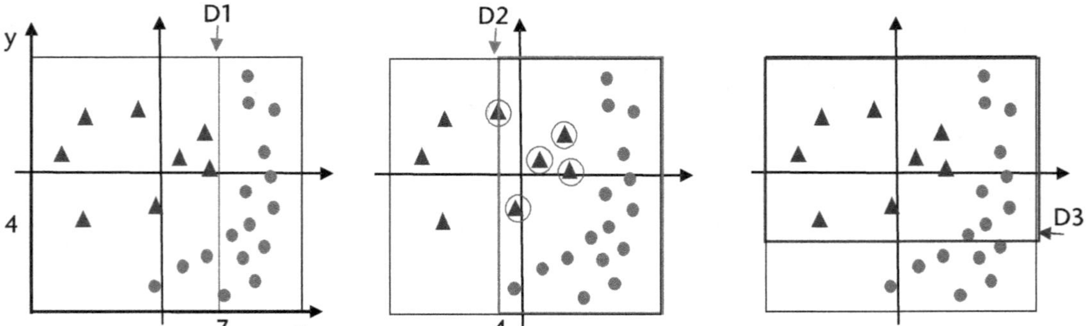

Figure 3.18 Example of AdaBoost decisions (weak classifiers).

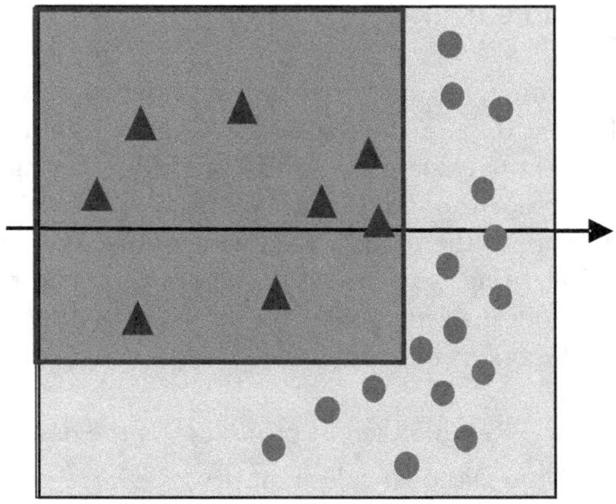

Figure 3.19 Example of AdaBoost decision (strong classifiers).

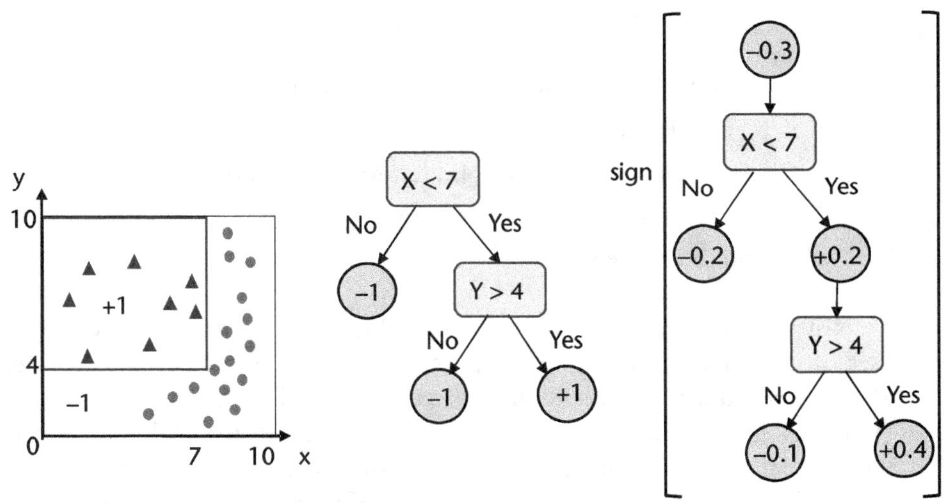

Figure 3.20 Example of AdaBoost decision (strong classifiers).

3.2.2.4 Neural Networks

Neural networks extend the perceptron for nonlinear classification. Bishop [25] provides a classic introduction to NNs for pattern recognition covering the MLP and RBF. Many of the popular techniques extending from the basic NN include MLP, PCNN, RNN, CNN, and SNN.

The emergence of these techniques within the framework of deep learning has exploded the design, development, and deployment. The two reasons for this growth include advances in hardware capability (e.g., graphical processing unit) and software availability (e.g., Python). While the family of DNN continues to evolve, Chapter 4 is dedicated to these methods so that the reader can appreciate the DNN approaches. Many of the NN approaches are developed as semisupervised methods, so the next section reviews elements of unsupervised methods.

3.3 Unsupervised Learning

In machine learning, unsupervised learning tries to find a hidden structure or pattern in unlabeled data. Since the examples given are unlabeled, there is no error or reward signal to evaluate a potential solution. During training, there is only input data and no corresponding output variables (labels), as opposed to supervised learning where both the input and labeled output data are available. The goal of unsupervised learning is to model the underlying structure or distribution of the data. The methods are left to their own devices to discover interesting structure in the data. Key functions of unsupervised learning include:

- *Clustering:* Reveal inherent groupings in the data;
- *Association:* Discover rules that describe large portions of the data;
- *Dimension reduction:* Reducing the number of features, or inputs, in a set of data.

Some classical unsupervised learning algorithms [26]:

- PCA, independent components analysis (ICA);
- K-means, K-medoids;
- Gaussian mixtures models, hdden Markov models;
- Expectation-eximization (EM) algorithm;
- Graphical models.

In unsupervised learning, the algorithm builds a mathematical model from a set of data that contains only inputs and no output labels of the desired targets. Unsupervised learning algorithms are used to find structure in the data to discover patterns, group the inputs into categories as in feature learning, or determine clusters of data points. The algorithms therefore learn from test data that has not been labeled, classified, or categorized. Instead of responding to feedback, unsupervised learning algorithms identify commonalities in the data and react based on the presence or absence of such commonalities in each new piece of data. A central application of unsupervised learning includes statistical analysis called *density estimation.* Density estimation is the process of discovering the distribution of the data.

Cluster analysis is the assignment of a set of observations into subsets (called *clusters*) so that observations within the same cluster are similar according to one or more predesignated criteria, while observations drawn from different clusters are dissimilar. Different clustering techniques make different assumptions on the structure of the data, often defined by some *similarity metric.* Similarity *internal compactness* is the similarity between members of the same cluster, and *separation* is the difference between clusters.

3.3.1 K-Means Clustering

To discern patterns from data, clustering is used to organize the information as shown in Figure 3.21. The goal of the clustering is to identify separate regions of

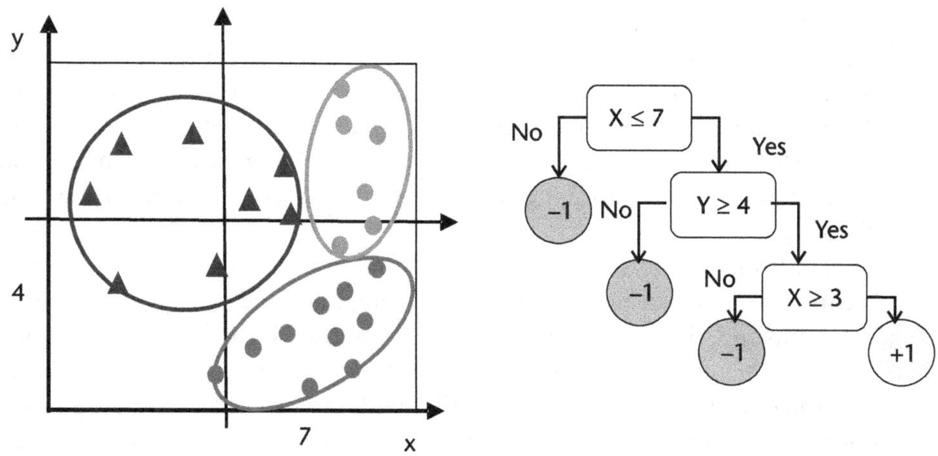

Figure 3.21 Example of K-means (N = 24, K = 3).

the parameter space in which the data points are concentrated, and then group the points accordingly. Hence, a data set with N objects can be grouped into any number of clusters K between 1 and N. However, there is no guarantee a globally optimal solution will be reached, as it depends on the initial seeding of the cluster centers (i.e., centroids).

K-means aims to partition N objects into K clusters, in which each observation belongs to the cluster with the nearest mean and the number of clusters is determined by minimizing the average squared Euclidean distance of objects from their centroids. The measure of how well the centroids represent the members of their clusters is the residual sum of squares (RSS), which is the sum of the squared distances from each observation to its centroid. The K-means algorithm is proven to converge to a local optimum.

Given a set of observations $\mathbf{X} = \{x_1, x_2, ..., x_n\}$, where each observation is a d-dimensional real vector $\mathbb{R}^D$, thus each x_n is a d-dimensional vector. K-means clustering aims to partition the n observations into $K(\leq n)$ sets $\mathbf{S} = \{S_1, S_2, ..., S_k\}$ to minimize the within-cluster sum of squares (WCSS) (i.e., variance). The union of all clusters should be $\mathbf{X}$ where the intersection of any two clusters should be zero.

The K-means splits the data halfway between the cluster means, which can lead to suboptimal splits. The EM algorithm (can be viewed as a generalization of K-means) using Gaussian models is more flexible by having both variances and covariances. The EM algorithm accommodates clusters of variable size much better than K-means, as well as correlated clusters. Another related approach to K-means is nonparametric Bayesian modeling.

SAR Examples

In [27], Richardson et al. seek the optimal K classes for polarimetric SAR using the feature density. They compared the Wishart H/A/alpha classifier and with Gaussian densities implemented a K-means that resulted in 16 classes as compared to a KNN for (K = 55, 22 classes) and (K = 35, 51 classes), revealing better segmentation.

The first methods for the MSTAR included the SVM (reported above) and K-means. Kaplan et al. [28] implemented the K-means to achieve probability of correct classification (PCC) = 80%. They further investigated a supervised template method of linear vector quantization (LVQ) of codewords (similar to Figure 3.21) with 90% accuracy [29]. As the k-means is a common method, the results from the MSTAR data set have been widely analyzed. Aitnouri et al. [30] used the K-means as part of the elimination of false cluster (EFC) algorithm with a linear solution for threshold computation between the two adjacent densities of the target and the shadow.

Additional efforts included manifold methods [31] for clustering using K-means with reasonable results of 95% using a limited number of training examples as an unsupervised method. In another example, [32] the K-means was compared to a level set method for SAR ATR segmentation of the MSTAR images. Using a generalized gamma distribution (GΓD) with the method of log-cumulants (MOLC) compared to the Gaussian K-means, fewer outliers were observed in segmenting the images.

3.3.2 K-Medoid Clustering

K-medoids is similar to k-means, but rather than calculating centroids as the cluster centers, it uses medoids that are the most centrally located objects in the cluster, not just points in space. The benefit of *K-medoids* is that it is less sensitive to outliers.

In K-medoid clustering, there is a data set *S* and F is a distance measure that evaluates distances between any two elements in *S*. Then the medoid m_s of *S* is the smallest sum.

$$m_s = \arg\min_{S} \sum_{x_n \in S_k}^{k} F(S_i, x_n) \tag{3.36}$$

In each successive iteration, the centroid of each of the *k* clusters becomes the new mean, as shown for three iterations in Figure 3.22.

The difference between the K-means and the K-medoid is that the k-means may not make sense for nonvector data. The K-medoid clustering can be applied on nonvector data with non-Euclidean distance measures. For example, the K-medoid can be used to cluster a set of time series objects, using a distance measure between time series data. Finally, the K-medoid clustering is more robust to outliers. Since a single outlier can dominate the mean of a cluster, it typically has only small

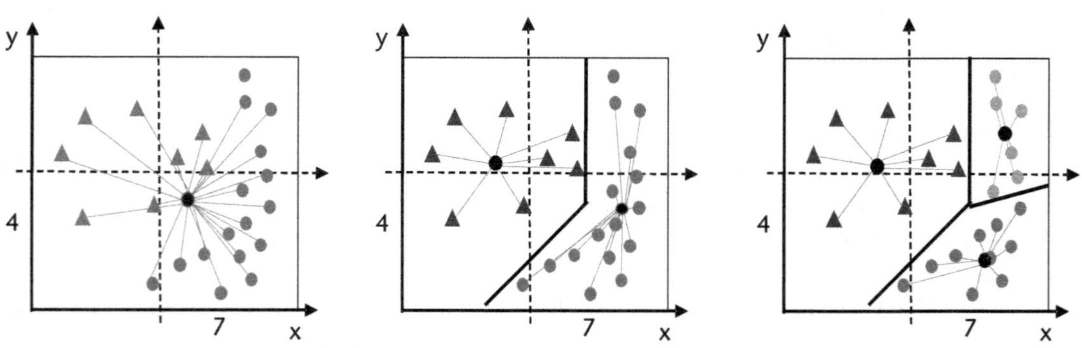

Figure 3.22 Example of K-medoids (N = 24, K = 3).

influence on the medoid. As with the K-means, the K-medoid algorithm may not converge. Another choice is the EM for clustering, which can be found elsewhere and applied to SAR classification.

SAR Example

While a few papers emphasized the k-medoid approach, only a handful of papers sought to utilize the approach. One example is for road extraction in SAR imagery [33].

3.3.3 Random Forest

The random forest (RF) approach is an ensemble learning method that derives its name from splitting decision trees where oblique hyperplanes gain accuracy as they grow without suffering from overtraining. If the splitting method is randomly forced to be insensitive to some feature dimensions or randomly restricted to be sensitive to only selected feature dimensions, precision is gained with each tree split in the forest. Note that when decision trees are designed to be deep, they learn irregular patterns and overfit (i.e., low bias, high variance) the data. The RF output is the mode of classes for classification or mean of individual trees for regressions. RF(s) average multiple deep decision trees using the same training set reducing the variance with small increase in bias and boosting the final performance. Constructing a forest of uncorrelated trees extends a classification and regression trees (CART) decision-tree-like procedure with randomized node optimization and bagging. Bagging is a method of bootstrap aggregation to reduce variance through averaging. Hence, bootstrap sampling is a way of decorrelating the trees by showing them different training sets. RF estimates the generalization error, measures variable importance through permutation, and assesses the tree correlations (i.e., want trees to be uncorrelated).

There are three types:

- *Bagging:* Train learners in parallel on different samples of the data, then combine by voting (discrete output) or by averaging (continuous output) (see [34]);
- *Stacking:* Combine model outputs using a second-stage learner like linear regression;
- *Boosting:* Train learners on the filtered output of other learners.

A simple random forest tree is shown in Figure 3.23 with $N = 3$.

SAR Example:

Since the RF is a popular method with many software routines available, it has been applied to many different SAR image classifications, especially POLSAR. Borodinov, V. V. Myasnikov [35] applied the RF to the MSTAR problem and compared the RF with the decision tree; support vector machine, and AdaBoost. For the RF, they used the CART method. They first used the Canny operator to separate the target from the scene and then a PCA for dimension reduction. With the available

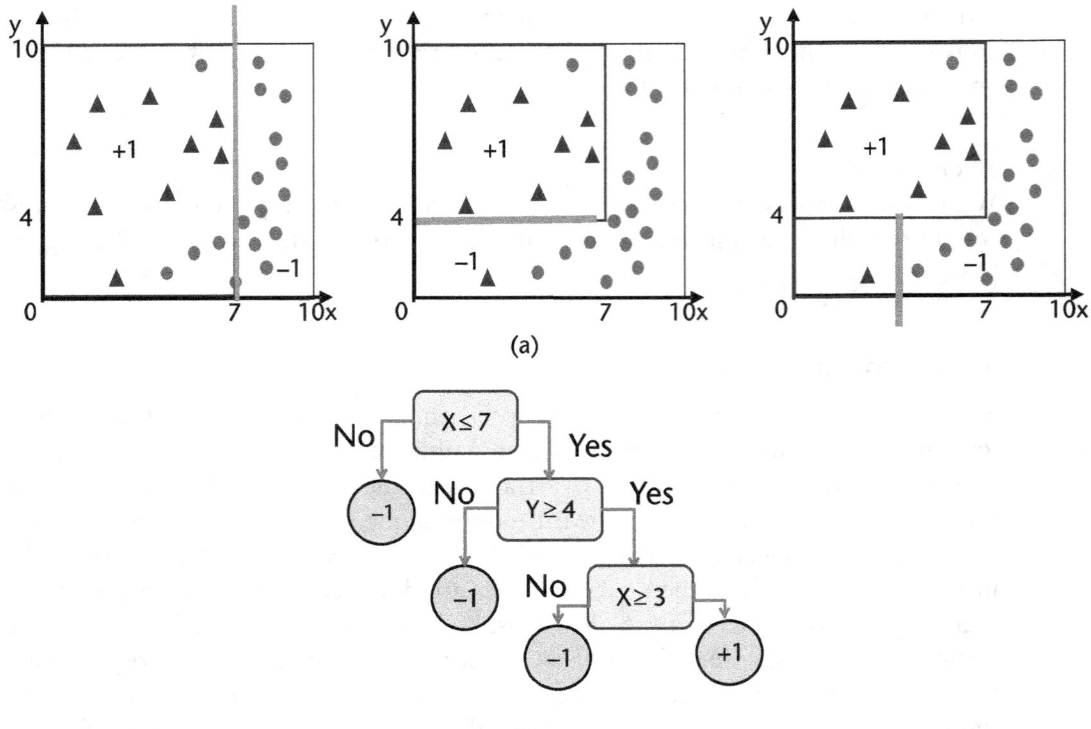

Figure 3.23 RF (tree) diagram from classification partitions.

features, they compared the methods for classification. Interestingly, their results demonstrated that the AdaBoost and RF outperformed the CART (decision tree) method; however, all four methods showed inferior performance to the SVM and KNN.

In another example, Wu et al. [36] compared KNN, SVM, RF, and stochastic gradient boosting (SGB) for LandSAT imagery. SGB outperformed the other methods by obtaining a much smaller mean value bias, followed by SVM, and RF where the RF had the largest bias. For classification accuracy, the RF was better than SGB, which both outperformed SVM and KNN.

As highlighted, the RF can be used for multimodal imagery, and general results [37] have been applied toward classification, though it was not compared to other methods. Given that the RF has been around for 20 years, it appears it has not been widely adopted for SAR methods, which results from the choice of sample selection, feature analysis, and comparative results.

3.3.4 Gaussian Mixture Models

The Gaussian mixture models (GMM) approach is used for data clustering where GMM parameters are typically estimated from training data by using EM or maximum a posteriori (MAP) algorithms.

The GMM method develops a hierarchy of models forming a finite-dimensional mixture model consisting of N observed random variables each distributed according to a mixture of K components. The observations and mixture components belong to the same parametric set of distributions (e.g., Gaussian with a mean μ and variance σ) but with different parameters. The N random latent variables specifying the identity of the mixture component of each observation is distributed according to a K-dimensional categorical distribution.

A Bayesian GMM is shown in Figure 3.24 where $[K]$ indicates a vector of size K.

Typical parameter estimation includes mixture decomposition using the ML methods such as EM or MAP. Additionally, system identification separately determines the number and functional form of components within a mixture while parameter estimation distinguishes the corresponding values. For example, graphical methods seek both system identification and parameter estimation in a GMM. Given a data set $X = \{x_1, \ldots, x_N\}$, GMM $f(x;\Theta)$ can be used for density estimation (e.g., histograms) for a cluster membership function $z_i^{(n)}$. The likelihood is:

$$p(X;\Theta) = p(x_1,\ldots,x_N;\Theta) = \prod_{i=1}^{N} f(x_i;\Theta) \tag{3.37}$$

The GMM training with the log-likelihood maximization

$$\Theta^* = \arg\min_{\Theta} \sum_{i=1}^{N} \ln p(x_i;\Theta) \tag{3.38}$$

The EM algorithm is an iterative method for the mixture model distribution in two steps: an *expectation step* and a *maximization step*.

- *E-step*: compute expectation of hidden variables given the observations;
- *M-step:* Maximize the expected complete likelihood using the MAP.

SAR Example

The MSTAR data was compiled from HRR profiles [38]. Various ML methods were used to classify the MSTAR data including the SVM, Bayesian (ML and MAP), and ANN. Using the 360° profiles of degree 1 of the targets and a bin length of 10, a GMM was trained for $S = 360$ to develop an HMM and results demonstrated 95% PCC [39]. Using track information to estimate the pose based on ANN classification, Blasch [40] demonstrated 98% PCC for the 10-target set. Another example

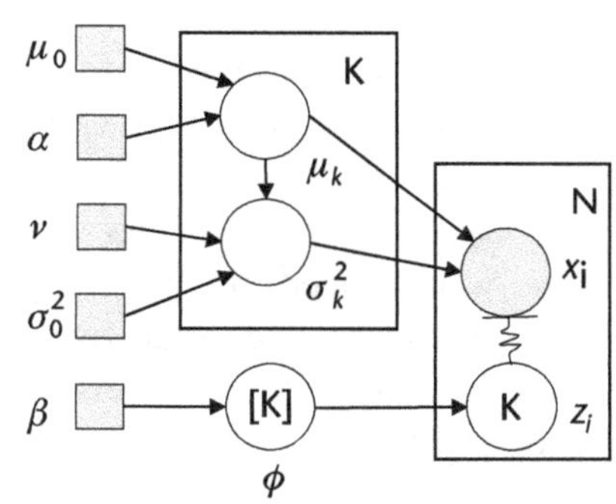

Figure 3.24 Factor graph representation with links for belief propagation.

[41] is for through-the-wall radar image analysis. Both a GMM and ML ratio test (MLRT) were analyzed with the EM using 22 feature vectors for target and clutter. Results improved using the GMM-KNN > MLRT > K-NN to designate a target and determine clutter.

Recent efforts [42] highlight change detection in SAR imagery using a multivariate Gaussian mixture model (MGMM) modeling and partitioning. Other methods compared include the EM-Markov random field and Bayes MGMM. Using the total error (TE) determined by a comparison: $P_{TE} = (MD+FA)/(N_u + N_c) \times 100\%$, where N_u and N_c are the total numbers of unchanged pixels and changed pixels in the validation images, respectively. The variational inference for MGMM showed the least TE with an average of 3% TE or 97% accuracy. Finally, the next section explores semisupervised methods.

3.4 Semisupervised Learning

Semisupervised learning algorithms develop mathematical models from incomplete training data, where a portion of the sample input doesn't have labels [43]. Given that some of the data is labeled, there is an interest to utilize supervised methods where the data includes the inputs and labeled desired outputs. For example, determining whether a SAR image contained a certain object, the training exemplars would contain chips from the SAR image with labels on whether there was an object present or not. For much of the SAR image, the pixels (or set of pixels as in a chip) would be unlabeled, yet possibly could contain an object. The labeled data can be used with the unlabeled data for clustering to classify the data. Note that most of the radar sensor data is unlabeled and even if labeled, there is only a small fraction that has most details associated with comprehensive supervised learning. Unsupervised and semisupervised learning techniques can be applied to much larger, unlabeled data sets, by labeling a minimum set of representative examples, which makes them very appealing to researchers, developers, and practitioners. Hence, semisupervised is a mixture of supervised and unsupervised learning techniques that constitutes most of the available products. To begin, the labeling of some of the data results from generative models.

3.4.1 Generative Approaches

A typical generative model-based kernel approach seeks a kernel to encode prior information. For example, one strategy uses convolution kernels to recursively define the data structure as developing knowledge for a global kernel from local kernel. Another approach is to combine a mix of labeled and unlabeled data when training the kernel. As will be shown in Chapter 4 with the generative adversarial network (GAN), there are two types of models: generative and discriminative (Table 3.9). Both approaches have been utilized in supervised, unsupervised, and in coordination for semisupervised methods. The generative model learns the joint probability distribution $p(x,y)$ for prediction; while a discriminative model learns a conditional probability distribution $p(x|y)$.

Table 3.9 Generative versus Discriminative Models

Model	*Generative*	*Discriminative*
Goal	Probability estimates	Classification rule
Performance measure	Likelihood	Misclassification rate
Mismatch problems	Outliers	Misclassifications

SAR Example

In [44], Tabti et al. developed a generative GMM with the EM based on patch learning. For each patch, the method determined the frequency of each pixel label $P(a_k|\mathcal{L})$ and radiometry label $P(v_q|\mathcal{L})$. The pixel was assigned a label by: $\log P(a^i = a_k, v^i = v_q|\mathcal{L}) = \log P(a_k|\mathcal{L}) + \log P(v_q|\mathcal{L})$. Finally, a Bayesian network was used to determine the a posteriori value:

$$P\left(L|a^i,v^i\right)=-\sum_i P\left(a^i,v^i|L^i\right)+\beta\sum_{i\neq j}\delta\left(L^i,L^j\right) \tag{3.39}$$

where $\beta > 0$, $i \sim j$ refers to the neighboring pixels and $\delta(\mathcal{L}^i, \mathcal{L}^j) = 1$ if $\mathcal{L}^i = \mathcal{L}^j$, and 0 otherwise.

Using conditional independence, graphical models are widely used for generative and discriminative methods.

3.4.2 Graph-Based Methods

A graphical model a probabilistic model where a graph provides the conditional structure of the random variables defining the mixture model. Two common examples are Bayesian Networks (BNs) and MRFs. A conditional random field is a discriminative model over an undirected graph.

Suppose you decompose the NN into a graphical format, noting that you have some knowledge of the relationships of the output variables (e.g., Figure 3.25): where the graphical inference relation is:

$$p\left(x_1,x_2,x_3,x_4,x_5\right)=p\left(x_1\right)p\left(x_2\right)p\left(x_3|x_1,x_2\right)p\left(x_4|x_2,x_3\right)p\left(x_5|x_3,x_4\right) \tag{3.40}$$

There are two methods: the naïve Bayes and belief propagation methods:

Suppose the inference desired is $p(x_1|\, x_3 = x)$ in Figure 3.25 and that the choice is $x = [0, 1]$, then

Naïve:

$$p\left(x_1|x_3=x\right)=\frac{p\left(x_1,x_3=x\right)}{p\left(x_3=x\right)} \quad \text{[2 Terms]} \tag{3.41}$$

$$p\left(x_1|x_3=x\right)=\sum_{x_2,x_4,x_5} p\left(x_1,x_2,x_3=x,x_4,x_5\right) \quad \text{[16 Terms]} = 2 \text{ times } 4 \text{ links} \tag{3.42}$$

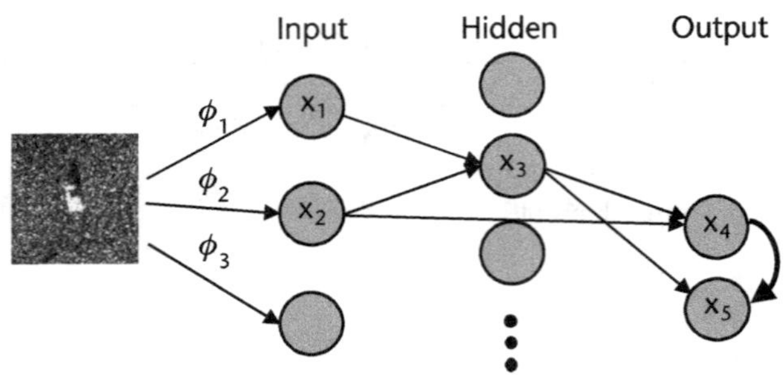

Figure 3.25 Graphical representation of a neural network.

$$p(x_3 = x) = \sum_{x_1} p(x_1, x_3 = x) \quad [2\ \text{Terms}] = 0 \text{ and } 1 \tag{3.43}$$

which results in 20 terms to be computed and summed.

Belief propagation: using the conditional independence relationships of the graph

$$\begin{aligned} p(x_1, x_3 = x) &= \sum_{x_2, x_4, x_5} \begin{matrix} p(x_1)p(x_2)p(x_3 = x|x_1, x_2) \\ p(x_4|x_2, x_3 = x)p(x_5|x_3 = x, x_4) \end{matrix} \\ &= \sum_{x_2} p(x_1)p(x_2)p(x_3 = x|x_1, x_2) \sum_{x_2} p(x_5|x_3 = x, x_4) \\ &\sum_{x_2} p(x_4|x_2, x_3 = x)p(x_5|x_3 = x, x_4) \end{aligned} \tag{3.44}$$

and knowing that the conditional relations

	x_5	
x_4	0	1
0	0·0 = 0	0·1 = 0
1	0·1 = 0	1·1 = 1

Then there only needs four terms as from:

$$p(x_1, x_3 = x) = \sum_{x_2} p(x_1)p(x_2)p(x_3 = x|x_1, x_2) \tag{3.45}$$

To implement a graphical model, it is more efficient to do *belief factor graph propagation.* In the factor graph, the variables $\mathrm{x} = \{x_1, \ldots, x_n\}$ and links $\mathcal{L} = \{\ell_1, \ldots, \ell_j\}$ are used to determine the joint probability distribution as a product of factors/links. A representation is shown in Figure 3.26, which is calculated as:

$$p(x) = p(x_1, \ldots, x_n) = \frac{1}{Z} \prod_j \ell_j \left(\mathrm{x}_{S_j}\right) \tag{3.46}$$

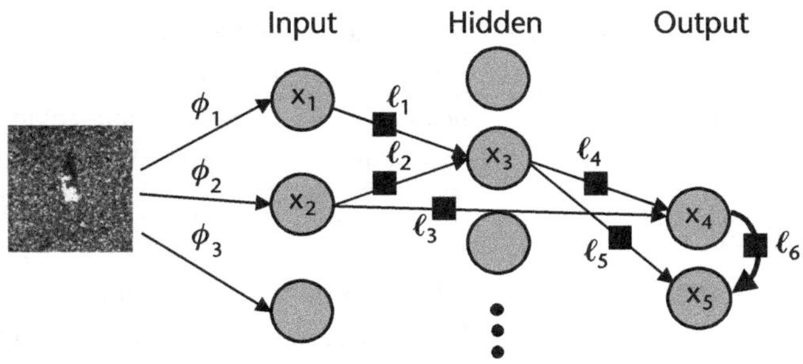

Figure 3.26 Factor fraph representation with links for belief propagation.

where Z is a normalization constant and S_j denotes the subset of {1, ... ,n} which contribute in link ℓ_j and $x_{S_j} = \{x_i : i \in S_j\}$.

As with conditional independence of belief propagation, the method extends to the Hidden Markov Model (HMM) as a directed acyclic graph. Figure 3.27 presents a HMM where each feature f is a pixel, and connection is determined over space (linking the pixels in a single image) or over time (the pixel change if the radar imager is fixed). For the HMM, the states x_t for $t = 1, \ldots, T$, are discrete evolving over time t, and there is a direct link between the input and the output. Hence

$$p(x_{1,\ldots T}, y_{1,\ldots T}) = p(x_1)p(y_1|x_1)\prod_{t=2}^{T} p(x_t|x_{t-1})p(y_t|x_t) \tag{3.47}$$

In building the graphical model from the HMM, there is added value by coordinating the development with the processing of the observed data. The methods associated with observable and hidden data are listed in Table 3.10.

In the common latent Dirichlet allocation (LDA) topic/features model, observations are a set of labeled features x_i and x_j drawn from I images and the K mixture components representing common topics shared in the images. Each image has a different set of mixture weights, which specify the features prevalent in that image. All sets of mixture weights share common hyperparameters. Extensions include a Markov chain to connect latent variables h_i and h_j to define the mixture component

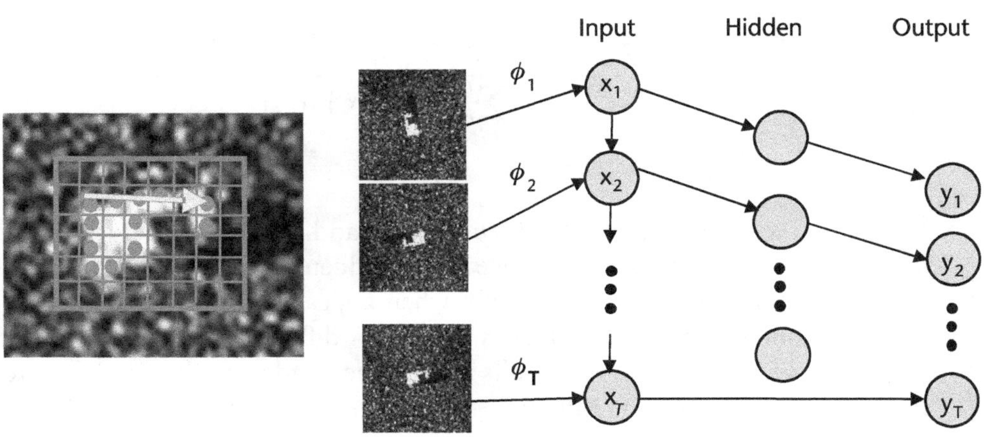

Figure 3.27 Hidden Markov Model

Table 3.10 Methods for Observable and Hidden Data

	Complete (Fully Observed) Data	*Incomplete (Hidden/Missing) Data*
HMM	Calculate frequencies	Expectation-maximization
Bayesian	Update Dirichlet distributions	Markov chain Monte Carlo (MCMC)

identities that do not have to be independent and identically distributed random variables, a HMM. The HMM is a sequential hierarchical model that extends the GMM, where the HMM can be used for tagging. The HMM assumes there is a set of visible variables $x_1, \ldots x_N$, and hidden variables $h_1, \ldots h_N$, and the goal is to impose a joint kernel $K_z((x_i, h_i),(x_j, h_j))$, where $K_z = 0$ if $h_i \neq h_j$. Marginalizing over the latent variables gives:

$$K(x_i, x_j) = \sum_{i=1}^{I} \sum_{j=1}^{J} p(h_i|x_i) p(h_j|x_j) K_z((x_i, h_i),(x_j, h_j)) \tag{3.48}$$

The posterior $p(h|x)$ are given by a generative model using Bayes rule. Hence, the use of the joint kernel $K_z((x_i, h_i),(x_j, h_j))$ couples generative models and kernel methods that afford the use of domain (user) knowledge. Learning with kernels works well if the class variable is contained in the latent variable of the probability model (MAP). For example, computing the similarity between two images, the label ($\mathbf{x}$) can be weighted differently based on visible variables (or features) (x) and context variables (h) [resolution, intensity, shape].

Three examples include:

1. *Gaussian mixture*: Choose :$K_z((x_i, h_i),(x_j, h_j)) = (x_i)^T A_h(x_j)$ if $h_i = h_j$.; else 0 if $h_i h_j$. which is the local Mahalanobis distance
   The mixture model is:

$$p(x_i) = \sum p(h_i|x_i) N(x_i|h_i, m_h, A_h) \tag{3.49}$$

   and

$$K(x_i x_j) = \frac{\sum_{i=1}^{I} \sum_{j=1}^{J} p(x_i|h_i) p(x_j|h_i) p^2(h_i)(x_i)^2 A_h(x_j)}{\sum_{i=1}^{I} \sum_{j=1}^{J} p(x_i) p(x_j)} \tag{3.50}$$

2. *Bag-of-words (BOW) kernel*: Let x_i and x_j be two label feature sequences corresponding to sequence of the hidden states h_i and h_j. Count the common label occurrences only when appearing in the same image (given h_i and h_j sequences), which is the standard BOW kernel restricted to common states, and then the results are summed and weighted by the posteriors.
3. *Fisher kernel*: Let $P(x|\theta_0)$ be the generative model (θ_0 is found by the maximum likelihood). Two objects appear similar if the adaptation of the parameters moves the model in the same direction [45]:

$$K(x_i, x_j) = \left(\nabla_\theta \log P(x_i|\theta)\big|_{\theta=\theta_0}\right) I_M^{-1} \left(\nabla_\theta \log P(x_j|\theta)\big|_{\theta=\theta_0}\right) \tag{3.51}$$

where I_M is the Fisher information matrix

$$I_M = E\left[\left(\nabla_\theta \log P(x_i|\theta)\big|_{\theta=\theta_0}\right)\left(\nabla_\theta \log P(x_i|\theta)\big|_{\theta=\theta_0}\right)^T\right] \tag{3.23}$$

Extensions to the Fisher kernel include a generalized k-stage Markov process with a decay factor λ.

Defining kernels for structures of paired sequences for tagging and trees for parsing can be extended to weighted stochastic structures of probabilities, such as the HMM. Hence, the goal is to utilize a few training examples and assumed kernels to discover the most plausible correct labels from a learning/classifier method without building a generative model.

SAR Examples

Multiple examples of HMM for SAR and HRR are referenced in [46]. Papson and Narayanan [47] developed an HMM likelihood chain code for a connected patch of pixels in the image to determine the target shape. Using the MSTAR data, the generated target models utilized the EM method to determine the path state:

$$P\left(O^{(t)}\middle|\lambda_k^1, \lambda_k^2, \lambda_k^3\right) = \max_n \sum_{j=1}^{3} \log P\left(O^{(t)}\middle|\lambda_k^j\right) \tag{3.52}$$

where $\lambda = \{A, B, \Pi\}$; where the path state $A = \{a_{ij}\}$ and $a_{ij} = P[n_{t+1} = S_j \mid n_t = S_i]$ and the emission matrix $B = \{b_j(k)\}$ and $b_j(k) = P[V_k$ at time $t \mid n_t = S_j]$. The results highlight that the shadow-only classifier did not perform as well as that of the scattering-center classifier. Additional results from the HMM on the MSTAR data presented promising results [48]. Other approaches that seek methods of labeling include active learning (few training examples), a restricted Boltzmann machine (RBM) [49] as well as contemporary approaches in Generative adversarial network (GANs) [50] featured in the next chapter.

3.5 Summary

This chapter reviewed methods for supervised, semisupervised, and unsupervised learning. The ML methods differ based on the type of training with structured, semistructured, and unstructured data (e.g., labeling); respectively. Many of the methods discussed have been available for quite some time and researched over available data sets (e.g., MSTAR and PoLSAR). The instantiations of supervised and unsupervised methods explore variations of semisupervised methods. It appears that the SVM was the proven method for SAR analysis as results were efficient, effective, and generally satisfiable. However, as soon as deep learning became a mainstay in

2012, there are many researchers who revisited these baseline ML techniques with adoptions of the most popular DL methods. The three highlights from the chapter include (1) importance of the SVM as the standard ML technique, (2) the value of semisupervised methods when there is not enough labeled data, and (3) that, as will be shown in Chapter 4, the baseline ML performances were improved with DL methods. Future chapters will highlight advances from recent efforts in DNN, reward learning approaches, and GANs.

References

[1] Bishop, C. M., *Pattern Recognition and Machine Learning*, Springer Science and Business Media, 2006.

[2] Blasch, E., C. Yang, and I. Kadar, "Summary of Tracking and Identification Methods," *Proc. SPIE*, Vol. 9091, 2014.

[3] Scarborough, S. M., C. H. Casteel Jr., L. Gorham, et al., "A Challenge Problem for SAR-Based GMTI in Urban Environments" *Proc. of SPIE*, Vol. 7337, 2009.

[4] Priddy, K., and P. E. Keller, *Artificial Neural Networks: An Introduction*, SPIE Press, 2005.

[5] Brownlee. J., *Mastering Machine Learning Algorithms,* Machine Learning Mastery, 2016.

[6] J. C., Principe, M. Kim, and M. Fisher, "Target Discrimination in Synthetic Aperture Radar using Artificial Neural Networks," *IEEE Transactions on Image Processing,* Vol. 7, No. 8, 1998, pp. 1136–1149.

[7] Bryant, M. L., and F. Garber. "SVM Classifier Applied to the MSTAR Public Data Set," *Proc. SPIE*, Vol. 3721, 1999.

[8] Zhao, Q., J. C. Principe, V. L. Brennan, D. Xu, and Z. Wang, "Synthetic Aperture Radar Automatic Target Recognition with Three Strategies of Learning and Representation," *Optical Engineering,* Vol. 39, No. 5, 2000, pp. 1230–1244.

[9] Blasch, E. P., J. J. Westerkamp, J. R. Layne, L. Hong, F. D. Garber, and A. Shaw, "Identifying Moving HRR Signatures with an ATR Belief Filter," *Proc. SPIE*, Vol. 4053, 2000.

[10] Fransson, J. E. S., F. Walter, and L. M. H. Ulander, "Estimation of Forest Parameters Using CARABAS-IIVHF SAR Data," *IEEE Transactions on Geoscience and Remote Sensing,*" Vol. 38, No. 2, 2000, pp. 720–727.

[11] Deng, H., and D. A. Clausi, "Unsupervised Segmentation of Synthetic Aperture Radar Sea Ice Imagery Using a Novel Markov Random Field Model," *IEEE Transactions on Geoscience and Remote Sensing*, Vol. 43, No. 3, 2005, pp. 528–538.

[12] Zheng. Y., E. Blasch, and Z. Liu, *Multispectral Image Fusion and Colorization*, SPIE Press, 2018.

[13] Lin, C., F. Peng, B-H. Wang, W-F. Sun, and X-J. Kong, "Research on PCA and KPCA Self-Fusion Based MSTAR SAR Automatic Target Recognition Algorithm," *Journal of Electronic Science and Technology*, Vol. 10, No. 4, 2012, pp. 352–357.

[14] Crammer, K., O. Dekel, J. Keshet, S. Shalev-Shwartz, and Y. Singer, "Online Passive-Aggressive Algorithms," *Journal of Machine Learning Research,* Vol. 7, 2006, pp. 551–585.

[15] Nie, X., S. Ding, B. Zhang, H. Qiao, and X. Huang, "POLSAR Data Online Classification Based on Multi-View Learning," IEEE *International Conference on Image Processing* (ICIP), 2017.

[16] Freund, Y., and R. E. Schapire, "Large Margin Classification Using the Perceptron Algorithm," *Machine Learning,* Vol. 37, No. 3, 1999, pp. 277–296.

[17] Kivinen, J., A. J. Smola, and R. C. Williamson, "Online Learning with Kernels," *IEEE Transactions on Signal Processing,* Vol. 52, No. 8, 2004, pp. 2165–2176.

[18] Zhao, Q., and J. C. Principe, "Support Vector Machines for SAR Automatic Target Recognition," *IEEE Transactions on Aerospace and Electronic Systems,* Vol. 37, No. 2, 2001, pp. 643–654.

[19] Dong, G., N. Wang, G. Kuang, and Y. Zhang, "Kernel Linear Representation: Application to Target Recognition in Synthetic Aperture Radar Images, "*Journal of Applied Remote Sensing*, Vol. 8, 2014, pp. 083613–083613–13.

[20] Falcon, R., R. Abielmona, and E. Blasch, "Behavioral Learning of Vessel Types with Fuzzy-Rough Decision Trees," *Internnational Conference on Information Fusion*, 2014.

[21] Rajamanickam, S., and S. M. M. Roomi, "SAR Image Target Classification: A Feature Fusion Approach," *International Journal of Computer Science and Network (IJCSN),*Vol. 6, No. 6, 2017, pp. 689–693.

[22] Ding, B., G. Wen, C. Ma, and X. Yang, "Target Recognition in Synthetic Aperture Radar Images Using Binary Morphological Operations," *Journal of Applied Remote Sensing,* Vol. 10, No. 4, 2016, pp. 1–14, 046006.

[23] Sun, Y., Z. Liu, S. Todorovic, and J. Li, "Adaptive Boosting for Synthetic Aperture Radar Automatic Target Recognition," *IEEE Transactions on Aerospace and Electronics Systems,* Vol. 43, No. 1, 2008, pp. 112–125.

[24] Gou, S., Y. Wang, and L. Jiao, "Transfer Network Learning-Based Remote Sensing Target Recognition," *Proc. SPIE,* Vol. 7495, 2009.

[25] Bishop, C. M., *Neural Networks for Pattern Recognition*, Oxford University Press, 2003.

[26] Ghahramani, Z., "*Unsupervised Learning,*" in *Advanced Lectures on Machine Learning* (O. Bousquet, G. Raetsch, and U. von Luxburg, eds.), Berlin: Springer-Verlag, 2004.

[27] Richardson, A., D. G. Goodenough, H. Chen, B. Moa, G. Hobart, and W. Myrvold, "Unsupervised Nonparametric Classification of Polarimetric SAR Data Using the K-nearest Neighbor Graph," *IEEE IGARSS*, 2010.

[28] Kaplan, L. M., R. Murenzi, K. R. Namuduri, and M. N. Cohen, "Improved Template-Based SAR ATR Performance Using Learning Vector Quantization," *Proc. SPIE,* Vol. 3462, 1998.

[29] Marinelli, A. M. P., L. M. Kaplan, and N. M. Nasrabadi, "SAR ATR Using a Modified Learning Vector Quantization Algorithm," *Proc. SPIE,* Vol. 3721, 1999.

[30] Aitnouri, E., S. Wang, and D. Ziou, "Segmentation of Small Vehicle Targets in SAR Images," *Proc. SPIE,* Vol. 4726, 2002.

[31] Chen, K., R. Stanfill, and A. Mahalanobis, "Aided Target Recognition Using Hyperdimensional Manifolds," *Proc. SPIE,* Vol. 9476, 2015.

[32] Wang, B., and G. Wang, "A GΓD Based Level Set Method for Target Segmentation in High Resolution SAR Images," *International Conference on Frontiers of Sensors Technologies*, 2017, pp. 256–259.

[33] Zhang, Q., and I. Couloigner, "Comparing Different Localization Approaches of the Radon Transform for Road Centerline Extraction from Classified Satellite Imagery," *International Conference on Pattern Recognition* 2, 2006, pp. 138–141.

[34] Breiman, L., "Bagging Predictors," *Machine Learning,* Vol. 24, No. 2, 1996, pp. 123–140.

[35] Borodinov, A. A., and V. V. Myasnikov, "Comparison of Classification Algorithms for Various Methods of Preprocessing Radar Images of the MSTAR Base," *Proc. SPIE 10696, Tenth International Conference on Machine Vision* (ICMV 2017), 2018.

[36] Wu, C., H. Shen, A. Shen, et al., "Comparison of Machine-Learning Methods for Above-Ground Biomass Estimation-Based on Landsat Imagery," *J. Appl. Remote Sens.,* Vol. 10, No. 3, 2016, p. 035010.

[37] Al-Ruzouq, R., A. Shanableh , M. B. Gibril, and B. Kalantar, "Multi-Scale Correlation-Based Feature Sselection and Random Forest Cclassification for LULC Mapping from the Integration of SAR and Optical Sentinel Images," *Proc. SPIE,* Vol. 11157, 2019.

[38] Williams, R., J. Westerkamp, D. Gross, and A. Palomino, "Automatic Target Recognition of Time Critical Moving Targets Using 1D High Range Resolution (HRR) Radar," *IEEE Aerospace and Electronic Systems Magazine,* Vol. 15, No. 4, 2000, pp. 37–43.

[39] Gudnason, J., J. Cui, and M. Brookes, "HRR Automatic Target Recognition from Super-resolution Scattering Center Features," *IEEE Transactions on Aerospace and Electronic Systems,* Vol. 45 , No. 4, 2009, pp. 1512–1584.

[40] Blasch, E., *Derivation of a Belief Filter for Simultaneous High Range Resolution Radar Tracking and Identification,* Ph.D. thesis, Wright State University, 1999.

[41] Kilaru, V., M. Amin, F. Ahmad, P. Sévigny, and D. DiFilippo, "Gaussian Mixture Modeling Approach for Stationary Human Identification in Through-the-Wall Radar Imagery," *Journal of Electronic Imaging,* Vol. 24, No. 1, 2015, 013028, pp. 1–11.

[42] Yang, G., H-C. Li, W. Yang, K. Fu, Y-J. Sun, and W. J. Emery, "Unsupervised Change Detection of SAR Images Based on Variational Multivariate Gaussian Mixture Model and Shannon Entropy," *IEEE Geoscience and Remote Sensing Letters*, Vol. 16, No. 5, 2019, pp. 826–830.

[43] Zhu, X., "Semi-Supervised Learning Literature Survey," Computer Sciences, University of Wisconsin, 2008.

[44] Tabti, S., C-A Deledall2, L. Denis, F. Tupin, "Patch-Based SAR Image Classification: The potential of modeling the Statistical Distribution of Patches with Gaussian Mixtures," *IGARSS*, 2015.

[45] Goutte, C., H. Dejean E. Gaussier, N. Cancedda, and J.- M. Renders, "Combing Labeled and Unlabeled Data: a Case Study on Fisher Kernels and Transductive Inference for Biological Entity Recognition," in *CoNLL*, 2002, pp. 1–7.

[46] El-Darymli, K., E. W. Gill, P. McGuire, D. Power, and C. Moloney, "Automatic Target Recognition in Synthetic Aperture Radar Imagery: A State-of-the-Art Review," *IEEE Access*, Vol. 4, 2016, pp. 6014–6058.

[47] Papson, S., and R. M. Narayanan, "Applying Target Shadow Models for SAR ATR," *Proc. SPIE,* Vol. 6567, 2007.

[48] Friend, M., *Combat Identification with Synthetic Aperture Rradar, Out-of-Library Identification, and Non-Declarations*, Ph.D. dissertation, Air Force Inst. Technol., Wright-Patterson AFB, OH, 2007, [Online], http://gradworks.umi.com/32/88/3288578.html.

[49] Luttrell, S., "The Implications of Boltzmann-Type Machines for SAR (Synthetic Aperture Radar) Data Processing: A Preliminary Survey," Royal Signals Radar Establishment, Malvern, U.K., *Tech. Rep. AD-A163 749*, 1985, [Online], http://tinyurl.com/nkrgn43.

[50] Schwegmann, C. P., W. Kleynhans, B. P. Salmon, L. W. Mdakane, and R. G. V. Meyer, "Synthetic Aperture Radar Ship Discrimination, Generation, and Latent Variable Extraction using Information Maximizing Generative Adversarial Networks," *IGARSS*, 2017.

CHAPTER 4

A Review of Deep Learning Algorithms

4.1 Introduction

Machine learning development begins with expert training and statistical learning, then moves toward context awareness. The methods for expert training and statistical analysis were developed in Chapter 3. The methods in Chapter 4 expand on statistical learning resulting from advances in computer hardware, data availability, and software reuse. For example, with faster computers and more data, advanced processing allows for more effective and efficient techniques with deep learning (DL). To begin, the chapter starts with DL approaches. DL uses a set of labeled data and progressive data processing architectures to extract more structure from the data through a deep neural network (DNN). Essentially, DL is machine learning with multiple layers of perceptrons to go from a coarse analysis that determines the data structure to a fine assessment of the nuances for decision discrimination. The advanced DL methods afford more accurate decision making, conditioned on the sensor, target, and environment context. One prominent DL technique is the *autoencoder* (AE) that encodes the data to features to be fed into a DNN, while a decoder maps the feature space to the decision output.

We also discuss popular DL methods to process spatially organized and time-dependent data. For the spatial 2-dimensional data, such as imagery, the pulse coupled neural network (PCNN) and convolutional NN (CNN) build multiple NN layers based on the feature processing at each layer. Basically, each layer learns another structure. For example, a NN needs one layer to segment the data with a line. Then, the next layer learns how to segment with two lines. The next layer is able to determine a box inside the image. Progressively, spatial data structure is revealed through multiple layers. For time-dependent 1-dimensional data, the sequence pattern is analyzed with the recurrent NN (RNN) that learns the dependency of 1D data. If the first layer learns the trends between two points, then the next layer learns across multiple points. Determining an adequate number of layers is a function of the changes in the data. For example, a RNN quickly learns that a heartbeat has a repetitive pattern. The RNN leads to the long short-term memory (LSTM) approach that learns long-term relationships by remembering short-term information for long periods of time. To address the deterministic assumptions in the CNN and RNN, methods such as the stochastic neural network (SNN) address the data uncertainty. SNNs including the Boltzmann machine and the spiking neural network, are used to get the pattern from noisy data through conditional contextual relationships between the NN layers.

The third section focuses on reward-based DL when labeled data is not explicit, such as in RL, TL, and GAN—all of which use another source of information to guide the learning process. Reinforcement learning uses goals to augment the labeled truth data with rewards in training the system. If the goal changes, the system adapts to the new goal. Transfer learning uses another model on a related domain to seed the developments in the desired source. For example, a model trained for ATR in visual imagery can be used to initiate training in radar imagery. TL, while difficult, can be useful in going from one situation to another that is slightly changed—such as altering tasks from target recognition to irregular pattern learning of targets. The final method, called GANs, is one of the most promising methods for robust performance. A GAN has a generative function and a discriminator function. The generator develops new data (as from a model), while the discriminator determines if the new information should update the statistical model or is an imposter. The power of GANs results from the fact that the generated data moves beyond simulated data based on the domain physics to that of detecting the unknown such as fake data. Currently there are thousands of GANs as it emerges as a promising contribution to science. The current state of the art for GANs with regard to ATR on the MSTAR data set is around 99.98% (as compared to a result of 90% on the MSTAR data set for the initial ML-based methods).

All of these DL techniques have been developed and implemented using RF data, especially CNNs for SAR and RNNs for emitted radar signals. These methods are important developments to enhance RF processing, especially when incorporated with AEs and GANs. The chapter begins with the AE to demonstrate the power of labeled data for learning; whereas the chapter ends with RL, TL, and GANs that seek to learn from situations in which no labeled data is available. Together, these methods provide powerful semisupervised approaches that are slightly different than unsupervised methods, as the auxiliary information (goal, related model, and synthetic data) has some labeling that is inherent in the domain application. Thus, AE and GAN methods have promise for increasing the robustness of AI/ML systems in real-world application situations. Likewise, with the known information, there is a better chance for interpretability as the new information can shape the learning function and then use the knowledge as a source of understanding. Thus, contextual awareness is developed to align the new situational knowledge for decision making from the unknown of unforeseen with practical predictions. Together the situation assessment and awareness lead to situation understanding and contextual knowledge.

4.1.1 Deep Neural Networks

A DNN is a NN with many deep hidden layers that is used for nonlinear data classification, as shown in Figure 4.1. In each successive layer, the NN learns more descriptive features. Eventually, the successive feature learning from each layer supports a more discriminative classification.

With the many arrows passing data from one node to another in another layer as in Figure 4.1, the DNN is typically shown as in Figure 4.2. The DNN is fully

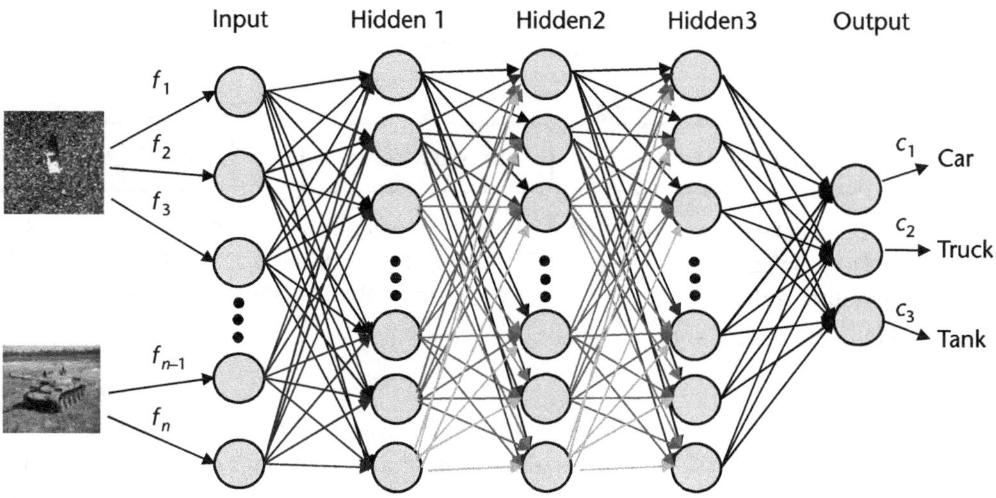

Figure 4.1 DNN model.

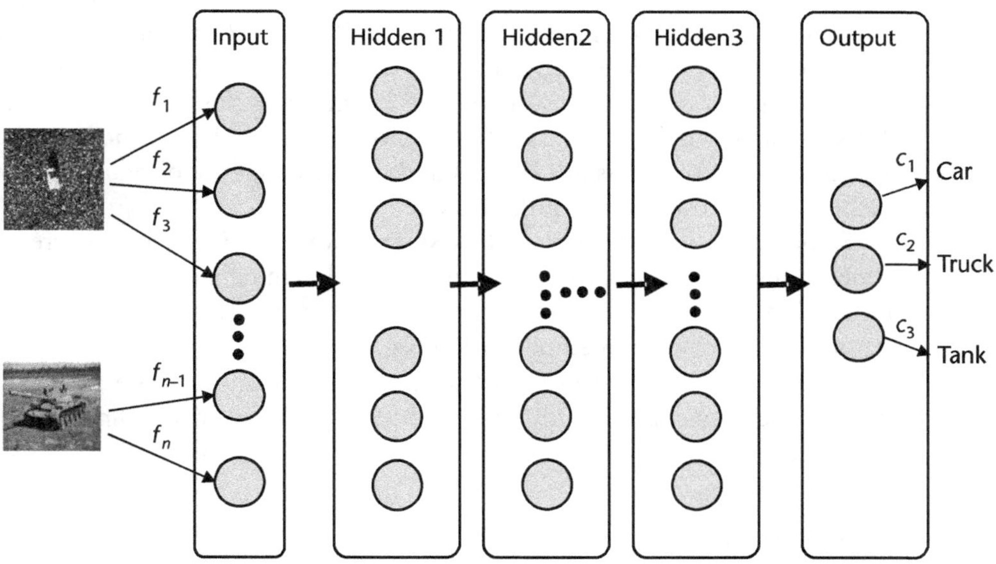

Figure 4.2 DNN model representation.

connected, with input, hidden, and output layers. The input layer learns to perform preprocessing on the input from a set of features. The data is then sent to a series of hidden layers, where the number of layers can vary and is determined by design. As the data propagates through hidden layers, more complex features are extracted and learned. The output layer performs classification and determines the label of the input. For classification, the last layer usually employs a *Softmax* function (described in Section 4.2.1.3) that normalizes the probability distribution to a fixed number of probabilities.

Figure 4.3 shows various DNN methods. One of the most prominent DNN methods is AE that both encodes and decodes data features as parameters for

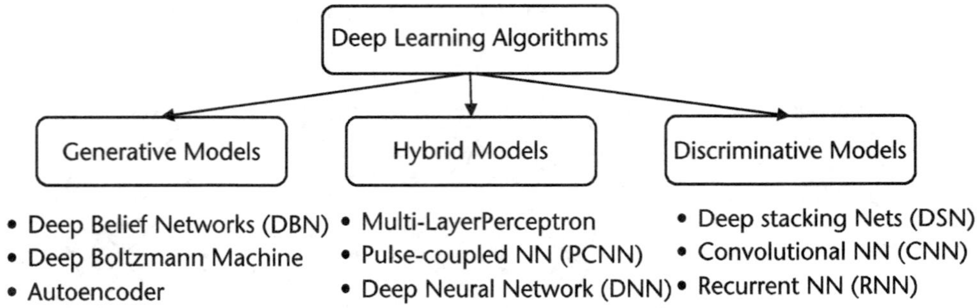

Figure 4.3 Deep learning methods.

learning. Within the *generative methods* category are deterministic deep belief networks (DBNs) and stochastic deep Boltzmann machines (DBMs). To afford robust analysis and adaptation to real-world conditions, stochastic analysis enhances DL methods. The chapter will end with GAN that can incorporate ideas from my DL techniques.

On the other spectrum of generative methods are *discriminative methods* such as deep stacking networks (DSNs). Two common approaches applied to RF analysis include the CNN for spatial data (e.g., SAR target classification) as well as the RNN for time-dependent (e.g., radar emitter detection). Hybrid methods that are somewhat generative and discriminative include the classic MLP as well as the DNN for spatial data and the PCNN to introduce time dependencies in the analysis. Each DL method has advantages and disadvantages and many contemporary methods combine elements of these techniques.

Some of the important early DNNs are shown in Table 4.1, where improvements in the number of layers and theoretical design enhance performance. Many of these DNNs use convolution filters for feature extraction as well as for discrimination. To describe the methods of DNNs, it is important to start with one of the common methods, called autoencoders, which seek to generate reduced models through successive layers and then remodel for effective discrimination for information retrieval.

4.1.2 Autoencoder

A simple autoencoder generates a model with an FNN. A single layer perceptron consists of input, hidden, and output layers. The single layer perceptron can lead to a deep instantiation through an MLP, which includes many connected hidden layers to decompose the input x and reconstruct the inputs for an output x'. An

Table 4.1 Classical Deep Neural Networks

Method	*Convolutional Layers*	*Reference*	
AlexNet	5	Krizhevsky et al. 2012	[1]
VGG (Visual Geometry Group)	19	Simonyan et al. 2014	[2]
GoogleNet	22	Szegedy et al. 2015	[3]
ResNet (Residual Network)	34	He et al. 2015	[4]

autoencoder is an unsupervised learning model that does not require labels to predict $y = f(x)$, but rather minimizes the difference between the input **x** and the output x'.

As shown in Figure 4.4, an autoencoder seeks an encoder mapping between the input space X and the feature space F as (θ: X →F). Similarly, the decoder maps the feature space to a remodeled output (ϕ: F → X). The combined encoder-decoder minimizes the error:

$$\theta,\phi = \underset{\theta,\phi}{\arg\min}\left\|X-(\phi\cdot\theta\ )X\right\|^2 \quad (4.1)$$

Using a single hidden layer **h** (called the code or *latent* variables representation), the encoder maps the input $\mathbf{x} \in \mathbb{R}^d = X$ to the hidden layer:

$$\mathrm{h} = f_\theta(\tilde{\mathrm{x}}) = \sigma(\mathbf{W}\mathbf{x} + \mathbf{b}) \quad (4.2)$$

where where **W** is a weight matrix, **b** is a bias vector, and σ is an element-wise activation function (e.g., rectified linear unit (ReLU), hyperbolic, or sigmoid function):

- sigmoid: $\sigma(\mathrm{x}) = 1/\ [1 + \exp(-x)]$
- hyperbolic tangent (tanh) $\sigma(\mathrm{x}) = [\exp(x) - \exp(-x)]\ /\ [\exp(x) + \exp(-x)]$
- ReLU: $\sigma(\mathrm{x}) = \max(x,0)$.

Backpropagation training iteratively updates the randomly initialized weights and biases. The decoder stage of the maps **h** to the reconstructed model **x′** as similar to **x**:

$$\mathrm{x}' = \mathrm{z} = g_\phi(\mathrm{h}) = \sigma'(\mathrm{W}'\mathrm{h} + \mathrm{b}') \quad (4.3)$$

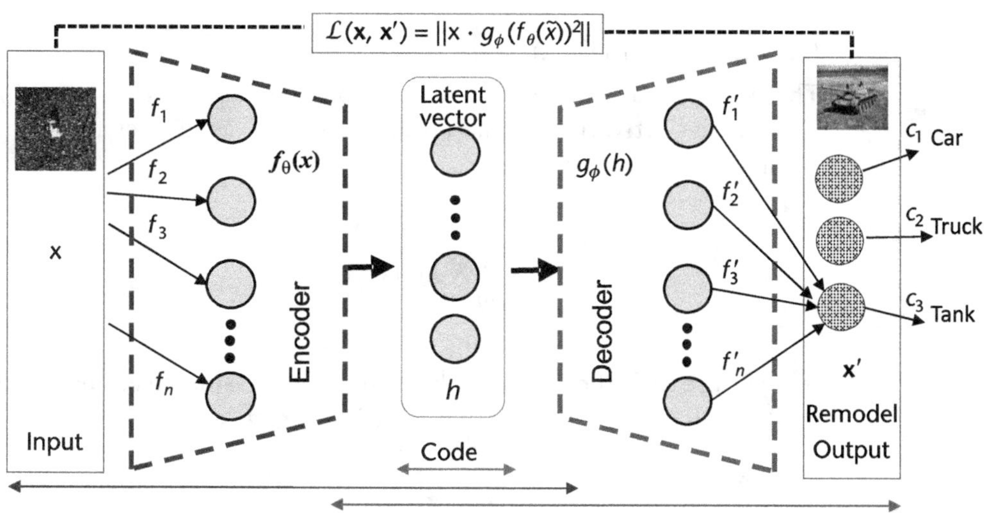

Figure 4.4 Autoencoder.

where for the decoder may be unrelated to the corresponding σ', $\mathbf{W}'$, and $\mathbf{b}'$ for the encoder. The decoder minimizes the reconstruction errors from the loss function by averaging $\mathbf{x}$ over some input training set:

$$\begin{aligned} L(\mathbf{x},\mathbf{x}') &= \|\mathbf{x},\mathbf{x}'\|^2 \\ &= \left\|\mathbf{x} - \sigma'\left(\mathbf{W}'\left(\left[\sigma(\mathbf{W}\mathbf{x}+\mathbf{b})\right]\right) + \mathbf{b}'\right)\right\|^2 \\ &= \left\|\mathbf{x} - g_\phi\left(f_\theta(\tilde{\mathbf{x}})\right)\right\|^2 \end{aligned} \tag{4.4}$$

To effectively design an AE, the selection of the code dimension d based on the data distribution can enhance the results. Caution should be exercised when the code dimension is small (e.g., $F^d << X^d$) from which the feature vector $\theta(x)$ is a compressed representation, leading to *undercomplete* AEs. Likewise, if $F^d >> X^d$, then the use of too many hidden layers leads to an *overcomplete* AE. The balance is realized through *regularized autoencoders*. Regularization implies the addition of information to solve an ill-posed optimization problem, to prevent overfitting for the subject ML applications. The regularized AE leverages *sparsity* in allowing only a small number of the hidden units to be active. Sparsity analysis is a concept for *compressive sensing,* which is not discussed here but is evident in many ML methods. A sparsity constraint (or penalty $\Omega(\boldsymbol{h})$) forces the model to respond to the unique statistical features of the input data used for training. Hence, a sparsity penalty $\boldsymbol{h}$ is employed:

$$L(\mathbf{x},\mathbf{x}') == \left\|\mathbf{x} - \sigma'\left(\mathbf{W}'\left(\left[\sigma(\Omega(h))\right]\right) + \mathbf{b}'\right)\right\|^2 \tag{4.5}$$

where $h = f(Wx + b)$. The penalty encourages the model to activate (i.e., yield an output value that is close to 1) some specific areas of the network as shown in Figure 4.5 on the basis of the input data, while forcing all other neurons to be inactive (i.e., to have an output value close to 0).

Some sparsity of activation penalty terms includes:

- *Kullback-Leibler (KL) divergence.*

  The sparsity parameter ρ determined from the hidden variables $h_j(x_i)$, where the average activation of the hidden unit averaged over the training examples is

$$\hat{\rho}_j = \frac{1}{m}\sum_{i=1}^{m}\left[h_j(x_i)\right] \tag{4.6}$$

  If $\hat{\rho}_j$ is close to 0, then most of the neurons will be inactive. The penalty term $\Omega(\boldsymbol{h})$ penalizes $\hat{\rho}_j = \rho$, exploiting the KL divergence:

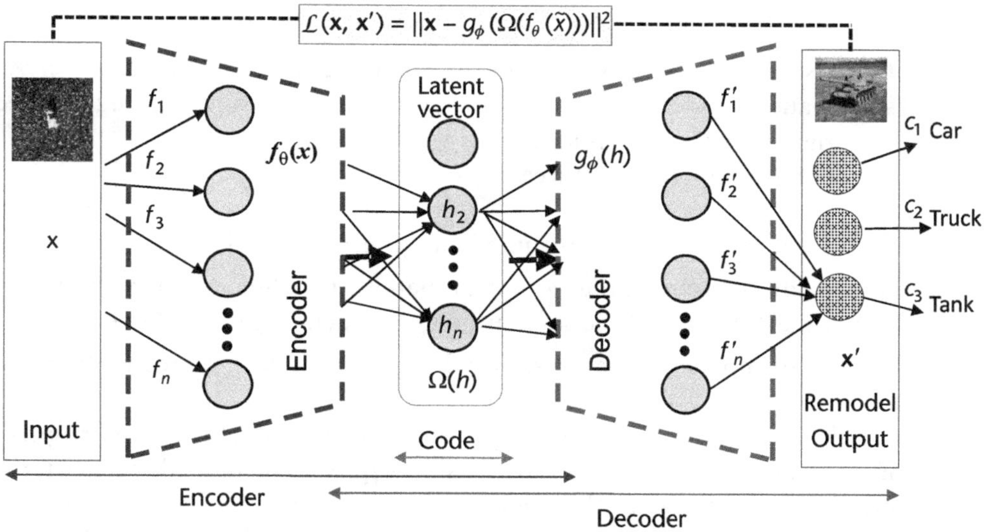

Figure 4.5 Sparse autoencoder (only the filled hidden variables are activated).

$$\Omega(\boldsymbol{h}) = \sum_{j=1}^{s} \mathrm{KL}(\rho \| \hat{\rho}_j) = \sum_{j=1}^{s} \left[\rho \log\left(\frac{\rho}{\hat{\rho}_j}\right) + (1-\rho)\log\left(\frac{1-\rho}{1-\hat{\rho}_j}\right) \right] \tag{4.7}$$

where the summation index is applied over the s hidden nodes in the hidden layer, and $\mathrm{KL}(\rho\|\hat{\rho}_j)$ is the KL-divergence between a Bernoulli random variable with mean ρ and a Bernoulli random variable with mean $\hat{\rho}_j$. Select the $\hat{\rho}_j$ that is closest to ρ.

- *ℓ_1 or ℓ_2 regularization*

  In the case of ℓ_1, the loss function for a scaling parameter λ is:

$$\Omega(\mathbf{h}) = \lambda \sum_{i=1} |h_i| \tag{4.8}$$

- *K-sparse autoencoder:*

  Based on a linear autoencoder (i.e., with linear activation function) and weights w, the strongest activations are retained by ranking the activities and keeping only the first k values, or by using hidden units that are comprised of ReLUs with thresholds that are adaptively adjusted until the k largest activities are identified.

$$\hat{x}_i = \arg\min_{z} \|x_i - \mathbf{W}\mathbf{x}\|_2^2 \quad s.t. \ \|\mathbf{x}\|_0 < k \tag{4.9}$$

Autoencoder adaptation strategies include many forms, such as the denoising AE (DAE), contractive AE (CAE), and variational (AE).

4.1.2.1 Denoising Autoencoder

DAEs seek a good representation by changing the *reconstruction criterion* from sparse autoencoders to clean a corrupted input such as denoising and recovering the original *undistorted input.*

DAE training follows three steps:

- The initial input is corrupted into $\tilde{x}$ through stochastic mapping $\tilde{x} \sim q_D(\tilde{x}|x)$;
- The corrupted input $\tilde{x}$ is then mapped to a hidden representation with the same process of the standard autoencoder, $\boldsymbol{h} = f(\tilde{x}) = s(\mathbf{W}\tilde{x} + \mathbf{b})$;
- The model reconstructs $z = g\ (\mathbf{h})$ from the hidden representation.

For each random example$\boldsymbol{x}$ presented to the model, a new corrupted version is generated stochastically on the basis of $q_D(|\boldsymbol{x})$. The model's parameters θ and θ' are trained to minimize the average reconstruction error over the training data by minimizing the difference between z and the original uncorrupted input$\boldsymbol{x}$. Many examples of noisy images are generated, such as Gaussian, salt-and-pepper, uniform, and shot noise. Gaussian noise is used with the RGB of color cameras, while Johnson-Nyquist noise occurs for thermal cameras. Salt-and-pepper (or spike noise) models the fat-tail impulsive noise associated with analog-to-digital bit error transmissions. Uniform noise models the dithering of pixel quantization noise. Shot noise can be the most dominant resulting from photo quantization for a given exposure and is modeled as a Poisson distribution.

4.1.2.2 Contractive Autoencoder

To enhance the ML robustness to variations on the input values, the CAE adds an explicit regularizer to the input activations of the objective function corresponding to the *Frobenius norm* of the Jacobian matrix. The regularizer penalizes learning the noise distribution during training by mapping the inputs to a smaller number of outputs as:

$$L(\mathbf{x}, \mathbf{x}') + \lambda \sum_{i=1} \left\| \nabla_s h_i \right\|^2 \tag{4.10}$$

where L is the loss function, λ is the scaling parameter, and $\|\nabla_s h_i\|^2$ is the Frobenius noise. Note, in the limit of small Gaussian input noise, DAE reconstruction resists small finite-sized input perturbations, while CAE feature extraction resists infinitesimal input perturbations.

4.1.2.3 Variational Autoencoder

Typically, there are not enough training samples to represent all variations of the objects to classify, Hence, VAEs generate models (akin to GANs) using directed probabilistic graphical models (DPGM). The VAE NN *approximates the posterior* by learning the generalization of the causal relations of how the data is simulated and the grouping of *unobserved variables*. Later in the chapter, the GAN offers a different approach with generative modeling to learn a predictor given the observation. The variation approach uses representational learning to model the latent variables distributions. The VAE adds a loss component of a Stochastic Gradient Variational Bayes (SGVB) estimator.

SAR Examples

There are numerous examples of autoencoders that have become popular in the late 2010s. One excellent summary by Dong et al. [5] presents methods for information fusion, convolutional neural networks, and transfer learning. Using the MSTAR data and multiclass logistic regression, the AE's accuracy improves as the window size increases while the accuracy is relatively stable using the cross-entropy for any number of hidden units between 400 and 1600 at 89%. However, the accuracy improves for the number of hidden nodes using the methods for an average of sigmoid (89%), tanh (90%), and ReLU (91%). Also, a stacked AE improves on the SVM and sparse representation-based classification with comparable results with the CNN, yet the CNN achieves a maximum of 94%. Another example is that of Lv et al. [6] who develop a stacked contractive autoencoder (CAE) with a simple linear iterative clustering (SLIC) superpixel algorithm that better reconstructs images. Finally, as a third case, to get multiple instances for real world analysis, a SAR image generation extension of a Wasserstein autoencoder (WAE) is proposed by Wang et al. [7]. Using the MSTAR imagery, the deep learning WAE-GAN achieves better recognition accuracy than the VAE and VAE-GAN methods with 21,000 generated images for VAE (96%), VAE-GA (97%), and WAE-GAN (98.5%). Given the popularity of VAE and GANs and their impressive results, the rest of the chapter will describe the DNN methods that lead to the GAN-VAE for deterministic classification and the RBM for stochastic classification.

4.2 Neural Networks

This section reviews the PCNN as a baseline method, the CNN as a conventional method, the RNN for dynamic assessment, and the RBM for stochastic variations. There are two important takeaway messages that can be inferenced by reviewing and comparing these DNN methods:

- Improvements in accuracy are obtained on the same MSTAR data set for standard operating conditions (SOC) to the point of achieving 99% accuracy in progressing from PCNN to CNN and then to RNN (LSTM);
- Reduced performance on the MSTAR data for EOC with stochastic variations is obtained at 96%, which represents real-world robustness analysis.

4.2.1 Feed Forward Neural Networks

A simple form of a FFNN is the ANN (see Chapter 3), as shown in Figure 4.6. The FNN is enhanced through use of the MLP. Various FFNNs include radial basis function networks (RBFN) and CNNs.

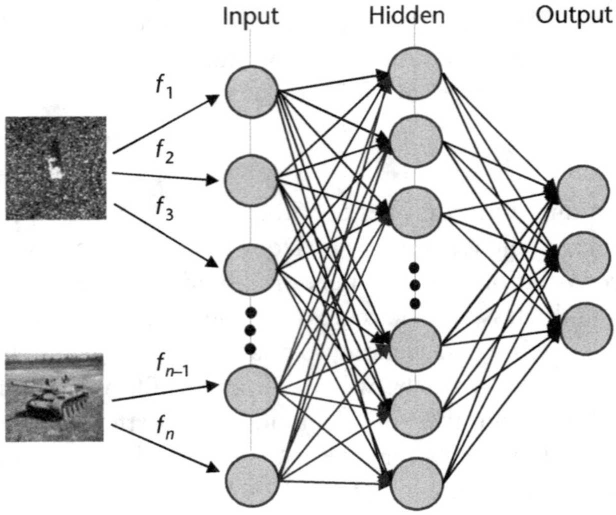

Figure 4.6 Feedforward neural network.

4.2.1.1 Multilayer Perceptron

The MLP is the simplest form of FNN based on linear perceptron, as shown in Figure 4.7. Generally, the MLP consists of three or more layers of nonlinear activation nodes wherein the MLP network learns from backpropagation estimation processing as the layers are fully connected. CNNs are regularized versions of MLPs.

SAR Example

An ANN MLP has been developed for the analysis of the MSTAR data sets. Specifically, in 1997, Catlin et al. [8] demonstrated the MLP as an approximation to the Bayes optimal discriminant function [9] for the two feature outputs of aspect (z_1) and depression (z_2) angles. To determine the probability of target identification (PID), the output pair of (z_1, z_2) was used, wherein z_1 is the output and z_2

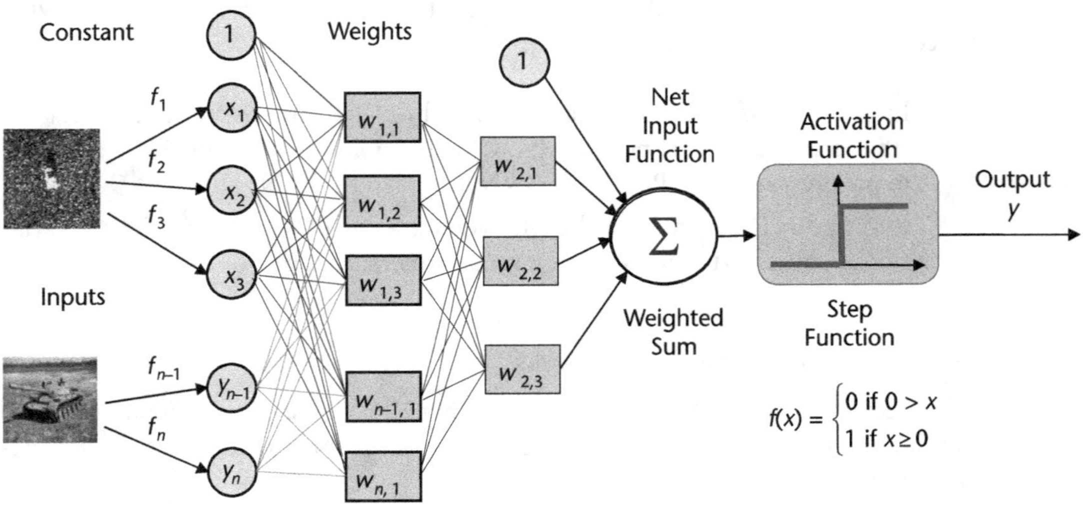

Figure 4.7 Multilayer perceptron.

= (1 – z_1) for training. The ANN MLP contained a single hidden layer with five nodes. Results demonstrated that concentrating on 90° and 180° aspect angles for any depression angle results in the best performance, while the other directions had minimal PID results. The ANN investigation provided a baseline for performance evaluation using the MSTAR data.

4.2.1.2 Pulse-Coupled Neural Networks

An extension to the MLP is the PCNN. The PCNN is a flexible biologically inspired filter. The PCNN, as developed by Johnson [10], models the neuronal pulse height, duration (repetition rate), and the interneural linking observed in the primate visual cortex. Figure 4.8 presents a PCNN model.

SAR Example

Blasch [11, 12] developed a PCNN model for the MSTAR data. The PCNN code processes features from the image set (i.e., pixel intensity and color). The feature extraction algorithm determines the grouped feature information, such as orientation, entropy, and size. Using the PCNN for feature extraction, a belief filter fuses features and discerns target type by maintaining a belief in one of the possible targets. When the accumulated belief for a target is confirmed, the decision-making process declares the target type.

The MSTAR radar image data set is shown in Figure 1.2. Three types of targets are presented to the algorithm, including the standard BMP2, BTR70, and T72 tanks selected for recognition. Figure 4.9 shows a PCNN-processed image.

Given the PCNN image, various features were extracted, such as length/width ratios, orientation, entropy, turrets, and wheels, as revealed in Figure 4.10. It is assumed that the system has a priori information pertaining to the expected

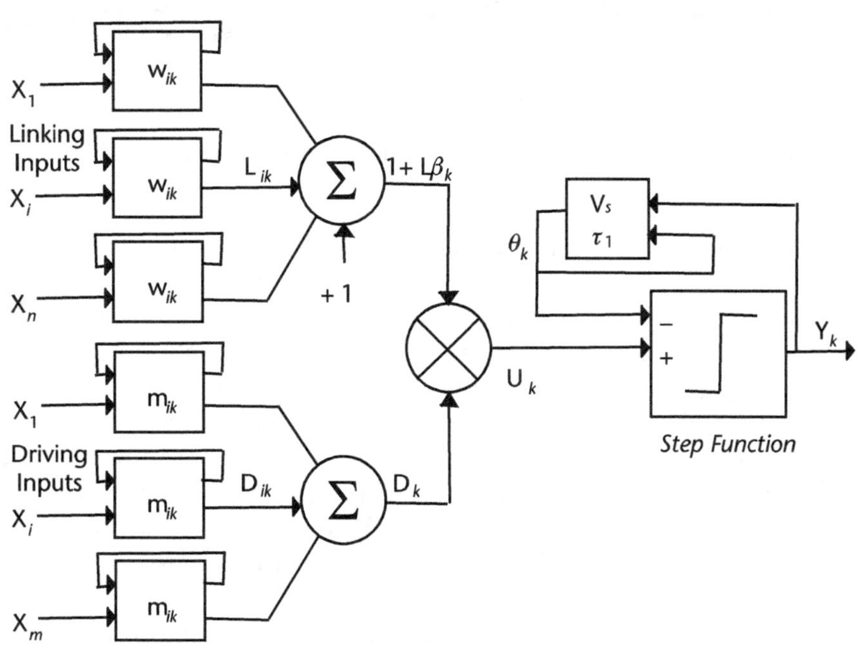

Figure 4.8 PCNN model.

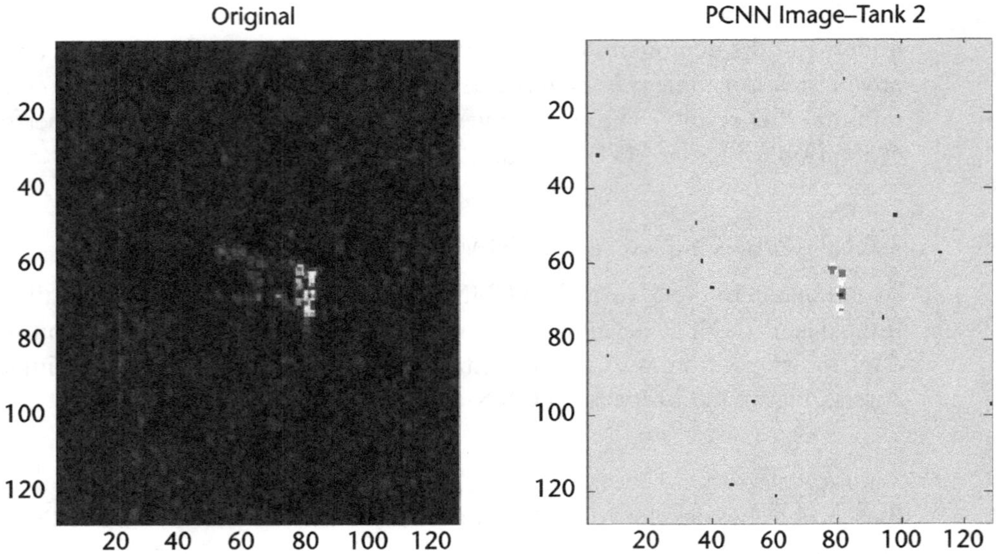

Figure 4.9 PCNN from MSTAR target.

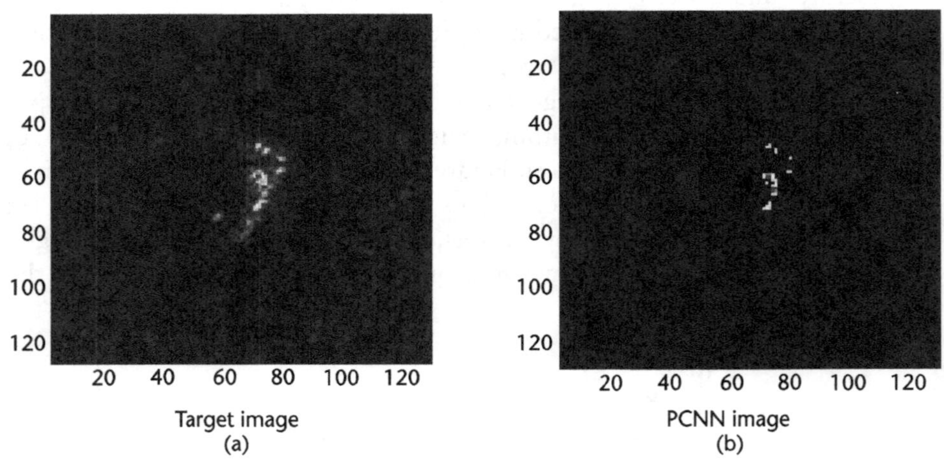

Figure 4.10 PCNN analysis: (a)target image, and (b) PCNN image.

orientation, size, and general features of the target of interest. The algorithm robustness is dependent on the measuring system and the expected learned-feature associations. Using the extracted features, the belief-confirming process propagates weighted features, fuses feature information over time and space, and declares the target type. Before confirming the target type, a conflict measure is performed to assess the plausibility of the target given previously accumulated information. Two runs were made. The first addressed the time delay of the PCNN for belief confirming, where the computational time step was increased with each run until a target was confirmed (Table 4.2, column 3). Assuming a fixed PCNN processing rate, the second run assessed sequential images with a changing orientation (Table 4.2, column 4).

From the MSTAR data set, 12 targets with 15° depression angles were used for training over 360° aspect angles. The 17° depression targets were sequestered for

Table 4.2 Estimated Time for Decision Making

Target	*Radius*	*Test 1*	*Test 2*
BMP2	100	(110, 5)	(50,9*)
BTR70	100	(100, 5)	(50,6)
T72	100	(50, 5)	(50,2)

I = # images; TD = time delay.

detection testing. In this study, the performance of the PCNN for target recognition was assessed from which the gray scale images were used. Figure 4.11 shows PCNN results from the training image, and Figure 4.12 shows the matched image (for automatic target recognition) of the resulting test image.

Researchers continue to expand on the basic PCNN concept [13]. For example, Figure 4.13 presents results from the 12-target MSTAR target database (as opposed to the standard three-target publically released target set). These results reveal that the PCNN improves the PCC by using multiple passes from the PCNN (or multiple looks). The final results include an average PCC of 80%. Given the development of the multilook classifier decision-level fusion results for SAR ATR before the explosion of deep learning methods, then the method is a baseline for comparison shown in Table 4.3.

4.2.1.3 Convolutional Neural Networks

CNNs are regularized versions of MLPs. The CNN follows from deep learning ideas rooted in *connectionist* neural networks from which a single perceptron (1 layer) could not perform the XOR function. The challenge was quickly solved by adding another layer in the network (2 layers) to achieve an MLP so that the middle layer learned the XOR function, while the final layer did the classification. An

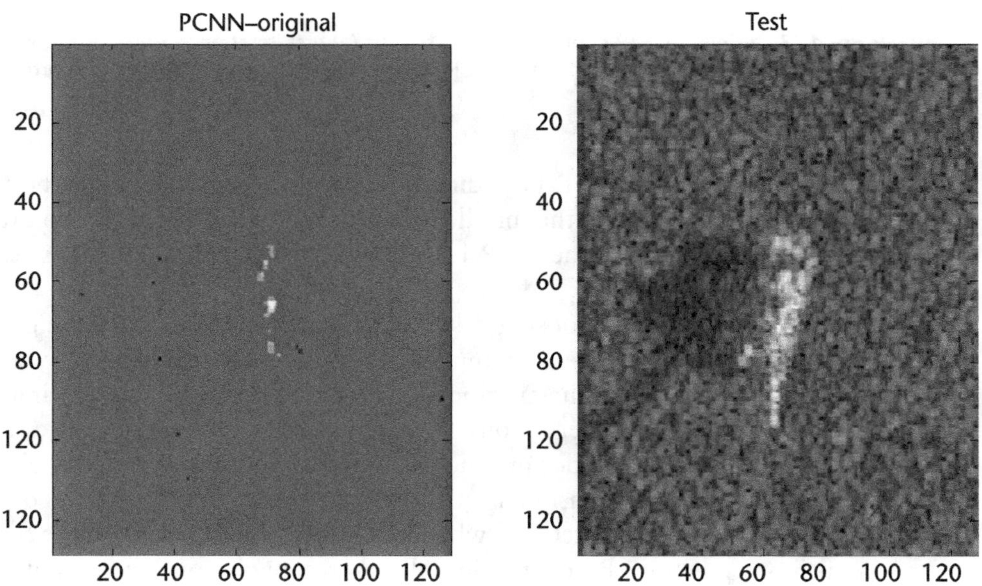

Figure 4.11 PCNN results.

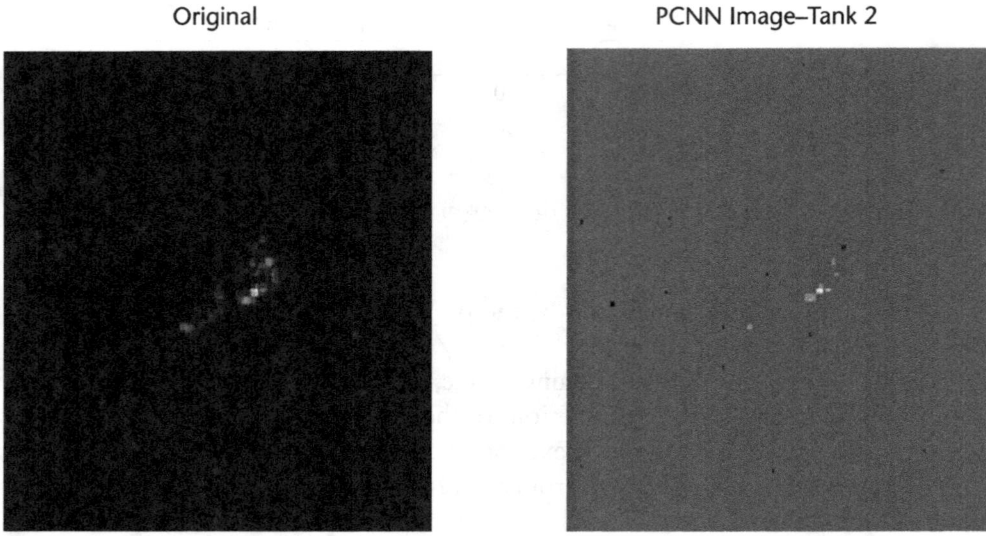

Figure 4.12 PCNN match comparisons to images.

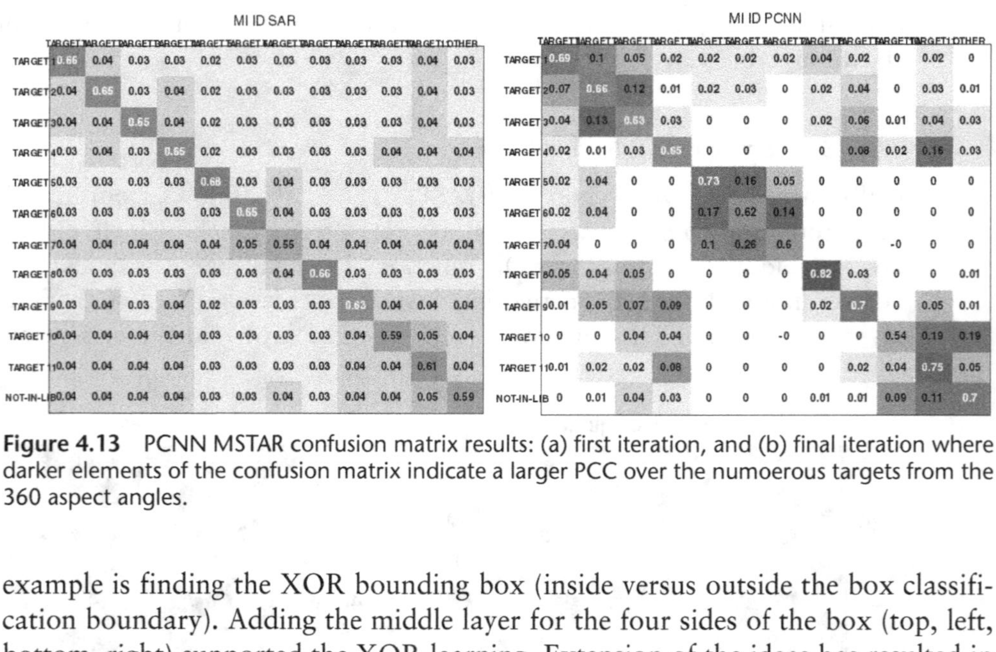

Figure 4.13 PCNN MSTAR confusion matrix results: (a) first iteration, and (b) final iteration where darker elements of the confusion matrix indicate a larger PCC over the numoerous targets from the 360 aspect angles.

example is finding the XOR bounding box (inside versus outside the box classification boundary). Adding the middle layer for the four sides of the box (top, left, bottom, right) supported the XOR learning. Extension of the ideas has resulted in the CNN.

The CNN includes choosing a kernel as a receptive field filter (e.g., a 5 × 5 bounding box in an image), learning the classification, and typically employing backpropagation with stochastic gradient descent to learn one layer at time. During the forward pass, each filter is convolved across the width and height of the input volume by computing the dot product between the input and the filter elements in order to produce a 2-dimensional activation map. Essentially, *network convolution* learns the best filters that activate when it detects a specific feature type for a given spatial input position. By connecting all other convolution layer results, reasonable classifications are obtained subject to the appropriate choice of the number of

Table 4.3 MSTAR Classification

Method	*Performance (%)*			*Reference*	
	SOC	*EOC-1*	*EOC-2*		
Pulse Couple Neural Network (PCNN)	87.0	75	67	Blasch, 1998	[11]
Conditional False Alarm Rate (CFAR)	89.0	91	85	Mossing, *et al.*, 1998	[16]
Extended maximum average (EMACH)	88.0	77	68	Singh, *et al.*,2002	[17]
Support Vector Machine (SVM)	90.0	81	75	Zhao, *et al.*, 2001	[18]
Conditional Gaussian	92.0	80	79	O'Sullivan, *et al.*, 2001	[19]
AdaBoost	92.0	82	78	Sun, *et al.*, 2007	[20]
Bayesian compressive sensing (BCS)	92.0	–	–	Zhang, *et al.*, 2013	[21]
Sparse Representation of Monogenic Signal (MSRC)	93.6	88.4	–	Dong, *et al.* 2014	[22]
Iterative graph thickening (IGT)	95.0	85	80	Srinivas, *et al.*, 2014	[23]
Modified Polar Mapping Classifier (M-PMC)	98.8	–	97.3	Park, *et al.*, 2014	[24]
Monogenic scale space (MSS)	96.6	98.2	–	Dong, *et al.*, 2015	[25]
A-ConvNets	99.1	96.1	98.9	Chen, *et al.*, 2016	[26]
AFRLeNet	99.4	–	–	Profeta *et al.*, 2016	[27]

Adapted from Srinivas et al. 2014 and Chen et al. 2016.

layers. The availability of more exemplar data increases the likelihood the system would be able to perform well with increasingly complex situations. The advantage of the CNN is that the kernels do not need the full transformation equations, and with an AE as an ANN, neighborhood functions can be used to filter the results for parametric reduction. The optimum number of layers in a CNN is part of ongoing work; for example, hyperparametric learning has replaced the sigmoid function in traditional NNs to include the ReLU for gradient analysis.

In CNNs, the convolution and fully connected layers perform linear transformations on their inputs. Figure 4.14 highlights the typical layers of a CNN architecture that begins with the convolution layer (C_0), followed by additional layers of sampling (S_i) and convolutions (C_i). Using the activation and pooling layers introduces nonlinearity that searches the input space for very specific features. Each filter creates a feature map that gets fed into the next layer to be filtered again. CNNs create very high-dimensional representations of inputs depending on the number of filters that are used and as a result, must be resized with pooling and

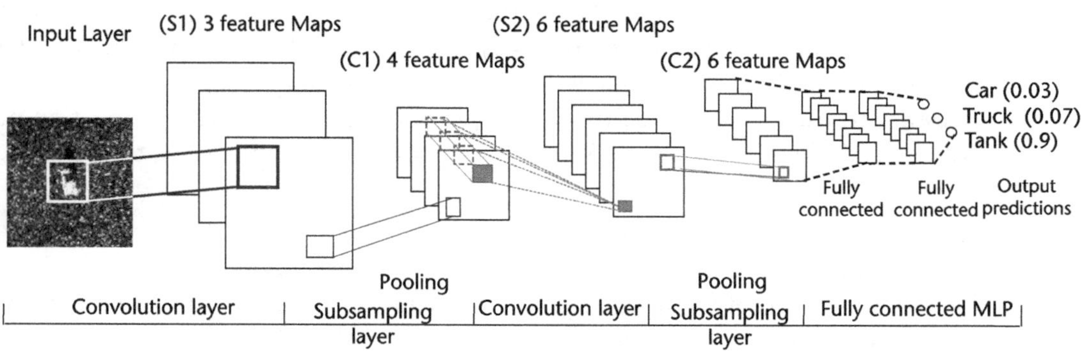

Figure 4.14 CNN concept.

dow-sampling layers. Finally, a fully connected MLP (e.g., Softmax) is typically applied in the final layer in order to determine the output training classifications. The CNN is widely accepted as a contemporary DL method showing great promise for radio frequency classification.

Convolution Layers

The data input is a tensor with shape $T = w \times h \times d \times I$, where w = image width, h = image height, d = image depth, and I = number of images. For example, $d = 1$ in processing magnitude imagery in SAR, $d = 2$ for complex-valued SAR imagery. However, for hyperspectral data, d = number of spectral bands. The convolutional layer abstracts a feature map consisting of three hyperparameters: $w \times h$ convolutional kernels, the number of input channels of the feature map $C_I = d$, and the number of output channels C_O. Three hyperparameters control the size of the output channel: the depth, stride, and zero padding.

- *Depth* controls the number of neurons K in a layer that connect to the same region of the input volume. These neurons learn to activate for different features in the input, such as oriented edges (e.g., target size) and intensity blobs (e. g., SAR shadows).
- *Stride* allocates the spatial dimensions (i.e., width and height). A stride of S causes the filter to be translated by S units at a time per output. $S = 1$ moves the filters one pixel at a time but results in overlapping receptive fields between the columns and large output volumes. When the stride is 2, then the filters jump 2 pixels. In practice, $S \geq 3$ is rare, as the receptive fields overlap less (for a 5×5 receptive field).
- *Zero padding* appends border zeros P to the input volume determining the output volume spatial size. The spatial size of the output volume can be computed as a function of the input volume size T, the kernel field size K, the applied stride S, and the amount of zero padding P. The formula for the number of neurons that fit into a given volume is

$$N = \frac{T - K + 2P}{S} + 1 \tag{4.11}$$

If N is not an integer, then the neurons cannot be tiled to symmetrically to fit across the input volume. Hence, for $S = 2$; then $P = (T + K - 2)/2$ ensures that the input volume and output volume will have the same spatial size. However, it is not necessary to use all of the neurons of the previous layer, from which a portion of padding is desired. In Figure 4.15, if $T = 9 \times 9 = 81$, $S = 2$, and $K = 8$, then $P = 31$.

Convolution Layers

A convolution layer convolves the result $C = I^*F$ where it is pointwise multiplication (row*column) of the image I pixel values and the filter F. For example, from Figure 4.15 for the second stride and the first filter (shown as element row 1, column 2 in the convolution matrix):

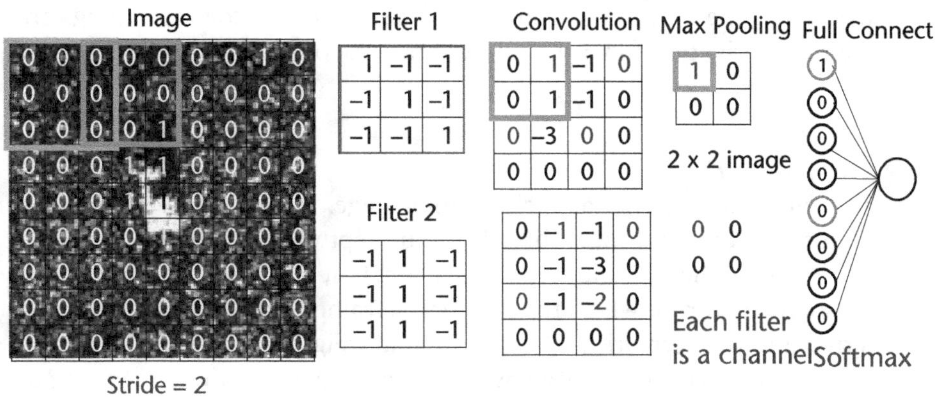

Figure 4.15 CNN example.

$$\begin{bmatrix} 0 & 0 & 0 \\ 0 & 0 & 0 \\ 0 & 0 & 1 \end{bmatrix} * \begin{bmatrix} 1 & -1 & -1 \\ -1 & 1 & -1 \\ -1 & -1 & 1 \end{bmatrix} =$$

$$\left[0(1)+0(-1)+0(-1)\right]+\left[0(-1)+0(1)+0(-1)\right]+\left[0(-1)+0(-1)+(1)(1)\right]=1$$

Pooling Layers

The pooling layer spatially resizes the input and operates independently on every input depth. The most common filters size is 2 × 2 where a stride of $S = 2$ downsamples at every input depth slice in the input by 2 along both width and height, discarding 75% of the activations. Using the ReLU, then the nonsaturating activation function $f(x) = \max(0, x)$ removes negative values from an activation map by setting them to zero. The max function increases the nonlinear properties of the overall network decision without affecting the receptive fields of the convolution layer. In Figure 4.15, each filter the pooling layer goes from 8 × 8 to 2 × 2. An example is shown in the boxes with thicker borders, where $\max \begin{bmatrix} 0 & 1 \\ 0 & -1 \end{bmatrix} = 1$.

Fully Connected Layers

Neurons in a fully connected layer have connections to all activations in the previous layer, as seen in the regular ANN (see Figure 4.7). The activations are given by a matrix multiplication followed by a vector addition of a bias offset. The final layer is the *loss layer*, which penalizes the deviation between the predicted (output) and the true labels. Many CNNs use the *Softmax* loss to predict a single class from K mutually exclusive classes:

$$P(c|y_i) = \frac{\exp(z_c(y_i))}{\sum_{k\in[1,\ldots,K]} \exp(z_k(y_i))} \tag{4.12}$$

where y_i is the output from the last hidden layer for corresponding target c, $z_c(\cdot)$ is the final activation function, and k is the one of target classes among the set [1, ..., K].

Dropout Layer

To prevent overfitting, at each training stage, individual nodes are either dropped out of the net with probability $(1 - p)$ or kept with probability p, reducing the network size. In the training stages, the probability that a hidden node will be dropped is usually 0.5, from which the network is trained. The removed nodes are then reinserted into the network with their original weights for testing. The use of a dropout layer improves training speed by reducing node interactions, allowing them to learn more robust features that better generalize to new data.

SAR Examples

Beginning in 2015, two papers presented results on the MSTAR data using CNN architectures. Malmgren-Hansen and Nobel-JØrgensen [14] developed a CNN using the features of annotated images that result in pixel intensity corresponding to the density among shadow and target regions. In the CNN, the architecture included convolutional, ReLU, max-pooling, and softmax FC-MLP layers. The CNN was able to segment the target from the radar shadow with a score of 83±6.5% with the SARBake code available on github. As another example, Wang, Chen et al. [15] developed a CNN for MSTAR classification and tested on 10 targets in the MSTAR data set. Their results presented in Table 4.3 demonstrate a superior approach of the All-Convolutional Networks (AConvNets) by comparing different methods, and these results have been replicated many times in the literature. Additionally, the results compared the SOC, as well as EOC of configuration variants (EOC-1) and version variants (EOC-2). The work in 2015 from the compiled results demonstrates significant improvement in the accuracy for SOCs.

Likewise, in 2016, Profeta et al. [27] achieved a PCC of 99.4% for the SOC for a seven-target MSTAR set. In addition to the standard CNN, Profeta et al. added a *dropout layer* to AlexNet, which they called AFRLeNet.

4.2.2 Sequential Neural Networks

The CNN is used for spatial and spectral data. For data that repeats over time, there are methods to learn dynamic sequences of patterns over time such as the RNN. Sequential networks are used in speech recognition, handwriting recognition, and recently for multimedia video data in combination of the CNN. In addition, sequential NNs are used for radar signals analysis, such as in communications, and could be adopted for moving SAR due to the nature of radio frequency signals. The most common sequential NN is the RNN.

4.2.2.1 Recurrent Neural Networks

RNNs have feedback loops in the learning process to determine sequential patterns. Some of the common approaches include:

- LSTM—locally-stored short-term results;
- Boltzmann machine—NN that receives stochastic inputs;
- Reservoir computing-(e.g., liquid state machine)—for low-power applications;
- Spiking neural network—randomly connected neurons.

RNNs allow a network to maintain information persistence via loops, as with GMMs. A given neural network A receives an input x_t and outputs a value y_t, according to a hidden (latent relationship) h_t. By passing data through iterative steps in the network, the loop is realized, and hence is recurrent. A RNN applies multiple copies of the same network, each passing a message to a successor as illustrated by unrolling the loop. The RNN chainlike architecture processes a sequential list of data, as shown in Figure 4.16.

RNNs can utilize previous information to determine current decisions. All RNNs utilize a chain of repeating modules of a standard NN, such as a single tanh layer, as shown in the middle functional block in Figure 4.17. RNNs are useful for learning a sequence of information over a short time interval over which the gap between the relevant information does not change much, as with the following: (1) a single sensor observing a vehicle conducting a stationary turning maneuver, or (2) a moving sensor observing a stationary target. As the gap between the relevant information timepoint and the time wherein it is needed becomes very large, RNNs become less accurate. One way to enhance performance is to exploit contextual information in order to mitigate such larger time intervals.

Many approaches to radio frequency data seek to develop methods based on DNN and CNNs, while the RNN is implemented with the LSTM or gated recurrent unit (GRU). LSTM architectures are developed in the next section. Note that a GRU has fewer gates than the LSTM and thus requires less memory. Since the RNN is designed for 1D signals analysis, the example presented here demonstrates the RNN for RF signals analysis, while a LSTM approach is presented in the next Section for SAR.

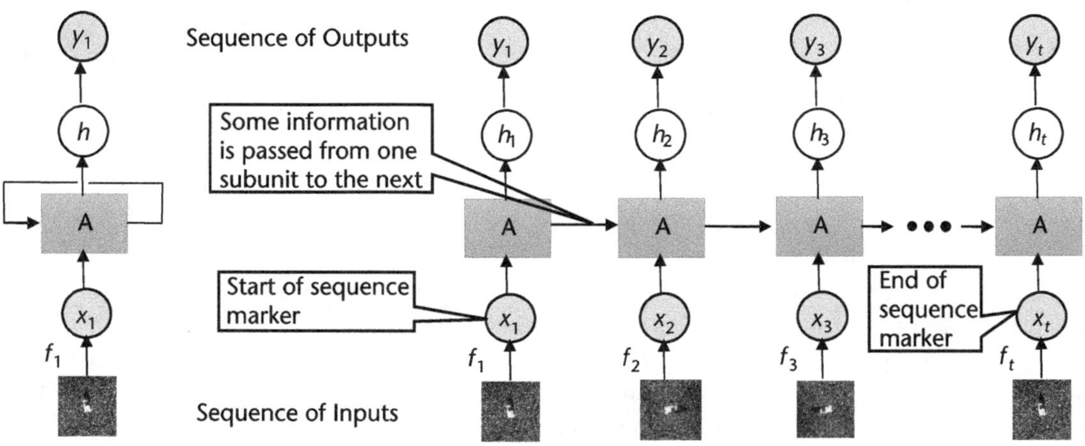

Figure 4.16 (a) RNN loop, and (B) RNN rolled out.

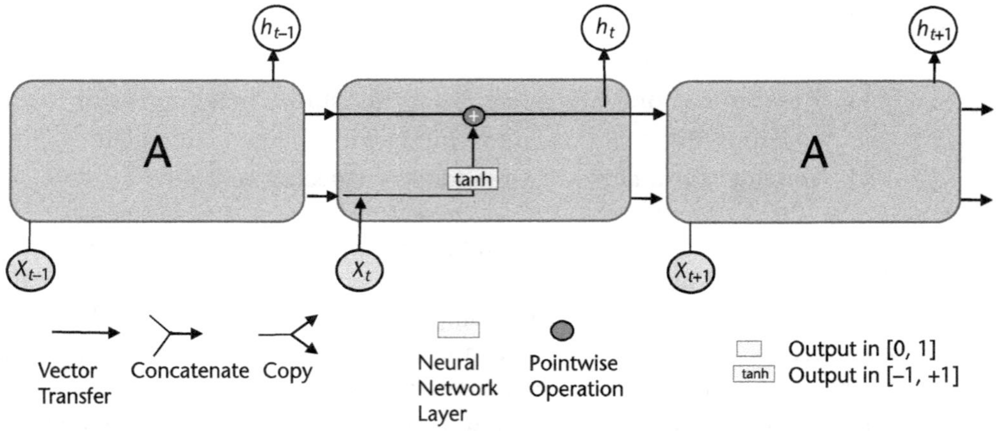

Figure 4.17 Standard RNN repeating module contains a single layer.

SAR Examples

In [28], Roy et al. designed a sequential RNN analysis with a GAN for RF transmitter identification. The radio frequency adversarial learning (RFAL) framework exploits transmitter specific signatures like the in-phase (I) and quadrature (Q) imbalance (i.e., the I/Q imbalance) present in all transmitters by learning feature representations using a DNN that uses the I/Q data from received signals. After elimination of the adversarial transmitters, the trusted transmitters are classified using a CNN, a fully connected DNN, and an RNN. The RNN is combined with LSTM and GRU discussed in the next section. Experimental results from collected data (Figure 4.18, Table 4.4) reveal that the CNN, DNN, and RNN (shown in Figure 4.19) can correctly distinguish between the eight trusted transmitters (Table 4.5). The parameters of the transmitters include: a gain of 45 dB, a frequency of 904 MHz (ISM), a bandwidth of 200 KHz, a sample of size 1024, and a value of the ratio of samples/transmitter to be 40,000 for 1 to 8 transmitters.

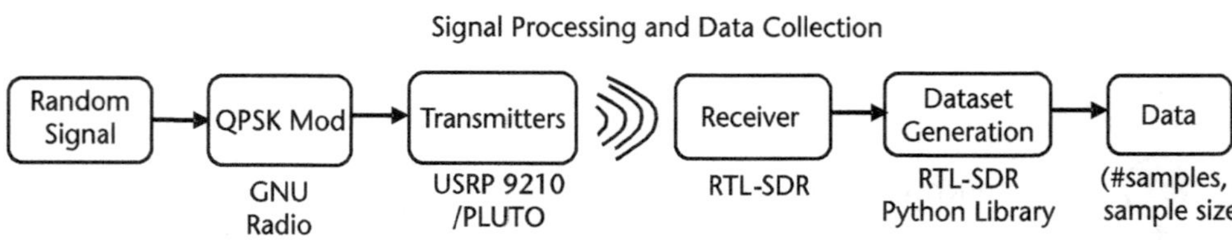

Figure 4.18 RF Signal data collection.

Table 4.4 Data Collection

	# Transmitters	*Type*	*Rows*	*Columns*	*SNR*	*Size*
Test 1	4	USRP B210 transmitters	160K	2048	30 dB	6.8 GB
Test 1	8	USRP B210 transmitters	320K	2048	30 dB	13.45 GB
Test 2	1	USRP B210 transmitter	80K	2048	30 dB	3.31 GB
Test 2	1	PLUTO transmitter	80K	2048	30 dB	2.85 GB
Test 3	8	USRP B210 transmitters	320K	2048	20, 10, 0 dB	13 GB

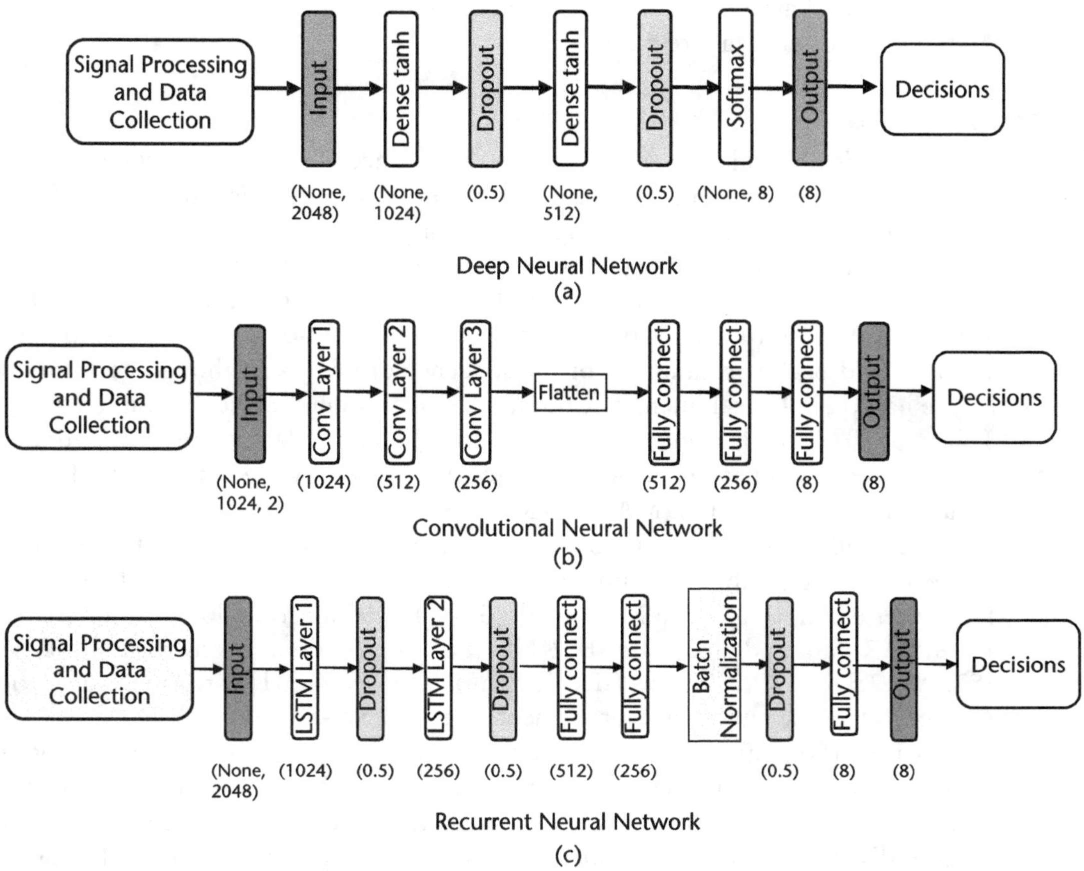

Figure 4.19 (a) DNN, (B) CNN, and (C) RNN.

Table 4.5 Accuracy Comparison of NN Approaches to Classify 4 and 8 Transmitters

#Trans	*Model*	*#Parameters*	*Acc (%)*	*Time (min)*
4	CNN	38 million	89.07	~ 25
4	DNN	6.8 million	96.49	~ 12
4	RNN-LSTM	14.2 million	97.40	~ 12
4	RNN-GRU	10.7 million	97.85	~ 12
8	CNN	38 million	81.59	~ 30
8	DNN	6.8 million	94.60	~ 15
8	RNN-LSTM	14.2 million	95.78	~ 16
8	RNN-GRU	10.7 million	97.06	~ 16

4.2.2.2 Long Short-Term Memory

The LSTM RNN learns long-term relationships between data by explicitly avoiding the long-term dependency problem. Hence, the LSTM remembers information for long periods of time. The LSTM chain structure employs a repeating module different than that of the standard single NN layer to include four layers (indicated by the four boxes in Figure 4.20).

SAR Examples

The LSTM is not widely used for SAR analysis, since the images are collected one at a time. If the image data is collected over multiple aspect articulations (e.g., 0° to 360°); then the images can be arranged according to a sequence as shown in Figure 4.21. Zhang et al. [29] sought to use a series of multiaspect images and design a bidirectional LSTM RNN to learn spatially varying scattering features. As with the CNN, MLP and Softmax layers are employed for classification after the LSTM does forward and backward analysis of the sequence of images. Although each target image has 16,384 pixel intensity values, the size of input features for bidirectional LSTM is 1024, the learning rate is 10^{-7}, and the LSTM layers are 512, 256 and 128, respectively. Zhang et al. developed the multiaspect aware bidirectional LSTM recurrent neural networks (MA-BLSTM).

The results for the volume of training data used are presented in Table 4.6. Table 4.7 updates Table 4.3 with contemporary results on the MSTAR data.

Other examples include Minh et al. [30] who demonstrated LSTM for space-based SAR data and compared the KNN (86.1%), random forest (87.1%), SVM (87.3%), LSTM (89.2%), and GRU (89.8%). Wang et al. [31] performed comparisons of POLSAR, revealing improvement on the LeNet-5 CNN.

As a final comparison on the MSTAR data set, Amrani and Jiang [32] proposed a combination of a pretrained VGG-S net, followed by deep feature extraction, and discriminant correlation analysis feature fusion. In addition, they demonstrated a LogDet divergence-based KNN metric classifier for target classification. The SOC and EOC of configuration variants (EOC-1) and version variants (EOC-2) are shown in Table 4.8.

The DNN of CNN and RNN operate on deterministic data. However, there are uncertainties associated with the data collection that require stochastic approaches.

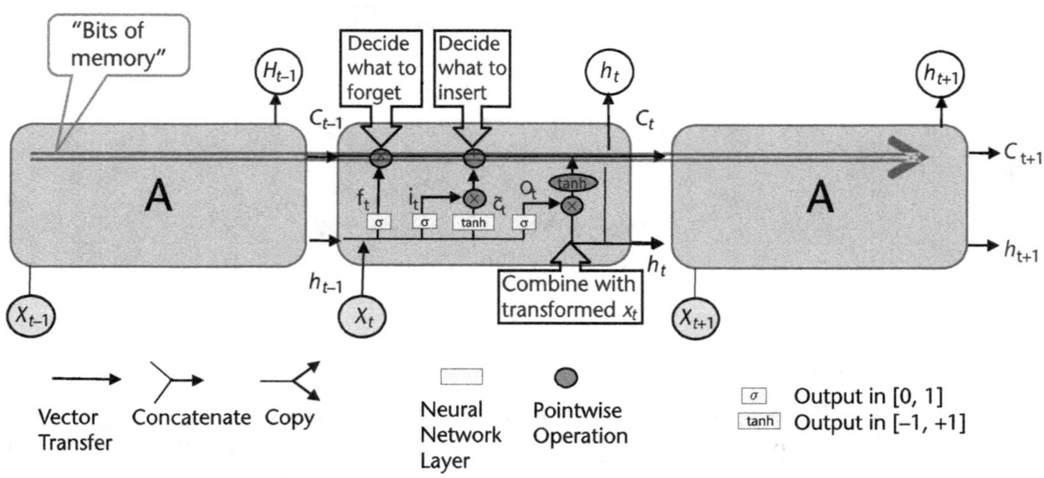

Figure 4.20 LSTM RNN repeating module contains four layers.

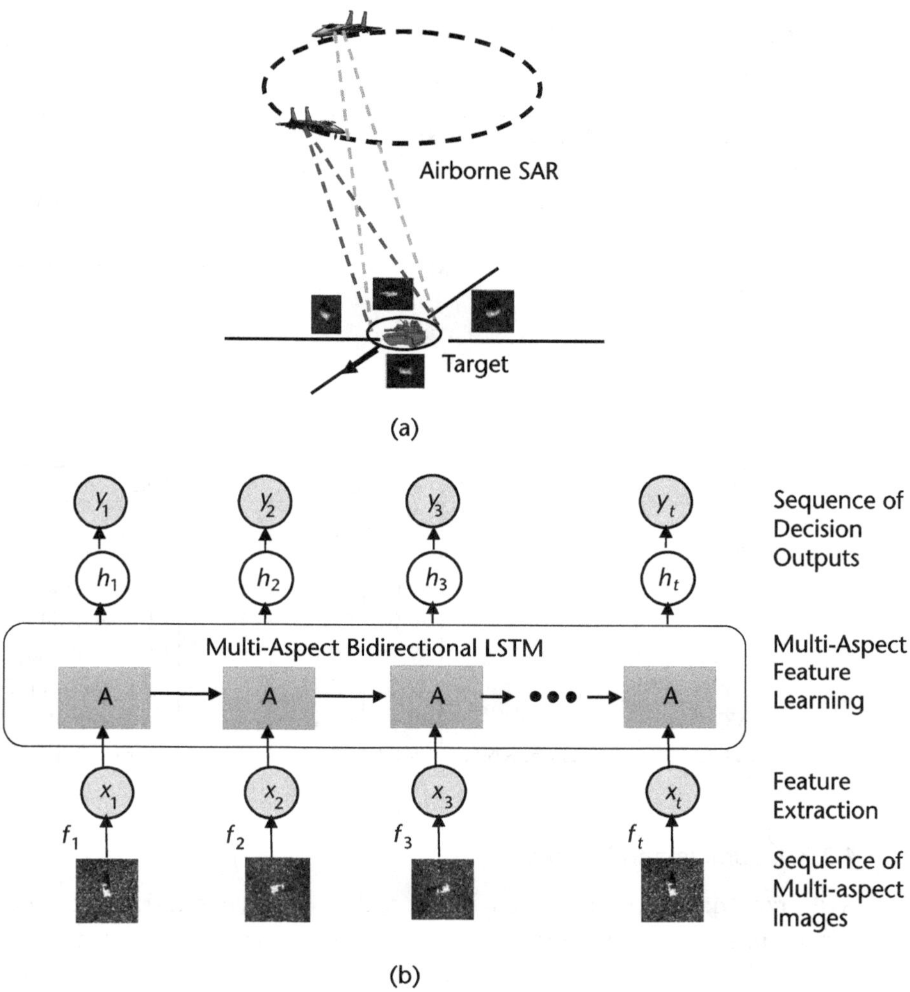

Figurde 4.21 (a) Multiaspect SAR collection, and (b) multiaspect RNN.

Table 4.6 Results from Multiaspect RNN

Training Percentage	100%	74%	47%	24%
Training Samples per Class	232	171	108	55
Training Epochs	150	191	287	750
Classification Accuracy	99.90%	98.36%	97.74%	95.79%

4.2.3 Stochastic Neural Networks

With many of the deep learning methods, especially for real-world applications, there is a need to understand the stochastic nature (or uncertainty) associated with the sensor data used to make decisions. The uncertainty comes from the sensor electronics, data transmission, and environmental noise. One popular method is the Boltzmann machine. In addition, these SNN methods support generative approaches for which uncertainty can enter into the simulation for generating exemplar data.

Table 4.7 MSTAR Classification

Method	*Performance (%)*			*Reference*
	SOC	*EOC-1*	*EOC-2*	
Pulse Coupled Neural Network (PCNN)	87.0	75	67	Blasch, 1998
Conditional False Alarm Rate (CFAR)	89.0	91	85	Mossing, *et al.*, 1998
Support Vector Machine (SVM)	90.0	81	75	Zhao, *et al.*, 2001
A-ConvNets	99.1	96.1	98.9	Chen, *et al.*, 2016
Multi-aspect aware bidirectional LSTM RNN (MA-BLSTM)	99.90	–	99.59	Zhang, *et al.* 2017

Adapted from Srinivas et al. 2014 and Chen et al. 2016.

Table 4.8 MSTAR Classification from PCNN to VGG

Method	*Performance (%)*			*Reference*
	SOC	*EOC-1*	*EOC-2*	
PCNN	87.0	75	67	Blasch [11]
CFAR	89.0	91	85	Mossing et al. [16]
SVM	90.0	81	75	Zhao et al. [18]
A-ConvNets	99.1	96.1	98.9	Chen et al. [26]
CNN-SVM	99.5	95.75		Amrani et al. [33]
VGG-S1	98.8	94.15		Amrani et al.
VGG-S, feature fusion, KNN	99.82	96.16		Amrani et al.

4.2.3.1 Boltzmann Machine

A Boltzmann machine, which is also called a stochastic Hopfield network with hidden units, is a type of stochastic recurrent neural network that employs a Markov random field. Boltzmann machines are a generative counterpart of Hopfield networks and one of the first NNs that are capable of learning internal representations over combinatoric problems. The difference in the global energy results from a single unit i equaling 0 (off) versus 1 (on) and is written as ΔE_i. The value of ΔE_i assumes a symmetric matrix of weights and is given by:

$$\Delta E_i = \sum_{i<j} w_{ij} s_j + \sum_{i>j} w_{ji} s_j + \theta_i \tag{4.13}$$

where:

- w_{ij} is the connection strength between unit j and unit j;
- s_i is the state, $s_i \in \{0,1\}$ of unit i;
- θ_i is the bias of unit i in the global energy function, and $-\theta$ is the activation threshold for the unit.

Often the weights w_{ij} are represented as a symmetric matrix $W = [w_{ij}]$ with zeros along the diagonal.

4.2.3.2 Restricted Boltzmann Machine

Restricted Boltzmann machines are probabilistic, wherein the model assigns probabilities as opposed to assigning discrete values. At each point in time, the RBM lies in a particular state. The units in the Boltzmann machine are divided into *visible* units **v** and *hidden* units **h**. The visible units are those that receive information from the *environment* (i.e., the training set) and is a set of binary vectors over the set **v**. The state refers to the values of neurons in the visible and hidden layers **v** and **h**, respectively. The probability that a certain state of **v** and **h** can be observed is given by the following joint distribution function:

$$p(\mathbf{v},\mathbf{h}) = \frac{1}{Z}e^{-E(\mathbf{v},\mathbf{h})} \tag{4.14}$$

where

$$\mathrm{Z} = \sum_{\mathbf{v},\mathbf{h}} e^{-E(\mathbf{v},\mathbf{h})} \tag{4.15}$$

Here, **Z** is called the partition function that is the summation over all possible pairs of stochastic binary visible vectors **v** and hidden stochastic binary vectors **h**. The update of the weight matrix occurs during the *contrastive divergence* step. The vectors $\mathbf{v}_0$ and $\mathbf{v}_\mathrm{k}$ are used to calculate the *activation probabilities* for the hidden values $\mathbf{h}_0$ and $\mathbf{h}_\mathrm{k}$. The difference between the outer products of those probabilities with the input vectors $\mathbf{v}_0$ and $\mathbf{v}_\mathrm{k}$ yields the update matrix:

$$\Delta W = \mathbf{v}_0 \otimes p(\mathbf{h}_0|\mathbf{v}_0) - \mathbf{v}_k \otimes p(\mathbf{h}_\mathrm{k}|\mathbf{v}_\mathrm{k}) \tag{4.16}$$

The energy function for the RBMs is defined as:

$$\begin{aligned} E(\mathbf{v},\mathbf{h};\theta) &= -\mathbf{v}^T W\mathbf{h} - \mathbf{b}^T\mathbf{v} - \mathbf{a}^\mathrm{T}\mathbf{h} \\ &= \sum_{i=1}^{D}\sum_{j=1}^{F} \mathrm{W}_{ij}v_i h_j - \sum_{t=1}^{D} b_i v_i - \sum_{j=1}^{F} a_j h_j \end{aligned} \tag{4.17}$$

where $\theta = \{W, \mathbf{b}, \mathbf{a}\}$ are the model parameters, W_{ij} represents the symmetric interaction term between the visible variable i and the hidden variable j, and b_i and a_j are bias terms. The architecture is shown in Figure 4.22, showing the visible and hidden nodes.

However, the structure of the hidden nodes for each layer are organized according to a *directed graph*. In Figure 4.22, for the first RBM, there are three visible $v \in \{0, 1\}^3$ and three hidden layers of two states each for $h^{(3)} \in \{0, 1\}^2$, so $W \in \mathrm{R}^{2\times 3}$.

4.2.3.3 Deep Boltzmann Machine

A DBM, shown in Figure 4.23, is a type of binary pairwise Markov random field (i.e., an undirected probabilistic graphical model) with multiple layers of hidden

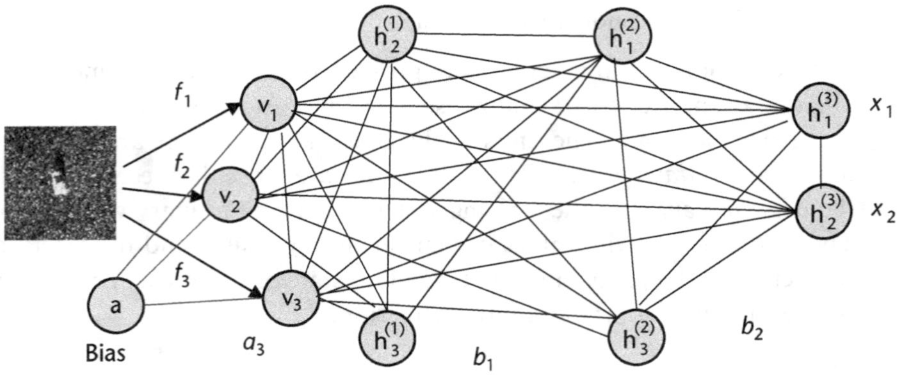

Figure 4.22 Boltzman machine.

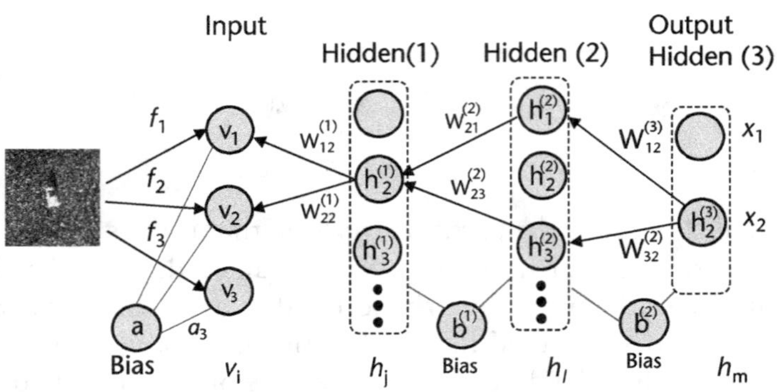

Figure 4.23 Deep Boltzman machine.

random variables. It is a network of symmetrically coupled stochastic binary units for explicit uncertainty analysis. It comprises a set of visible units $v \in \{0,1\}^D$ and layers of hidden units $h^{(1)} \in \{0,1\}$, $h^{(2)} \in \{0,1\}^{F_2}$, ..., $h^{(L)} \in \{0,1\}^{F_L}$. In the DBM, there is no connection linking units of the same layer (as with the RBM). For the DBM, the probability assigned to vector v is:

$$p(v)=\frac{1}{Z}\sum_{h} e^{\sum_{ij}W_{ij}^{(1)}v_i h_j^{(1)}+\sum_{jl}W_{jl}^{(2)}h_j^{(1)}h_l^{(2)}+\sum_{lm}W_{lm}^{(3)}h_l^{(2)}h_m^{(3)}} \tag{4.18}$$

where $h = \{h^{(1)}, h^{(2)}, h^{(3)}\}$ are the set of hidden units, and $\theta = \{W^{(1)}, W^{(2)}, W^{(3)}\}$are the model parameters, which represent visible-hidden and hidden-hidden interactions. In a DBN, only the top two layers form an RB). The RBM is an *undirected* graphical model, whereas the lower layers form a *directed* generative model. In a DBM, all layers are undirected and symmetric. The DBN is similar to the DBM.

Advantages

DBMs can learn complex and abstract internal representations of the input in tasks using *limited, labeled data* to fine-tune the representations built using a large supply of unlabeled sensory input data. However, unlike DBNs and deep convolutional neural networks, they adopt the inference and training procedures in both directions (bottom-up and top-down passes) that enable the DBM to better unveil the representations of the input structures. See Table 4.11 for the comparison analysis with GANs.

Disadvantages

The slow speed of DBMs limits their performance and functionality. Because exact maximum likelihood learning is intractable for DBMs, only approximate maximum likelihood learning is possible, as with the expected sufficient statistics via the MCMC. The MCMC approximate inference, which must be done for each test input, is about 25 to 50 times slower than a single bottom-up pass, thereby making joint optimization impractical for large data sets.

SAR Example

The RBM has been investigated by various researchers on the MSTAR database. Like the PCCN based on the visual cortical systems, Ni and Zu [33] used a sparse autoencoder to learn features from unlabeled data with RBM in the training layers to extract features. Using a 3-target MSTAR data set, they achieved 96% accuracy. Likewise, Deng et al. [34] used the RBM in a stacked AE to obtain an accuracy of 91.3% in the 3-target case and an enhanced AE-SVM to yield 94%. Chen et al. [35] compared the RBM to the SVM by training the DBN layer by layer and then used backpropagation for fine-tuning with these results using different kernels (i.e., polynomial, RBF, sigmoid). From the results, it is not clear that the DBN design develops advantages over the CNN.

4.3 Reward-Based Learning

This section reviews reinforcement, active, and transfer learning, all of which benefit learning efficiency. Efficient learning seeks to train the classifier more rapidly, to augment ATR human tasks, and to adapt for different domains.

4.3.1 Reinforcement Learning

Reinforcement learning seeks to resolve unlabeled input-output pairs without suboptimal error correction. RL finds a balance between the exploration of unknown knowledge and the exploitation of current knowledge. Typically, RL approaches are based on a Markov decision process (MDP) and utilize *dynamic programming* techniques (e.g., approximate DP and neuro-DP). Reinforcement learning algorithms do not assume knowledge of an exact mathematical model of the MDP and are used where exact models are infeasible. However, RL yields better results for

Table 4.9 MSTAR Classification Results from RBM

Shallow	*Accuracy*	*Deep*	*Accuracy*
Polynomial-SVM	44.5%	3-layer DBN	93.5%
Sigmoid-SVM	74%	4-layer DBN	92.5%
RBF-SVM	83.5%	5-layer DBN	89.5%

situations in which a model of the environment is known, but an analytic solution is not available; or for cases in which a simulation model of the environment is given (e.g., simulation-based optimization), as in generative adversarial modeling. RL software agents take actions to maximize the cumulative reward applied in game theory, control theory, operations research, information theory, simulation-based optimization, and genetic algorithms. In machine learning, the environment is typically represented as a MDP.

A MDP for RL model includes:

- A set of environment and agent states, S;
- A set of actions A of the agent;
- $P_a(s, s') = \Pr(s_t+1 = s' \mid s_t = s, a_t = a)$ is the probability of transition (at time t) from state s to state s' under action a.
- $R_a(s, s')$ is the immediate reward after transition from s to state $s'$with action a:
- A set of rules (stochastic) that describe what the agent observes.

The only way to collect information about the environment is to interact with it. The observation, either full or partial, of the current environment typically involves a scalar immediate reward associated with the last transition. A RL agent interacts with its environment in discrete time steps. At each time t, the agent receives an observation o_t, which typically includes the reward r_t. It then chooses an action a_t from the set of available actions, which is subsequently sent to the environment. The environment moves to a new state s_{t+1} and the reward r_{t+1} associated with the *transition* (s_t, a_t, s_{t+1}) is determined. The goal of a reinforcement learning agent is to collect as much reward as possible.

When the agent's performance is compared to an agent that acts optimally, the difference in performance gives rise to the notion of *regret*. In order to act near optimally, the agent must reason about the long-term consequences of its actions (i.e., maximize future income), although the immediate reward associated with this might be negative.

Reinforcement learning requires clever exploration mechanisms, such as in multiarm bandits, where random choices are available. Randomly selecting actions, without reference to an estimated probability distribution, generally yields poor performance. The case of small finite MDPs using *-greedy* (myopic), where $0 < \varepsilon < 1$ is a parameter controlling the amount of exploration versus exploitation. Exploitation is chosen with probability $1 - \varepsilon$, and the agent chooses the action it believes has the best long-term effect. Alternatively, exploration is chosen with probability ε, and the action is chosen uniformly at random. The probability is usually a fixed parameter but can be adjusted either according to a schedule, thus making the agent explore progressively lesser adaptively based on the situation.

An example of RL ATR is shown in Figure 4.24. In model-based approaches, a model is generated to predict expected features, while the search engine uses image processing to extract features. RL, on the other hand, has models as well as ground truth labels to compare performance. While the match function in model-based approaches determines the decision output, the RL uses the Q-value to learn the optimal policy for extracting the parameters that result in the best decision.

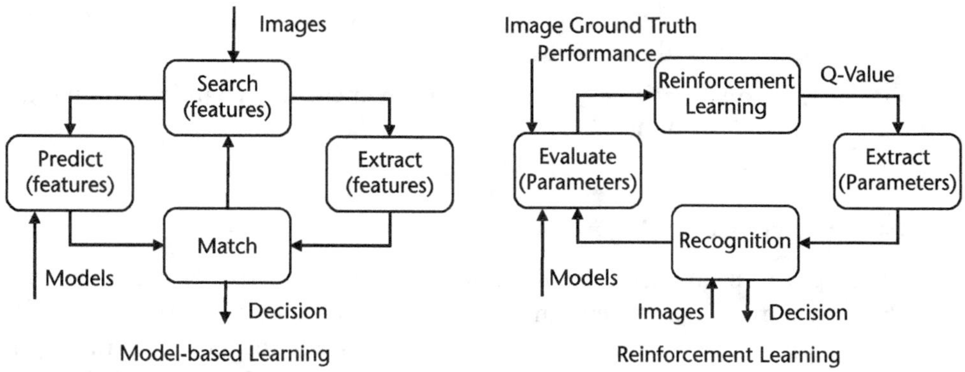

Figure 4.24 Model-based versus reinforcement learning ATR.

SAR Examples

RL was utilized in the 1990s for image segmentation and task learning, and yet was not directly implemented for target classification. Blasch and Bryant [36] developed a RL method based on a mutual information approach wherein the reinforcement is based on evidential reasoning in order to maximize mutual information. Using RL for the MSTAR data set, the system learns more rapidly toward optimal ATR but was not used to increase accuracy over the SVM. Using Q-learning, Bhanu et al. [37] utilized searching for performance results based on thresholds (T) and number of scatterers (N) as

$$Q(N,T) = Q(N,T) + \alpha\left[R - Q(N,T)\right]$$

where α is the rate ($\alpha = 0.1$) and R is the reinforcement. The trade-offs were shown that afford learning so an optimal set of parameters could be used for classification.

In 2003, Krawiec and Bhanu [38] utilize RL for data-driven synthesis of recognition procedures for real-world object recognition task. They used evolutionary computation, linear genetic programming (LGP), and cooperative coevolution for feature analysis. Using the 3-target MSTAR data, they achieved ~90% accuracy in 2,000 seconds. The elementary operations give the learner an access to complex, structural input data that otherwise could not be directly used and allows the learner to search for performance improvement by modifying the input representation. Building on these approaches, Ravichadran et al. [39] did an exhaustive study with results from a 5-target MSTAR data set with training/testing for SOC (99%, 90%) and then for EOCs (99.43%, 81.63%). A recent example [40] developed a spatial and polarimetric RL (SPRL) reward for polarimetric SAR data image segmentation comparing SVM, kNN, CNN, DRL and SPRL, with SPRL providing the characteristics of continuous learning and model-free.

RL is not applied to enhance accuracy, but for efficiency. Other current research topics include:

- *Active learning*: acting under partial information (e.g., using predictive state representation);

- *Transfer learning*: seeding knowledge from one domain to another (e.g., model-based to target-based ATR);
- *Lifelong learning*: continuous refinement in dynamic and changing conditions (e.g., EOCs).

4.3.2 Active Learning

RL algorithms are given feedback in the form of positive or negative reinforcement in a dynamic environment and are used in image segmentation. A similar concept is used in active learning by accessing the desired outputs (i.e., training labels) for a limited set of inputs based on a budget and optimizing the choice of inputs for which it will acquire training labels. When used interactively, these can be presented to a human user for labeling. Learning agents can provide labels by:

- Interacting with the environment (such as in a dynamic scenario with target variations);
- Assessing consequences of its actions, rather than from being taught explicitly and selects actions based on its past experiences (i.e., exploitation) and by new choices (i.e., exploration);
- Receiving a numerical reward for each of its actions that encodes a level of success and learns actions that maximize the reward.

SAR Examples

The idea of active learning for SAR ATR dates back to the development of radar imagery as humans were the classifiers. In the 1980s, with access to NN techniques, there were many approaches to align the machine classifier to augment human work classification tasks. In the 1990s, tests were conducted with machine analytics the operated over the observe-orient-decide-action (OODA) loop for SAR ATR with the human ATR tasks, as presented by Blasch [41]. Using the MSTAR examples, various approaches were reported that utilized active learning, such as incorporating negative examples for machine training [42], allowing the user to vary the classification thresholds to enhance man-machine performance [43], providing the user the ability to adjust the time, probability of hit rate, and probability of false alarm rate in multiview training [44], and incorporating context in the ATR cycle [45]. A second area of active learning focuses on the feature selection for which the MSTAR shadow analysis provides an example that leverages machine and human training for target segmentation [46–48]. Recent results incorporate active learning for image annotation [49] and dictionary-based label learning [50]. Since active learning enhances learning efficiency, it is also a future case that DNN methods could provide data efficiency methods in selecting the most intelligent generated images toward decision tasks.

4.3.3 Transfer Learning

DNN methods require a large training data set to produce accurate learning representation toward effective classification of the test data. Toward the end of the twentieth century, an ML approach leveraging the teaching of humans from educational psychology was presented as *transfer learning* (TL). For example, when a person learns to throw a baseball, then this skill can be transferred to that of throwing a football. The learning of the skill, throwing, is the task domain, while the learned object of a baseball is the source domain and the football is the target domain. Such a motor skill would be similar to image perception and mathematical comprehension, wherein an ATR operator learns to perform in different situations with only a few exemplars. In addition, there are different types of data that can be used: physical principles, measured data, and historical scenarios. Hence, TL seeks methods to learn from one task and to utilize that information for another task through demonstration, guidance, and experimentation. In Figure 4.25 the task of object detection has a given a source domain D_S and learning task T_S, a target domain D_T and learning task T_T, from which TL aims to help improve the learning of the target predictive function $f_T(x)$ in D_T using the knowledge in D_S and T_S, where $D_S \neq D_T$, or $T_S \neq T_T$. Since the radar model has similar attributes as if analyzing another frequency (e.g., electro-optical), the methods of training classifiers in visual imagery can be used as knowledge transferred to the SAR domain. The domain adaptation could leverage knowledge about variations of a source target and the ability to distinguish between two types of targets. Thus, when the situation changes, the TL can adapt to new situations and determine new target domains.

TL is useful when a new domain target data set is comparatively small. Table 4.10 provides some examples of data augmentation for the sensor, environment, and target (SET) domains for which operating conditions vary from one situation to the next. The idea is to use information that you already know and then to apply that knowledge to a new, similar data set. The application of functions already learned on data similar to a new dataset affords *domain adaptation* by selecting generalizable features from the training domains and specific features from the test domain. The domain D consists of: a feature space X and a marginal probability distribution $P(X)$, where $X = \{x_1, ..., x_n\}$. Given a specific domain $D = \{X, P(X)\}$, a task consists of two components: a label space y and an objective predictive function $f(\cdot)$ (denoted by $T = \{y, f(\cdot)\}$), which is learned from the training data consisting of pairs $\{x_i, y_i$, where $x_i X$ and $y_i Y$. The function $f(\cdot)$ can be used to predict

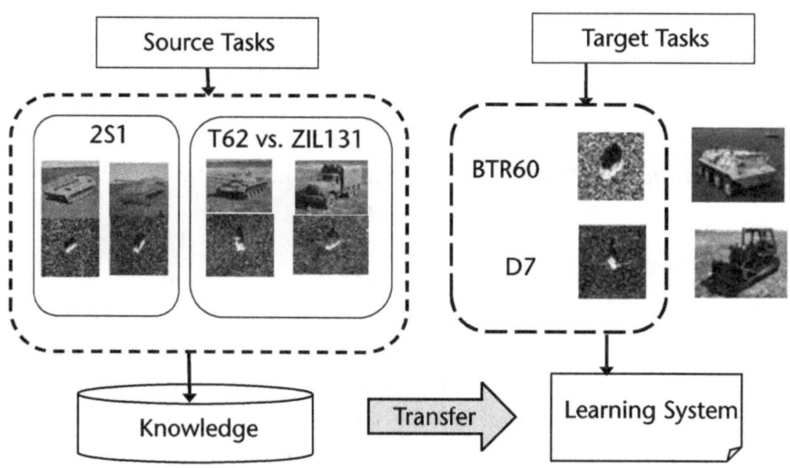

Figure 4.25 Transfer learning.

Table 4.10 Methods of Domain Transfer from First-Principle Physics

Domain Adaption	*Variation* Target	*Sensor*	*Environment*
Domain Transfer	Tanks to trucks	EO to SAR	High to low grazing angle
Model	Computer-aided design (CAD) models of SAR targets	Camera model as affine transfer applied to SAR image viewing	Electromagnetic code for SAR radar shadow representation in moving SAR
Representative Knowledge	Synthetic	Data collections	Historical scenarios

Table 4.11 GAN Comparisons

	Deep Directed Graphical Models	*Deep Undirected Graphical Models*	*Generative Autoencoders*	*Adversarial Models*
Model Design	Most models difficult	Careful design to ensure properties	Any differential function is theoretically permitted	Any differential function is theoretically permitted
Training	Inference needed during training	Inference needed. MCMC approximates partition function gradient	Enforced trade-off between mixing and power of reconstruction generations	Synchronizing the discriminator with the generator
Inference	Approximate inference	Variational inference	MCMC-based inference	Approximate inference
Sampling	Available	Requires Markov chain	Requires Markov chain	Available
Evaluation $p(z)$	Intractable, but can be approximated	Intractable, but can be approximated	Not explicitly represented, may be approximated with Parzen density estimation	Not explicitly represented, may be approximated with Parzen density estimation

Adapted from [55].

the corresponding label, $f(x)$, of a new instance x. Some target pair modeling techniques include:

- *Generative models*: Training and test data are generated from a mixture of different models, such as Dirichlet process prior to couple the parameters of several models from the same parameterized family of distributions;
- *Nonparametric models*: Boost the classifier with labeled examples that represent the true test distribution.

Three methods of TL include sampling, adaptation, and learning. As shown in Figure 4.26, the availability of labeled data and domain assumptions determine the TL approach. With labeled source data from a single domain, *sampling* can be used to assess the target classification task. If it is a different domain, such as target variations, *domain adaptation* is possible. Recent methods for inductive TL include multitask and self-taught learning, wherein some labeled data is in the target and source domains.

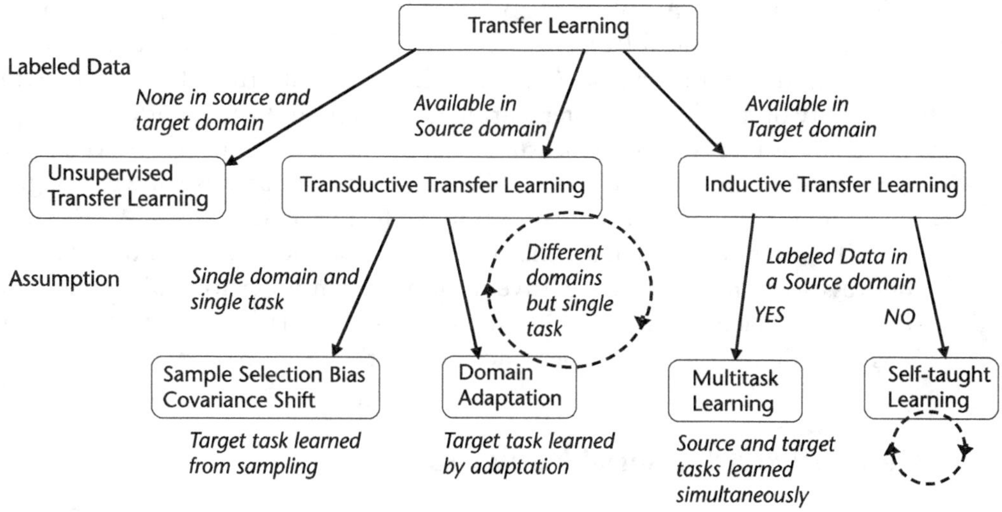

Figure 4.26 Transfer learning approaches.

Some of the research directions in TL include:

- *One-shot learning:* Rather than having thousands of training models per class, there is just one that takes advantage of knowledge from previously learned categories and the information from the training model to learn the new model. The goal is to apply the knowledge that has been gained so far, as opposed to starting from scratch each time to learn a new category/class.
- *Low-shot learning:* As with one-shot learning, having a few exemplar models of the source and target variations.
- *Bayesian networks:* Develop a directed acyclic graph of conditional probabilities, whose nodes correspond to random variables that can be discrete or continuous and which encodes X (to which the random variable is associated) has a direct influence on Y (with which the variable is associated). BN intuitively encodes causal and influence relationships, thus greatly facilitating the knowledge representation where domain expert knowledge must be formally represented.
- *Mark logic networks:* Combining logical and probabilistic reasoning for situations in which the knowledge base does not have strict first order logic pairs for data to labeling assignments.

SAR Examples

Recently, there have been some attempts at using TL for SAR ATR. Al Mufiti et al. [51] used three pretrained CNNs (AlexNet, VGG16, and GoogLeNet) for feature extraction versus image classification. The goal was to generalize knowledge about features in the source domain (15° data) and to transfer that knowledge for target domain (17° data), followed by an SVM, which demonstrated 99.27% accuracy with AlexNet. As it was a first attempt at developing a TL method, the source and target domain were relatively close. Subsequent efforts [52] used TL as a method of data augmentation to investigate the effects of transferring to a target domain with noisy images. Arnold et al. [53] developed another CNN pretrained Canadian Institute for Advanced Research (CIFAR) SAR source domain data set and created an adaptation of the original CIFAR-10 LeNet architecture. They demonstrated the feasibility of first-layer transfer learning, so that that diverse clutter and target types that are not represented in measured training data could be introduced in synthetic training data and later recognized in measured test data, as shown in Figure 4.27. Consistent approaches are used to transfer feature parameters [54].

The three reward approaches (reinforcement, active, and transfer learning) all address some notion of generative information in order to guide the learning process using machine, human, and data guidance. The interest in these approaches has been limited until the emergence of generative adversarial networks.

4.4 Generative Adversarial Networks

Generative adversarial networks are popular with many variations on the architecture, data sets, and training [62–64]. For more details, the GAN zoo (https://github.com/hindupuravinash/the-gan-zoo) provides a listing of available GANs. The fundamental design of GANs includes (data) generative models with (classifier) discriminative models. Discriminative algorithms (e.g., SVM correlation) attempt to classify input data as discussed in Chapter 3. That is, given the features of an instance of data, they predict a label or category to which that data belongs. To enhance discrimination, a generator is used to challenge the discriminator. Instead of predicting a label given certain features, the generators attempt to predict features given a certain label. A traditional generator model affords data augmentation, target variation, and corrupted exemplars. The generative model *G* in a GAN is different than the traditional data simulation model in that noise and bias is applied to enhance decision discrimination. The GAN, as its name implies, also provides generative malicious inputs designed to fool ML models. Injecting adversarial examples into the training set increases neural networks to robustly learn over incomplete, spurious, and manipulated data.

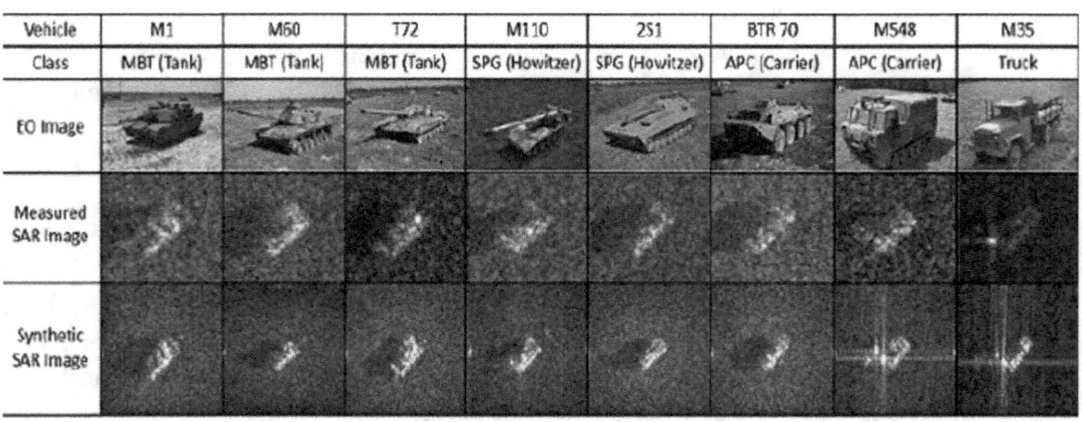

Figure 4.27 EO of each target with corresponding measured and synthetic SAR images. (From Arnold et al. [53]).

In GANs, the MLP generator creates new data instances, while the MLP discriminator evaluates them for authenticity. That is, the discriminator decides if each instance of data under review belongs to the actual training data set:

- *Discriminative models* learn the boundary between classes;
- *Generative models* model the distribution of individual classes.

Example 1: Object Recognition

Moving beyond ATR verification of collected data, the goal is to validate ATR in the real world. Since the ATR classifier does not know everything about the real world, model generation seeks exemplars of unknown possibilities. The goal of the discriminator, when shown an instance from the generator, is to recognize the data that is authentic. As the generator creates new, synthetic images that it passes to the discriminator, some are authentic (i.e., real instances) and some are imposters (i.e., fake instances). The generator seeks to pass plausible imposter target instances through the classifier, while the discriminator attempts to identify images coming from the generator as imposters.

Example 2: Information Retrieval

Moving beyond text classification (such as in a SAR intelligence product report), given all the words in a report, a discriminative algorithm predicts whether the message is factual (correct classifications: track, tank) or not-factual (imposter classifications: hovercraft). If not-factual data is one of the labels y, and the bag of words gathered from the report are the features that constitute the input data x, then $p(y|x)$ assesses the probability that a report is not-factual given the words it contains. A generator seeks to determine that if the report is not factual, how likely are these features? While discriminative models assess the relations between y and x, generative models care about how to get x. Generators allow you to capture $p(x|y)$, or the probability of features given a label or category $p(\text{features} \mid \text{nonfactual})$. Hence, GANs are useful for determining the likelihood of generating data that is not validated in the real world.

Generative Models

There are many deep generative models, such as graphical models, DBMs, and DBNs. Belief networks utilize Bayesian reasoning in a Markov chain to determine $p(c|y) = [p(y|c)\, p(c)]/p(y)$; where $p(y|c)$ is the likelihood and $p(c)$ is the prior instance probability of a class. A Boltzmann machine (with stochastic variables of the uncertainty and bias on c and y) is impossible to compute, so a DBN is used for the conditional of each layer as in Figure 4.30. Using the DBN, many layers of latent variables and millions of parameters can be learned efficiently, and high-level feature representations are successfully resolved (e.g., object recognition, information retrieval). The key attribute is conditional analysis among each layer.

$$p\left(y,h^1,h^2,\dots h^n,c\right) = p\left(y\middle|h^1\right)p\left(h^1\middle|h^2\right)\dots p\left(h^{n-2}\middle|h^{n-1}\right)p\left(h^{n-2}\middle|h^n\right)p\left(h^n\middle|c\right)p(c) \quad (4.19)$$

One of the limitations of the DBN is application in the real world. The DBN uses an explicit density that must be derived with the real-world classification priors

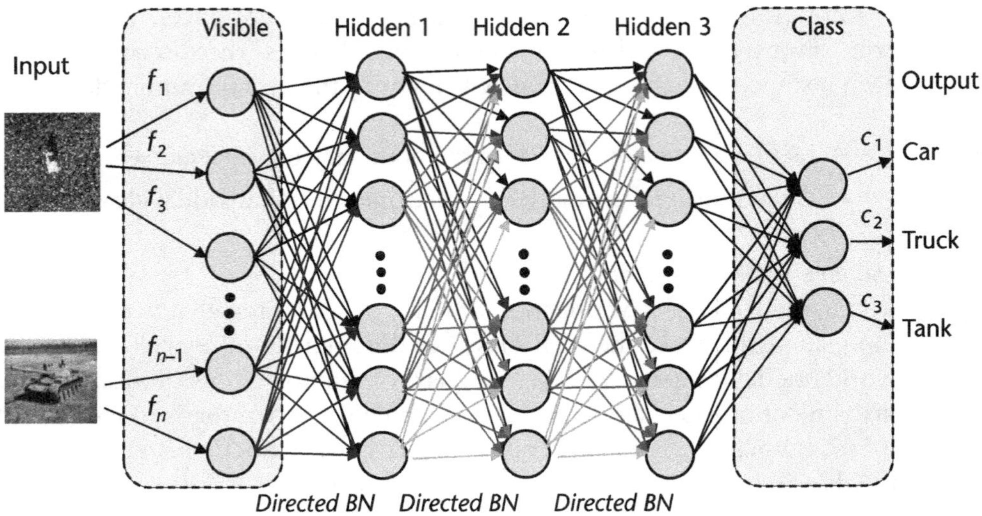

Figure 4.28 Deep belief network.

$p(c)$. In many situations, the real-world prior is not available, especially for data instances that have yet to be realized, such as the enormous potential of imposter data (e.g., synthetic images, emerging classifications). The DBN gives tractable results; however, other methods discussed provide approximate densities $p(y|c)$ using the maximum likelihood:

$$L(c|y) = \max_c \left\{ \prod_{i=1}^{m} p(y_i|c) \right\} \tag{4.20}$$

The advantage of the GAN is that it uses an implicit transition operator to determine the decision density. The different methods discussed in the chapter are shown in Figure 4.29. For example, instead of instances of the real observations y, generative instances are used z. The chapter highlights the current ML methods using the AE and the GAN, while other traditional methods employ a Markov chain such as MCMC sampling.

Generative algorithms align to these three types:

- Given some of the features, they predict the associated features (GMM, latent Dirichlet allocation);
- Given a label, they predict the associated features (naive Bayes, autoregressive);
- Given a hidden representation, they predict the associated features (VAE, GAN).

Generative Adversarial Networks

A GAN consists of two MLP models shown in Figure 4.30. The generative model G generates images, and the discriminative model D estimates the probability that, whatever data is received from G, it is from training data or it is from generator G itself. To understand the generator's probability distribution P_g over data (z), the

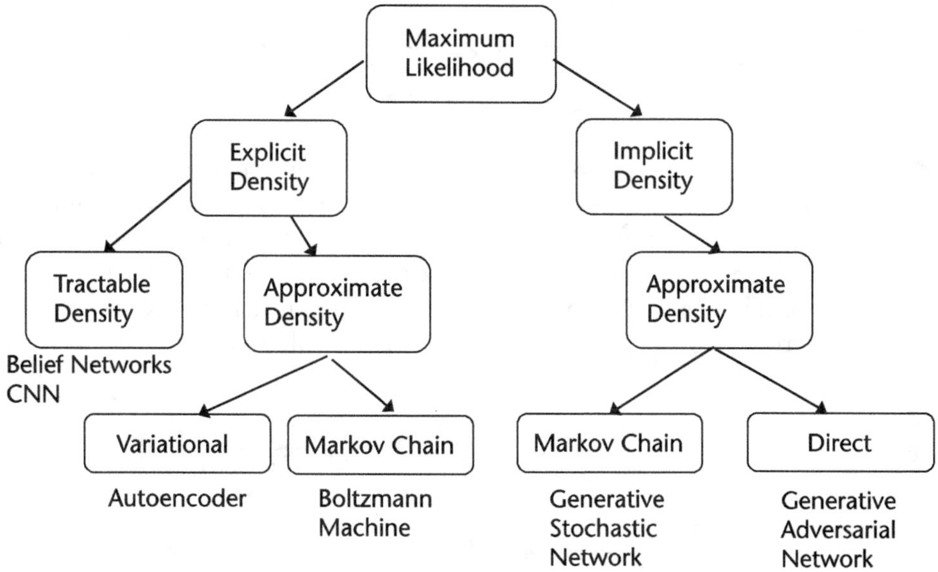

Figure 4.29 Explicit versus implicit densities.

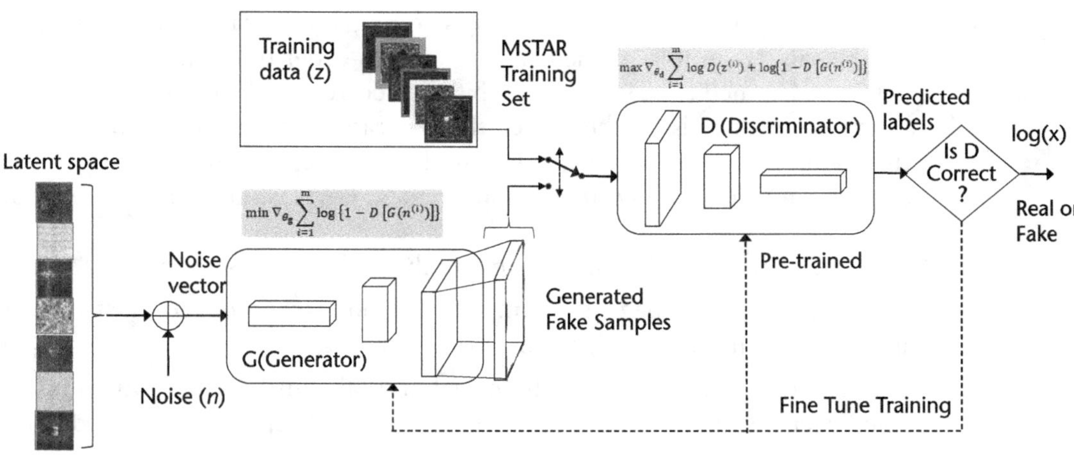

Figure 4.30 GAN method.

prior is defined on input noise variables $P_n(n)$, which represents a mapping to the image data space as $G(n; \theta_g)$, where G is a differentiable function represented by a MLP with parameters θ_g [55].

According to GAN design, a second MLP is defined as $D(y; \theta_d)$, which outputs a learned single scalar. The function $D(z)$ represents the images z that came from the data, not from the generator. The discriminator D was trained with real imagery in order to increase the probability of correctly recognizing z from the output y and the sample n as being produced from the generative model G. The loss or energy function of trained model G and D can be represented mathematically as:

$$\text{Generator} \quad \min \nabla_{\theta_g} \sum_{i=1}^{m} \log\left\{1 - D\left[G\left(n^{(i)}\right)\right]\right\} \tag{4.21}$$

$$\text{Discriminator} \quad \max \nabla_{\theta_d} \sum_{i=1}^{m} \log D\left(z^{(i)}\right) + \log\left\{1 - D\left[G\left(n^{(i)}\right)\right]\right\} \tag{4.22}$$

Both generator and discriminator models (or networks) try to optimize different and opposing objective functions, or loss functions, in a zero-sum game, which is essentially an actor-critic model. As the discriminator changes its behavior, so does the generator, and vice versa. The model losses push against each other. As the game progresses, each model player gets better at their respective action. Thus, the two-player zero-sum min-max game with the value function is:

$$\min_{\mathrm{G}} \max_{\mathrm{D}} V(\mathrm{D}, \mathrm{G}) = \sum_{i=1}^{m} \log D\left(z^{(i)}\right) + \log\left\{1 - D\left[n^{(i)}\right]\right\} \tag{4.23}$$

The generator is fed random data (which can be derived from a given image) as a seed and generates new imagery that, at the outset of training, is random noise. At the same time, the discriminator tries to distinguish between the collected data and the generated data. The generator G is trained with a loss metric that maximizes the probability D of making a mistake. By so doing, it learns to generate imagery that is on the same manifold as the original data. In order to obtain its objective, a loss metric is designed to maximize its ability to correctly identify generated and real images. Steady state is reached when the generator creates images that are no longer distinguishable from the target data set by the discriminator.

The processing required in order to implement at GAN is shown in Figure 4.30:

Step 1: The generator takes in random numbers and returns an image.

Step 2: The generated image is fed into the discriminator alongside a stream of images taken from the actual, ground-truth data set.

Step 3: The discriminator injects both real and fake images and returns probabilities, each being a number between 0 and 1, with 1 representing a prediction of authenticity and 0 representing the imposter.

Note the use of fine-tuning feedback loops, as the discriminator is in a feedback loop with the ground truth of the images (pretrained), and the generator is in a feedback loop with the discriminator. Hence, the system is dynamic. The discriminator down-samples (e.g., max-pooling) an image in order to produce a probability that categorizes images fed to it, such as a binomial classifier labeling images as real or imposter. The generator works as an inverse using a vector of random noise and upsamples it to an image. For image analysis, a generator seeks to create an increasingly realistic product. In contrast, the discriminator determines whether a given product is genuine or fabricated.

GANs are compared with other NN, such as AEs and VAEs, and improvements result when conditional relations are available such as $p(z|c)$ corresponding to knowledge about the sensor. Additionally, semisupervised GAN approaches utilize labeled features can improve efficiency and accuracy through intelligence sampling.

Autoencoders *encode* input data as vectors. They create a hidden, or compressed, representation of the raw data. They are useful in dimensionality reduction; that is, the vector serving as a hidden representation compresses the raw data into a smaller number of salient dimensions. Autoencoders can be paired with a decoder, which allows the machine to reconstruct input data based on its hidden representation, similar to a restricted Boltzmann machine.

Variational autoencoders are generative algorithms that add an additional constraint to encoding the input data—namely that the hidden representations are normalized. Variational autoencoders are capable of both compressing data, as with an autoencoder, and synthesizing data, like a GAN. However, while GANs generate data in fine, granular detail, images generated by VAEs tend to be more blurred. One example is the deep convolution GAN (DC-GAN) [56].

SAR Examples

Recent examples of GAN methods have been applied to the MSTAR data. Lewis et al. [57] noted that the synthetic images exhibit a higher contrast relative to the background and have a higher target-to-ground intensity ratio than that of the real images. They evaluated the DualGAN and SimGAN. The DualGAN calculates a correlated pixel loss function with a cycle-consistent constraint. SimGAN makes small changes and assesses consistency and similarity with a self-regularization term. The results provide realizable imagery with classifications around 95%. The SimGAN was sensitive to tuning parameters and produced good images while the DualGAN resulted in artifacts. Liu et al. [58] used self-similarity preserved GAN (SPGAN) consisting of an autoencoder and CycleGAN and a Siamese network (SiaNet). Using the SAR simulator OpenSARSim and the MSTAR data, the classifications with the synthetic data maintained robustness as the available MSTAR DL achieved 98.7%, the mixing of the real and synthetic achieved 97.7% for the SP-GAN-SAR. Zheng et al. [59, 60] used a GAN model based on multiple discriminators (MGAN) that employs a semisupervised CNN approach with label smoothing regularization (LSR) to strengthen the distinction of generated images and system robustness. They compare multiple approaches from which the accuracy relative to sampling is shown below where the generative images improved training. A summary of the performance of GAN methods applied to the MSTAR data set is shown in Table 4.12.

Table 4.12 GAN Results using MSTAR Imagery

Method	*Accuracy*			
	20 Samples	*40 Samples*	*80 Samples*	*All Samples*
PCA-SVM	76.43	87.95	92.48	94.32
SRC	79.61	88.07	93.16	95.04
DNN	79.39	87.73	93.76	96.50
CNN	81.80	88.35	93.88	97.03
GAN-CNN	84.39	90.13	94.29	97.53
MGAN-CNN	85.23	90.82	94.91	97.81

Adapted Zheng et al., 2019.

Finally, Gao et al. [61] implement a conditional generator (cG) as a semisupervised method for a 5-player model called integrated GANs (I-GAN), with a cG, unconditional generator (uG), classifier, conditional discriminator (cD), and unconditional discriminator (uD). With 600 labels out of ~2,600 images, they achieved for comparison C-GAN (99.77%), AC-GAN (99.82%), tripleGAN (99.96%), and I-GAN (99.93%). A key takeaway is that the quality of our generated image may still be inferior to that of the real SAR images. However, the experiment for the GAN-train and the GAN-test indicates that our generated images can fool a common DCNN classifier significantly.

4.5 Summary

This chapter reviewed the contemporary methods in DL. While GANs are the leading techniques for SAR ATR that were published recently at the time of writing, the chapter overviewed the development of methods and approaches that evolved toward the GAN techniques. Using the DNN techniques as a case history for the MSTAR data set, accuracies improved from 87% to 99.9%. The accuracy improvement also came with effectiveness, efficiency, and robustness. The many techniques of DL NN enhanced effectiveness of discerning features affording the techniques as effective in target recognition. The groups of approaches include (1) feedforward neural networks with the CNN as a fundamental approach for spatial analysis, (2) the sequential neural networks with the LSTM instantiation of the recurrent neural network as a method for temporal analysis, and (3) stochastic neural networks with the DBN to support frequency analysis. Other DL concepts are aligned to machine learning techniques that include reinforcement, active, and transfer learning. While these methods can improve accuracy over varying domains and tasks, their primary advantage is to increase efficiency in training. Data efficiency comes from providing the classifier the right subset of data to learn quickly. If the right data, goals, and domains are provided, computational efficiency is gained by processing less data. For example, methods of transfer learning seed the learner so that only the new situation must be learned.

The chapter began with the autoencoder and ended with the GAN. For real-world situations over various physical operating conditions (e.g., sensor, target, and environment variations) and human operating conditions (e.g., workload, reporting, and intent), the AE-GAN approach can provide enormous instances to train, test, verify, and validate DNN methods, providing robustness of DL methods. Future research questions still exist for the deployment of DL approaches, such as human-machine interaction, DL understanding and interpretation, as well as contextual awareness. At minimum, the chapter demonstrated from a common data set (MSTAR) that DL has shown, to near perfection, that classification over standard operating conditions (known target variations, known situations) and extended operation conditions (unknown target variations, known situations). On the other hand, DL techniques are supported for imposter operating conditions (known target variations, unknown situations). The final approach would be to combine physics-based (or model-based), and human-based (or situation-models) to deal with contextual operating conditions (unknown target variations, unknown situations).

Finally, DL RF analysis focused on RF imagery and RF transmitter detection, classification, recognition, and identification, from which the methods could be applied to RF communications (e.g., waveforms), electronics, medicine, appliances, and smart sensing. The next wave for DL RF analysis will emerge with autonomous cars to support sensor management, information fusion, and social data analytics.

References

[1] Krizhevsky, A., I. Sutskever, and G. E. Hinton, "Imagenet Classification with Deep Convolutional Neural Networks," *Advances in Neural Information Processing Systems*, 2012, pp. 1097–1105.

[2] Simonyan, K., and A. Zisserman, "Very Deep Convolutional Networks for Large-scale Recognition," arXiv preprint, *arXiv.1409.1556*, 2014.

[3] Szegedy, C., W. Liu, Y. Jia, et al.,, "Going Deeper with Convolutions.," *IEEE Conference on Computer Vision and Pattern Recognition*, 2015.

[4] He, K., X. Zhang, S. Ren, and J. Sun, "Deep Residual Learning for Image Recognition," arXiv preprint, *arXiz:1512.03385*, 2015.

[5] Dong, G., G. Liao, H. Liu, and G. Kuang, "A Review of the Autoencoder and Its Variants," *IEEE Geoscience and Remote Sensing Magazine*, 2018, pp. 44–68.

[6] Lv, C. Chen, N., T. Qiu, and A. K. Sangaiah, "Deep Learning and Superpixel Feature Extraction Based on Contractive Autoencoder for Change Detection in SAR Images," *IEEE Transactions on Industrial Informatics*, Vol. 14, No. 12, 2018, pp. 5530–5538.

[7] Wang , K., G. Zhang, Y. Leng, and H. Leung , "Synthetic Aperture Radar Image Generation with Deep Generative Models," *IEEE Geoscience and Remote Sensing Letters*, Vol. 16, No. 6, 2019, pp. 912–916.

[8] Ruck, D. E., et al. "The Multilayer Perceptron as an Approximation to a Bayes Optimal Discriminant Function," *IEEE Transactions on Neural Networks,* Vol. 37, 1990, pp. 296–298.

[9] Catlin, A., L. Myers, K. Bauer, S. Rogers, and R. Broussard, "Performance Modeling for Automatic Target Recognition Systems," Proc. SPIE, Vol. 3070, *Algorithms for Synthetic Aperture Radar Imagery IV,* July 1997.

[10] Johnson, J. L., "Pulse-Coupled Neural Nets: Translation, Rotation, Scale, Distortion, and Intensity Signal Invariances for Images," *Applied Optics*, Vol. 33, No. 26, 1994, pp. 6239–6253.

[11] Blasch, E. P., "Biological Information Fusion Using PCNN and Belief Filters," *IEEE Intl Joint Conference on Neural Networks* ,Gaithersburg, MD, Vol. 4, 1999, pp. 2792–2795.

[12] Blasch, E., *Derivation of a Belief Filter for Simultaneous High Range Resolution Radar Tracking and Identification,* Ph.D. thesis, Wright State University, 1999.

[13] Kou, G., J. Yue, Y. Ma. Q. Wang, and Z. Zhang, "SAR Image Invariant Extraction by Anisotropic Diffusion and Multi-Gray Level Simplified PCNN," *IEEE Access,* Vol. 7, 2019, pp. 47135–47142.

[14] Malmgren-Hanse, D., M. Nobel-JØrgensen, "Convolutional Neural Networks for SAR Image Segmentation," *IEEE International Symposium on Signal Processing and Information Technology* (ISSPIT), 2015.

[15] Wang, H., S. Chen, F. Xu , and Y-Q. Jin, "Application of Deep-Learning Algorithms to MSTAR Data," *IEEE International Geoscience and Remote Sensing Symposium (IGARSS),* 2015.

[16] Mossing, J., T. Ross, and J. Bradley, "Evaluation of SAR ATR Algorithm Performance Sensitivity to MSTAR Extended Operating Conditions," *Proc. SPIE,* Vol. 3370, Algorithms for Synthetic Aperture Radar Imagery V, September 1998.

[17] Zhao, Q., and J. C. Principe, "Support Vector Machines for SAR Automatic Target Recognition," *IEEE Transactions on Aerospace and Electronic Systems,* Vol. 37, 2001, pp. 643–654.

[18] O'Sullivan, J. A., M. D. DeVore, V. Kedia, and M. I. Miller, "SAR ATR Performance Using a Conditionally Gaussian Model," *IEEE Transactions on Aerospace and Electronic Systems,* Vol. 37, 2001, pp. 91–108.

[19] Singh, R., and B. V. K. V. Kumar, "Performance of the Extended Maximum Average Correlation Height (EMACH) Filter and the Polynomial Distance Classifier Correlation Filter (PDCCF) for Multi-Class SAR Detection and Classification," *Algorithms for Synthetic Aperture Radar Imagery IX,* Vol. 4727, 2002, pp. 265–276.

[20] Sun, Y. J., Z. P. Liu, S. Todorovic, and J. Li, "Adaptive Boosting for SAR Automatic Target Recognition," *IEEE Transactions on Aerospace and Electronic Systems,* Vol. 43, 2007, pp. 112–125.

[21] Zhang, X. Z., J. H. Qin, and G. J. Li, "SAR Target Classification Using Bayesian Compressive Sensing with Scattering Centers Features," *Progress in Electromagnetics Research-Pier,* Vol. 136, 2013, pp. 385–407.

[22] Dong, G., N. Wang, and G. Kuang, "Sparse Representation of Monogenic Signal: With Application to Target Recognition in SAR Images," *IEEE Signal Processing Letter*, Vol. 21, No. 8, 2014, pp. 952–956.

[23] Srinivas, U., V. Monga, and R. G. Raj, "SAR Automatic Target Recognition Using Discriminative Graphical Models," *IEEE Transactions on Aerospace and Electronic Systems,* Vol. 50, 2014, pp. 591–606.

[24] Park, J. I., and K. T. Kim, "Modified Polar Mapping Classifier for SAR Automatic Target Recognition," *IEEE Transactions on Aerospace and Electronic Systems,* Vol. 50, 2014, pp. 1092–1107.

[25] Dong, G., and G. Kuang, "Classification on the Monogenic Scale Space: Application to Target Recognition in SAR Image," *IEEE Trans. Image Process.*, Vol. 24, No. 8, 2015, pp. 2527–2539.

[26] Chen, S., H. Wang, F. Xu, and Y.- Q.Jin, "Target Classification Using the Deep Convolutional Networks for SAR Images," *IEEE Transactions on Geoscience and Remote Sensing,* Vol. 54, No. 8, 2016.

[27] Profeta, A., A. Rodriguez, and H. Scott Clouse, "Convolutional Neural Networks for Synthetic Aperture Radar Classification," *Proc. SPIE,* Vol. 9843, Algorithms for Synthetic Aperture Radar Imagery XXIII, May 2016.

[28] Roy, D., T. Mukherjee, M. Chatterjee, E. Blasch, and E. Pasiliao, "RFAL: Adversarial Learning for RF Transmitter Identification and Classification," *IEEE Transactions on Cognitive Communications and Networking*, 2019.

[29] Zhang, F., C. Hu, Q. Yin, W. Li, H-C. Li, and W. Hong, "Multi-Aspect-Aware Bidirectional LSTM Networks for Synthetic Aperture Radar Target Recognition," *IEEE Access*, Vol. 5, 2017, pp. 26880–26891.

[30] Minh, D. H. T., D. Ienco, R. Gaetano, et al., "Deep Recurrent Neural Networks for Winter Vegetation Quality Mapping via Multitemporal SAR Sentinel-1," *IEEE Geoscience and Remote Sensing Letters*, Vol. 15, No. 3, 2018, pp. 464–468.

[31] Wang, L., X. Xu, H. Dong, R. Gui, R. Yang, and F. Pu, "Exploring Convolutional LSTM for Polsar Image Classification," IGARSS *IEEE International Geoscience and Remote Sensing Symposium*, 2018, pp. 8452–8455.

[32] Amrani, M., and F. Jiang, "Deep Feature Ee\xtraction and Combination for Synthetic a\ Aperture Radar Target Classification," *J. Appl. Remote Sensing,* Vol. 11, No. 4, 042616, 2017, doi: 10.1117/1.JRS.11.042616.

[33] Ni, J. C., and Y. L. Xu, "SAR Automatic Target Recognition Based on a Visual Cortical System," *International Congress on Image and Signal Processing* (CISP), 2013.

[34] Deng, S., L. Du, C. Li, J. Ding, and H. Liu, "SAR Automatic Target Recognition Based on Euclidean Distance Restricted Autoencoder," *IEEE Journal of Selected Topics in applied Earth Observations and Remote Sensing*, Vol. 10, No. 7, 2017, pp. 3323–3333.

[35] Chen, X-Y., X-Y. Peng, Y. Peng and J-B. Li, "The Classification of Synthetic Aperture Radar Image Target Based on Deep Learning," *Journal of Information Hiding and Multimedia Signal Processing*, Vol. 7, No. 6, 2016, pp. 1345–1353.

[36] Blasch, E., and M. Bryant, "Information Assessment of SAR data for ATR," *IEEE National Aerospace and Electronics Conference*, 1998.

[37] Bhanu, B., Y. Lin, G. Jones, and J. Peng, "Adaptive Target Recognition Using Reinforcement Learning," in *Proc. DARPA Image Understanding Workshop*, 1998, pp. 1177–1180.

[38] Krawiec, K., and B. Bhanu, "Visual Learning by Evolutionary Feature Synthesis," *International Conference on Machine Learning (ICML-2003),* Washington, DC, 2005.

[39] Ravichandran, B., A. Gandhe, R. Smith, and R. Mehra, "Robust Automatic Target Recognition Using Learning Classifier Systems," *Information Fusion,* Vol. 8, 2007, pp. 252–265.

[40] Wang, M., Z. Wang, C. Yang, S. Yang, and Y. Gao, "Polarimetric SAR Data Classification via Reinforcement Learning," *IEEE Access,* Vol. 7, 2019, pp. 137629–137637.

[41] Blasch, E., R. Breton, and P. Valin, "Information Fusion Measures of Effectiveness (MOE) for Decision Support," *Proc. SPIE,* Vol. 8050, April 2011.

[42] Zhao, Q., and J. C. Principe, "Improving ATR Performance by Incorporating Virtual Negative Examples," *International Joint Conference on Neural Networks*, Vol. 5, 1999, pp. 3198–3203.

[43] Blasch, E., and R. Broussard, "Physiologically Motivated Computational Visual Target Recognition Beta Selection," *Proc. SPIE*, Vol. 4055, April 2000

[44] Blasch, E., and S. Plano, "User Fusion to Constrain SAR Targeting for TSTs," *Proc. of SPIE*, Vol. 5095, April 2003.

[45] Blasch, E., and S. Plano, "Cognitive Fusion Analysis Based on Context," *Proc. of SPIE,* Vol. 5434, April 2004.

[46] Blasch, E., and U. Majumder, "Automatic Target Cueing Utilizing a SNAKE-Fusion Tracking Algorithm," *Proc. of SPIE,* Vol. 5810, April 2005.

[47] Papson, S., and R. M. Narayanan, "Classification via the Shadow Region in SAR Imagery," *IEEE Transactions on Aerospace and Electronic Systems*, Vol. 48, No. 2, 2012, pp. 969–980.

[48] Tu, S., Y. Su, and Y. Li, "Convex Active Contour Model for Target Detection in Synthetic Aperture Radar Images," *Journal of Applied Remote Sensing,* Vol. 9, No. 1, 2015, 095084, pp. 1–24.

[49] Babaee, M., S. Tsoukalas, G. Rigoll, and M. Datcu, "Visualization-Based Active Learning for the Annotation of SAR Images," *IEEE Journal of Selected Topics in Applied Earth Observations and Remote Sensing*, Vol. 8, No. 10, 2015, pp. 4687–4698.

[50] Ren, H., X. Yu, L. Zou, Y. Zhou, and X. Wang, "Joint Supervised Dictionary and Classifier Learning for Multi-View SAR Image Classification," *IEEE Access*, Vol. 7, 2019, pp. 165127–165142.

[51] Al Mufti, M., E. Al Hadhrami, B. Taha, and N. Werghi, "Automatic Target Recognition in SAR Images: Comparison between Pre-trained CNNs in a Transfer Learning-Based Approach," *International Conference on Artificial Intelligence and Big Data (ICAIBD),* 2018, pp. 160–164.

[52] Al Mufti, M., E. Al Hadhrami, B. Taha, and N. Werghi, "Using Transfer Learning Technique for SAR Automatic Target Recognition," *Proc. SPIE,* Vol. 11197, 2019.

[53] Arnold, J., L. Moore, and E. Zelnio. "Blending Synthetic and Measured Data Using Transfer Learning for Synthetic Aperture Radar (SAR) Target Classification," *Proc. SPIE,* Vol. 10647, 2018.

[54] Shang, R., J. Wang, L. Jiao, R. Stolkin, B. Hou, and Y. Li, "SAR Targets Classification Based on Deep Memory Convolution Neural Networks and Transfer Parameters," *IEEE Journal of Selected Topics in Applied Earth Observations and Remote Sensing*, Vol. 11, No. 8, 2018, pp. 2834–2846.

[55] Goodfellow, I. J., J. Pouget-Abadie, M. Mirza, et al.,"Generative Adversarial Nets," arXiv:1406.2661 [stat.ML], 2014.

[56] Radford, A., et. al., "Unsupervised Representation Learning with Deep Convolutional Generative Adversarial Networks," CVPR 2015.

[57] Lewis, B., J. Liu, and A. Wong, "Generative Adversarial Networks for SAR Image Realism," *Proc. SPIE,* Vol. 10647, Algorithms for Synthetic Aperture Radar Imagery XXV, April 2018.

[58] Liu, W., Y. Zhao, M. Liu, L. Dong, X. Liu, and M. Hui, "Generating Simulated SAR Images Using Generative Adversarial Network," *Proc. SPIE,* Vol. 10752, Applications of Digital Image Processing XLI, September 2018.

[59] Zheng, C., X. Jiang, and X. Liu, "Multi-Discriminator Generative Adversarial Network for Semi-Supervised SAR Target Recognition," *IEEE Radar Conference (RadarConf),* 2019.

[60] Zheng, C., X. Jiang, and X. Liu, "Semi-Supervised SAR ATR via Multi-Discriminator Generative Adversarial Network," *IEEE Sensors Journal*, Vol. 19, No. 17, 2019.

[61] GAO, F., Q. Liu, J. Sun, A. Hussain, and H. Zhou, "Integrated GANs: Semi-Supervised SAR Target Recognition," *IEEE Access,* Vol. 7, 2019, pp. 113999–114013.

[62] Lewis, Benjamin, Jennifer Liu, and Amy Wong, "Generative adversarial networks for SAR image realism," *Algorithms for Synthetic Aperture Radar Imagery XXV.* Vol. 10647, International Society for Optics and Photonics, 2018.

[63] Lewis, Benjamin, Jennifer Liu, and Amy Wong, "Generative adversarial networks for SAR image realism." *Algorithms for Synthetic Aperture Radar Imagery XXV.* Vol. 10647, International Society for Optics and Photonics, 2018.

[64] Lewis, Benjamin, et al., "Realistic SAR data augmentation using machine learning techniques," *Algorithms for Synthetic Aperture Radar Imagery XXVI.* Vol. 10987, International Society for Optics and Photonics, 2019.

CHAPTER 5

Radio Frequency Data for ML Research

5.1 Introduction

Data is pivotal for accurate ML algorithm development, robust ATR performance evaluation, and credible adversarial attack understanding. Unlike video imagery (mostly used by the commercial world), there is no abundance of RF data for ML research. However, with the growing importance of ML for RF exploitation (particularly intelligence, surveillance, and reconnaissance applications), several organizations are collecting RF data, and thus the library for RF data is getting bigger. A portion of these data sources are in the public domain, so that researchers can examine the technical challenges and solutions associated with RF data analytics. Hence, this chapter provides an overview of available public release RF data sources. The RF data sources support ML challenge problems emerging from ATR that include copious amounts of unlabeled data, nonuniform standards and metrics, as well as big variations in data collections. In a broader sense, the RF data includes 1D signals for communications, 1D HRR radar profiles for dynamic target tracking, 2D SAR imagery for target detection, 3D SAR volumetric cubes for target classification, and 4D video SAR for changes over time that results in *big data*. The term big data was introduced by IBM for multimedia and the use of IoT/cyberphysical systems (CPS) . RF applications align with the big data concepts for multisensory, multimodal, multisource, and multi-intelligence data for multidomain applications.

This chapter presents a multitude of RF data concepts for surveillance applications beginning with a review of big data in Section 5.2. Section 5.3 discusses the many variations of SAR data. Section 5.4 highlights the MSTAR and synthetic and measured paired labeled experiment (SAMPLE) data sets. Section 5.5 highlights communications RF data sets while Section 5.6 overviews challenge problems with RF data. Section 5.7 provides a summary.

5.2 Big Data

What is the definition of big data? Does big data require gigabytes, terabytes, or petabytes of data? While it is possible that a general consensus would define big data as being comprised of gigabytes and higher, it is perhaps more accurate to state that this delineation depends on the applications. For example, in medical imaging, big data could be somewhere between gigabytes and terabytes. On the other hand, for financial transactions, comprising texts, numbers, or recommender system data with big data could be smaller. In general, big data for image data in surveillance and security applications could be as large as petabytes to exabytes, perhaps even greater volumes.

One approach to big data analysis follows from the "V" metrics associated with the type of data from differences in velocity, volume, veracity, and variety. The emergence of new big data considerations include:

- *Data at rest* versus *data in motion* (static versus dynamic data analysis);
- *Data in open* versus *data of importance* (available versus valuable data analysis);
- *Data in collection* versus *data from simulation* (observed versus imagined data);
- *Data in use* versus *data as manipulated* (truthful versus fake data analysis).

Naturally, these definitions and organization follow from techniques used to process data such as the three waves of AI (see Chapter 1: expert logic, statistical learning, and contextual reasoning) [1]. For rest versus motion data, the goal utilizes an index to access a database using known rules. The open versus importance concept relates to machine learning that facilities a compression of results for decision making. The use versus manipulated delineation relates to advances in deep learning to learn from analysis over real versus fake data. Similar to the future of ML methods is the ability to simulate a large quantity of data that represents known and unknown scenarios.

5.2.1 Data at Rest versus Data in Motion

IBM [2] provided a definition of big data very thoroughly by considering various aspects of the data. According to IBM, big data exhibits four characteristics as the four common Vs: *volume, velocity, variety*, and *veracity* (Table 5.1). For RF data volume, it aligns with *data at rest*; that is, terabytes to exabytes of data (stored at disk) that need to be processed. For the collection of imagery data, the resolution has increased with the opportunity for ML to better resolve boundaries of targets. On the other hand, data velocity implies *data in motion* (i.e., processing of large volume streaming data within milliseconds to seconds). Data at rest and data in motion are elements for ML as data at rest is available for training while data in motion is related to real-time testing of the ML classifier to be deployed for new situations. While the velocity of the data could pose challenges, if the ML classifier is well-trained to similar scenarios, then the results would be sufficient.

Two other related concepts are variety and veracity, which relate to information fusion as shown in Figure 5.1 [3]. Data variety provides challenges and opportunities from various sources of data such as multimedia (i.e., text, voice, or videos), as well as intelligence data (i.e., SAR, radar, and sonar). A classic example is

Table 5.1 Attributes of Big Data

Big Data	*Metric*	*Analysis*
Data at rest	Volume	Terabytes to exabytes of existing data to process
Data in motion	Velocity	Streaming data, milliseconds to seconds to respond
Data in forms	Variety	Structured, unstructured, text, imagery, multimedia
Data in doubt	Veracity	Uncertainty due to data inconsistency and incompleteness, ambiguity, latency, deception model appropriateness
Data in the open	Visibility	Open data available to anyone with privacy, security, and provenance issues
Data of importance	Value	Large range of data with cost, importance, and appropriateness concerns
Data of clarity	Vividness	Data appearance enhancement for discerning between elements
Data in use	Virtue	Data determination of worthiness, credit, trust
Data as manipulated	Volatility	Data degradation from corruption, alteration, and loss

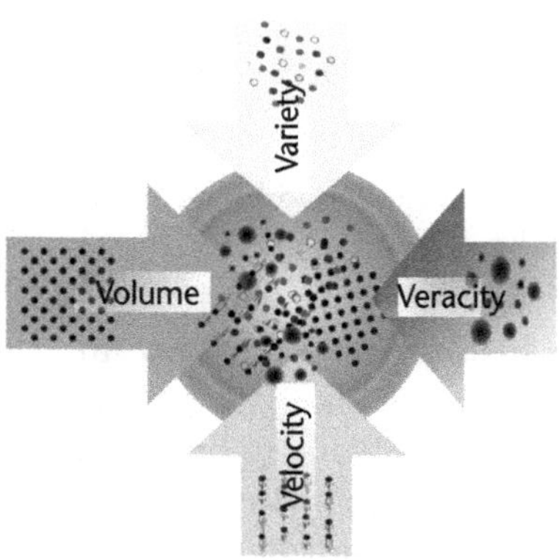

Figure 5.1 Four Vs of big data fused for information processing.

the variety of imagery, such as multispectral, infrared, and videos that can be fused to enhance vividness [4]. The challenges are that the data needs to be preprocessed such as registered, indexed, and labeled for meaningful information fusion. However, opportunities arise when the association of one data modality with another is able to reduce uncertainty. Hence, data in veracity includes a variety of uncertainty issues, such as ambiguous, compressed, incomplete, and inconsistent data; all of which require careful analytics to understand its contents.

5.2.2 Data in Open versus Data of Importance

IBM (2013) followed up with definitions of big data to include visibility and value. Figure 5.2 highlights the case in which the data, when amendable to machine learning, provides visibility and value. Visibility is the ability to observe the right data and/or results. For example, when the ML method processes the data, it can reduce noise to enhance visibility. Data presented in a meaningful way increases visibility. Note that visibility also is subject to privacy and security concerns. Many data collections are subject to security challenges, as while the data is collected, it is not necessarily available for learning and assessment. Likewise, not all data is available for

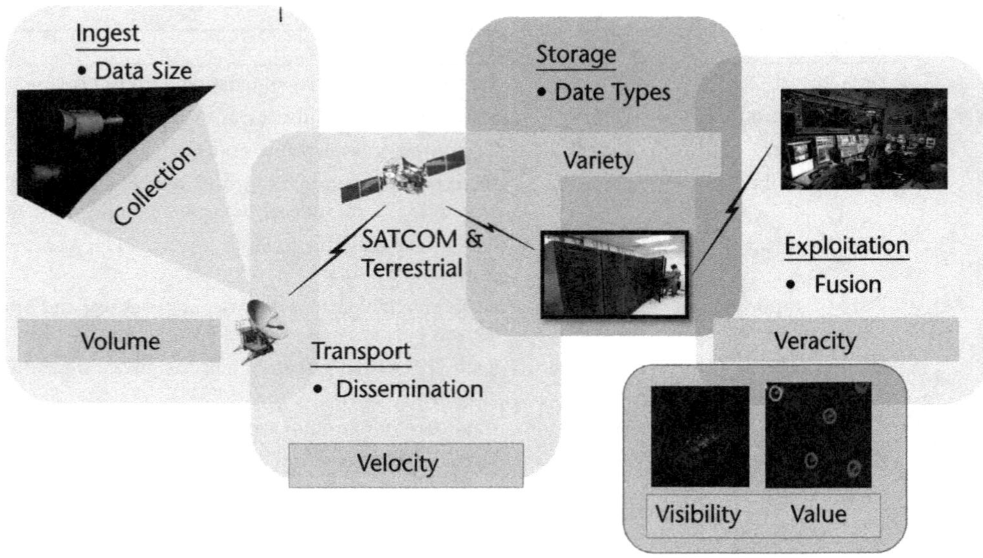

Figure 5.2 Visibility and value of big data such as discerning target types.

training given privacy concerns. Hence visibility includes concerns on data availability; especially when desiring access for training.

Value is a key element of big data. With such an enormous collection of data and methods to transform the data into information, an operator still needs to know what combination of the data and exploitation methods support decision making. Not all data is equivalent in value, especially for cases in which the data are incomplete, inconsistent, and atypical of the real world. While value is dependent on the mission and scenario, it has become the fifth V of most discussions on the big data topic. An example of data collection is that of Figure 5.2. If the ML method can extract and make a discerning analysis, the result can be presented to the user for verification. Likewise, as the exploitation of the information is visible, then the manner in which to present it could be of valuable importance, such as the identified detections with labels of known and unknown targets. Hence, when designing or selecting a ML method, the value to the user is required.

Value and visibility directly relate to the ML analysis, especially when considering the desired output. In many cases, it is pragmatic to offer the user the appropriate analysis, such as detection versus classification and results over standard and extended operating conditions. Classification, from the Greek word *taxonomy* (arrangement method), groups results based on shared attributes and a given naming convention. Most of ATR relies on a consistency of terminology that has unique labels so as to provide semantic results (e.g., car versus tank) reported in a dictionary. *Classification* answers the questions of how one group is related to another, while *identification* answers the specific name of the object being observed.

Within the various domains of mathematics, the word choices are well-defined in the analysis. *Pattern recognition* [5] is also a well-defined field of study that looks at data to find patterns through ML, advanced reasoning, or knowledge discovery such as data mining. The field of *system identification* uses statistical methods to build mathematical models of dynamical systems from measured data.

Using optimal design of experiments (DOEs), data is processed to generate information by fitting the measured data with mathematical models. Hence big data ML requires pragmatic approaches for efficient model learning. Data in use is the assessment of the data—whether raw, preprocessed, filter, exploited, or fused to support detection, recognition, classification, or identification. It is an appraisal of the information to make decisions for object and situation assessment and refinement [6]. Hence, visibility and value relate to the ATR process [7]:

- *Classification* is defined as (1) the act or process of putting *people or things into groups* based on ways that they are alike, or (2) an arrangement of people or things into groups based on ways that they are alike. Classifying is thus a systematic arrangement in groups or categories according to established criteria (e.g., taxonomy). The categories could be object behaviors (e.g., movement), appearances (e.g., size), and affiliations (e.g., group). In many cases, the determination of the information that supports a class discussion comes from signatures (1D, 2D, 3D), features, and combined attributes.
- *Recognition* is (1) the act of accepting that something is true or important or that it exists, or (2) the act of *knowing who or what someone or something is because of previous knowledge or experience*. Recognizing is an acknowledgment or feeling that something or someone is present alluding to a special notice or attention from the sensing and encoding by a machine. Given a structured approach to many man-made systems, examples include optical character *recognition*, ATR, or face recognition. The commonality of these definitions is that there is some defined template, model, or set of features that have been trained and linked to an object type (such as printed letters or characters). Recognition requires training, learning, or prior knowledge in order to match the sensed information to a known set of attributes.
- *Identification* is (1) the act of *finding out who someone is or what something is*, or (2) the act of identifying someone or something. We typically associate the term ID with material that confirms who or what something is (e.g., driver's license or car warranty). The ID information is linked to a specific entity that confirms their affiliation such as where they live, pedigree of activities, general history, name, and photograph. If the information is constant, then the ID is supposed to be a unique attribute that is not held by another entity. Identifying then includes verifying the evidence of the unique attributes associated with the entity.

Using these definitions, the relationship between recognition and classification is through the categories that have been defined: *recognize that something or someone belongs to a class*. Likewise, identification and recognition are related as to identify something requires the recognition. The confusion lies in the distinction between classification and identification. On the one hand, if you have all unique information, you can easily assess the class. The reverse is not true: when you classify something, it is merely the placement in a group, and it may not be unique within the group. There are special exceptions in which a man-made object has limited production, does not have extended versions, and classification is enough for a unique ID, such as a one-of-a kind car. The takeaway message is that identification is supposed to be definitive and unique, while a classification merely states a member of a group.

5.2.3 Data in Collection versus Data from Simulation

Big data processing follows from test and evaluation (T&E), which is explored in Chapter 9. While most of the ML methods focus on processing data at rest, there are other concerns, as shown in Figure 5.3. When designing, building, and deploying a ML system to operate on big data, it is important to consider other factors that can affect the performance of the ML system. For ease of discussion, these data properties are grouped together as (D-RUMCAT) [8]:

- *Data at rest*: Provide *structure* (i.e., translations) between data for integration, analysis, and storage;
- *Data in use:* Afford context-based, human-machine teaming based on dynamic mission priorities, users, and situations in order to balance needs, recommendations, and availabilities;
- *Data in motion:* Utilize *feedback control loops* to dynamically adapt to changing priorities, timescales, and mission scenarios;
- *Data in collect:* Leverage the power of *modeling* from which data is analyzed for information, delivered as knowledge, and supports prediction of data requirements;
- *Data in assumption:* Assess self-generated data, through supposition, error analysis, and presentation;
- *Data in transit:* Develop a *data as a service* architecture, which incorporates contextual, metadata, and archived information to support systems-of-systems design.

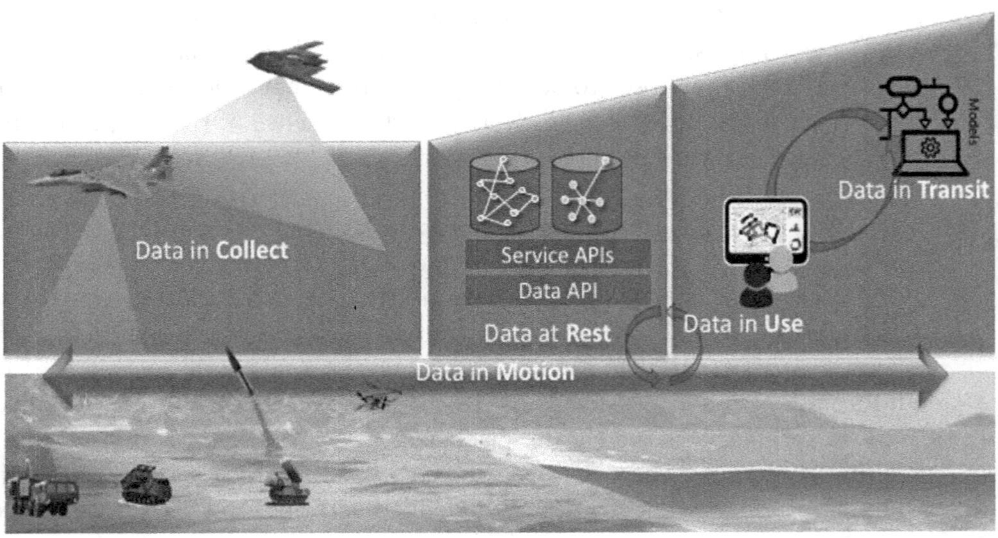

Figure 5.3 Types of big data, such as discerning target types.

The key distinction between these attributes of data for real-world systems follows two themes: physics-based and human-derived data. For the physics-based data, it is assumed that the data being collected is truthful (in the sense of veracity) and that the data being simulated (in the sense of four original Vs) represents the known world. Hence, all the data being used to train and test a ML system is verified. This concept is contrasted to that of fake data that represents manipulated data. Hence, the use of a generative adversarial network (GAN) provides both real and imposter data to train and test the ML classifier.

The development of traditional ML systems includes a variety of research areas in analytics, as illustrated by the dynamic data driven applications systems (DDDAS) framework (see Figure 5.4) [9]. Using the definitions of types of data, the goal of DDDAS is to leverage simulations to augment training and deployment of systems for scenarios in which data are not always available. Hence, context is utilized to determine the particular types of information to be simulated [10]. One simple example is that of using an ML target classifier for a tracking and classification system, in which the motion of the target can be used to simulate target features when the target is obscured.

- *Data at rest*: Techniques such as AI/ML that process *off-line* stored data typically involving information fusion to include historical context, physics models, and analytical displays;
- *Data in motion:* Techniques involving control that enable *online* analysis of dynamic data typically involving environment context, kinematic models, and signals display;
- *Data in use*: Methods for real-time performance assessment that bring together data at rest and data in motion for streaming *in-line* data typically involving situation context, data models, and user defined operating picture (UDOP) display.

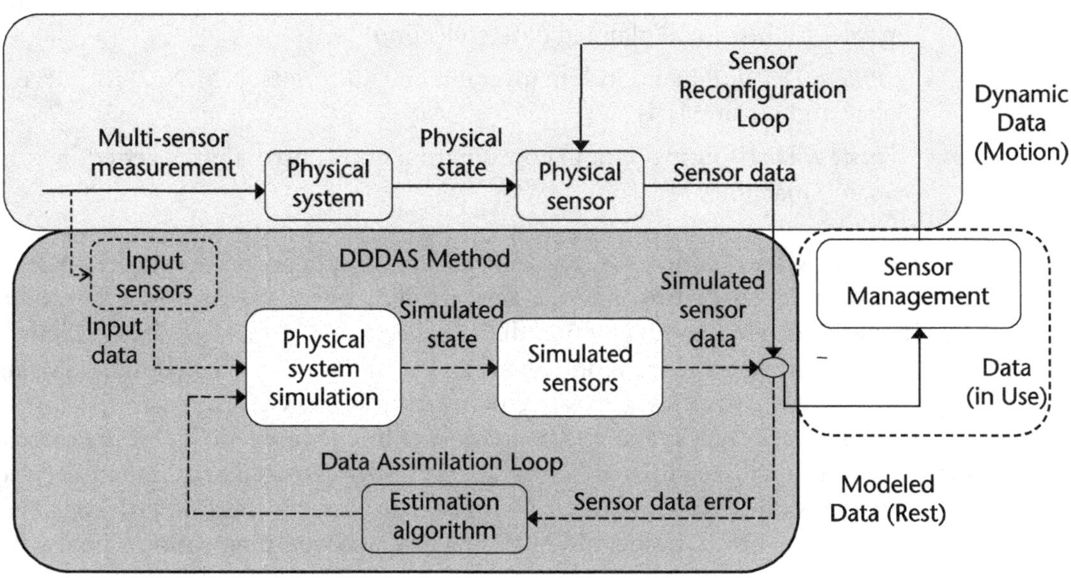

Figure 5.4 Dynamic data driven applications systems (DDDAS) framework.

The DDDAS framework brings together the sensor reconfiguration loop (SRL) (e.g., physical models) and the data assimilation loop (DAL) (e.g., trained models). The integration supports the deployment of ML systems, which focuses on reduced order modeling (ROM) approaches to combine the low-dimensional (i.e., SRL) and high-dimensional (i.e., DAL) state models.

The use of simulated data is well understood and is a current hallmark of ML systems, challenge problems, and data sets. However, the current state of the art in big data analysis is to leverage the simulation in order to imagine future data challenges in which the data has been altered by malicious intent (e.g., data attack, poisoning, and alteration).

5.2.4 Data in Use versus Data as Manipulated

The discussion and analysis of ML methods to process the copious amount of true and imposter data is a subject of debate and research. Much like game theory, the problem resides in the type of actors of malicious intent and in the variations of activities that can be imagined. Some ML methods can be conservative to train and utilize data only that conforms to the known situations. However, these systems would not readily adapt to changing conditions. If some new situation arises of an anomaly (such as a vehicle that has been changed), then the anomaly of the classification pattern should be able to adapt and recognize the new target. The challenge is that the observed image of such as vehicle modification could be a reality (e.g., mounting reflectors on a vehicle) or could be a manipulation of the data (e.g., extra scatterers in the image data from data poisoning). The key result is that future ML systems must be *trusted* and *robust* to these extra considerations.

To highlight these concepts, VAULT has been used to capture these ideas:

- *Visible*: Maintaining the UDOP visualization of each domain [11];
- *Accessible*: Providing the information to distributed locations [12];
- *Understandable*: Informing decision-making with context (to another domain with historical/planned data collection) [13];
- *Linked*: Developing multidomain command and control (MDC2) for surveillance and sensing [14];
- *Trusted:* Distributing data in coordination and control that is vetted by others and machines [15].

Emerging ML techniques—via an analysis of the data properties (D-RUMCAT) and attributes (VAULT) articulate the need for different scenarios toward evaluating ML methods. For example, if not all the collected data is available for training, how does that impact the training scenarios. Examples have been shown on using only a subset of the data for generative model training in GANs.

Of the DL techniques, Figure 5.5 provides a brief comparison. For the discussion, two additional scenarios are considered of physics-based and human-derived information fusion (PHIF) and natural language processing (NLP). PHIF considers the case in which RF data include both imagery and communication signals. An

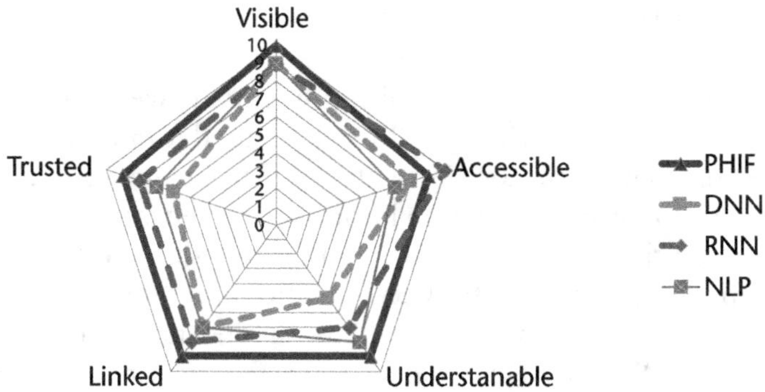

Figure 5.5 Normalized systems comparisons of PHIF, DNN, recurrent NN (RNN), NLP.

example corresponds to that of radar returns or emissions from a vehicle combined with radar observations from SAR. Additionally, another form of PHIF is when the target in an image has a semantic label. The NLP is able to determine the labels on imagery as microtext as well as any other supporting narrative that is aligned or associated with the SAR image.

From Figure 5.5, a DNN provides visual results, but is not well understood. Similarly, a DNN that is difficult to explain would not be as trusted as that of a semantic label. On the other hand, machine learning over multimodal data through PHIF supports both visibility and understandability.

In summary, big data is an important concept related to ML and classification analysis. Many ML methods exist and selecting the appropriate method is based on a variety of factors. Academic studies have the freedom to discuss ML methods usually derived from data at rest studies. However, when deploying the ML system to the real world, the results should capture the amount, type, maintenance, and sustainability of the classifier. Figure 5.6 shows that sensing measurements and applications modeling should consider the run-time architecture as well as the situation simulation approach. To aid in these concerns, many RF big data sets have been developed to support RF ML research that is highlighted in the next sections.

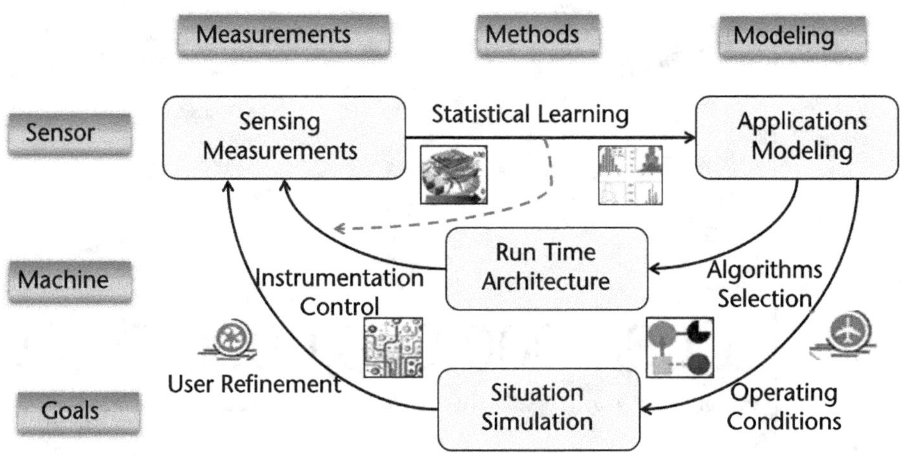

Figure 5.6 DDDAS framework.

5.3 Synthetic Aperture Radar Data

SAR is an important radar application for all weather surveillance operation. Among various applications, SAR has been used for monitoring areas of the earth's surface for unauthorized object entry, target recognition, and remote sensing for disaster planning and vegetation information. The quality of SAR imagery depends primarily on the radar frequency band, bandwidth, mode of operations (spotlight, stripmap, etc.), and the influence of clutter and interference. However, radar operating frequency and bandwidth plays significant role on the quality and volume of SAR data. With a higher signal frequency, the finer the resolution that can be achieved, also corresponds to a higher sampling rate and thus a high volume of data. Increasing bandwidth also adds more data samples and volume. Figure 5.7 shows that at UHF radar band, circular spotlight mode SAR data collection of a 20-km diameter ground spot corresponds to 50 gigabytes of data. However, at X-band, circular, spotlight mode SAR data collection of the same 20-km diameter ground spot yields 1.7 terabytes of data. This means that the act of simply changing the radar frequency from UHF to X-band increases the data volume roughly 30 orders of magnitude. The increased data size is necessary to obtain the image quality that is required to understand salient features and high accuracy object classification. Figure 5.7 also demonstrates that, as operational time increases, the magnitude of SAR data increases. It was demonstrated that at an X-band radar must collect roughly 100 petabytes of data in order to survey an area of 20-km diameter spot for 1 year. Thus, depending on the applications, RF systems collect extensive amount of data that it is impossible (or cost prohibitive) to exploit and understand by human analysts. Hence, RF systems generate big data that must use ML methods for efficient and effective exploitation.

As previously mentioned, there are very limited volumes of RF data (SAR imagery, communication signals) for DL/ML learning research, in comparison to visual imagery (electro-optical data). The reason for fewer RF data sets follows from the fact that RF data collection is expensive and requires authorization from multiple

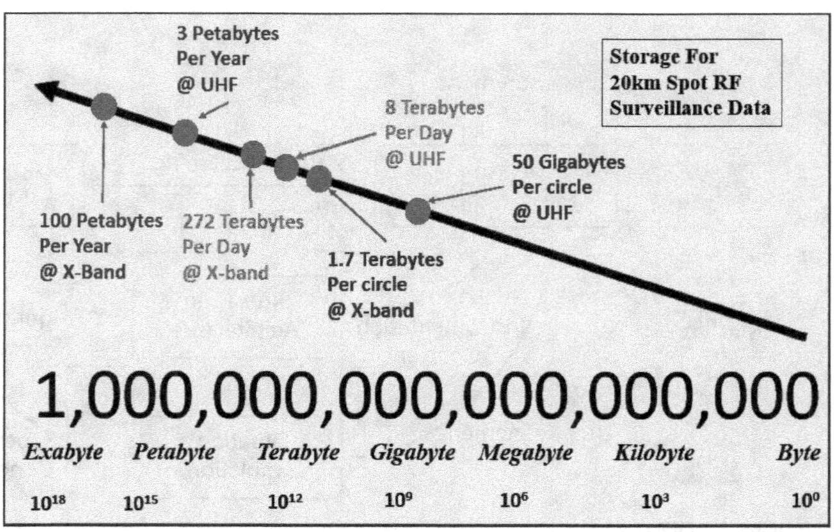

Figure 5.7 RF data exploitation as big-data problem. SAR data storage requirements as function of spot size, radar frequency, and interrogation time.

authorities (spectrum authorization, flight test, etc.). Once data is collected, preprocessing needs to be conducted to prepare the data for ML algorithms. For SAR data, the first step in preprocessing involves ensuring phase correction across radar pulses. *Phase correction* is also known as motion compensation—although with advanced inertial navigation systems (INS)/GPS system, radar pulse phase errors are not significant in many SAR data collection systems. The second step of preprocessing involves SAR image formation. The *image formation* step is significant, since the quality of SAR image depends on imaging algorithms used. Final steps include ground truthing and labeling the images for target localization and image normalization.

5.4 Public Release SAR Data for ML Research

This section describes several key SAR data sets available for ML research, shown in Table 5.2. The most popular SAR data used by the researcher is MSTAR public data released in 1996. The MSTAR data has been used extensively by the researcher across the globe. During the pre-DNN era, the MSTAR data set has been used for analyzing classification performance of various classical ML algorithms to include support vector machine (SVM), template-based and model-based among others. The second SAR data set with public release in 2010 was the civilian vehicles data set, also known as the CVDome data set. The civilian vehicle data dome (CVDome) data set is recent and has been published by the AFRL Sensors Directorate. Unlike the MSTAR data set (which is measured data), CVDome data is synthetically generated RF data to classify various civilian vehicles. The third SAR data set to introduce is the SAMPLE data set published by AFRL in 2019. The goal of the SAMPLE data is developing advanced algorithms to match synthetic RF data with measured RF data. Thus, as research progresses, the data sets have become larger and more challenging to emulate real-world conditions.

5.4.1 MSTAR: Moving and Stationary Target Acquisition and Recognition Dataset

The U.S. Air Force provided the MSTAR public released data set with 10 target types. The MSTAR data is the most used SAR image database for DL-based SAR object recognition research. The Defense Advanced Research Project Agency (DARPA) and Air Force Research Lab (AFRL) developed the MSTAR data set to support ML research. With the enormous use of the data, it has become a baseline

Table 5.2 List of Data Sets Described

Data Source	*RF Type*	*Published By*
MSTAR public release data	Measured SAR imagery	AFRL Sensors Directorate
CVDome data	Synthetic SAR imagery	AFRL Sensors Directorate
SAMPLE data	Measured and synthetic SAR imagery	AFRL Sensors Directorate
RF fingerprinting data set	RF signals (WiFi)	Northeastern University/ DARPA

data set of DL comparisons. Hence, MSTAR is presented throughout the book so the reader can appreciate the improvements in RF ATR.

In 1995, DARPA funded Sandia National Lab (SNL) to use their SAR sensor for the MSTAR RF data collection. The location of this data collection was Redstone Arsenal, Huntsville, AL. The MSTAR data set can be downloaded immediately and contains seven targets, which are shown in Table 5.3. The first data file named "MSTAR Public Target Chips" contains three different targets. It has three different variants of T-72 by changing articulation or changing several key features. These variations allow developing robust machine learning algorithms with added or deleted features of a target. Similarly, target type BMP-2 has three variations.

Another set of data available for download contains seven different targets shown in Table 5.4. The data file named "MSTAR Public Mixed Targets CD1" contains seven different targets at 15-degree elevation. The data file named "MSTAR Public Mixed Targets CD2" contains seven different targets at 17-degree elevation. Given the 10 targets, collections were competed at 1° intervals for aspect variations over 360°. During the data cleansing, not all aspect collections provided usable data, so of the 360 possible images for each target at two depression angles, the available number of images is presented in Table 5.5.

While the target visual images and representative SAR images (Figure 1.2) are presented and repeated in hundreds of papers, it is important to note that a large number of people contributed to the test design and data collection, imagery cleansing and analysis, as well as baseline results and analysis; much of which was not releasable at that time. Many of the introductory journal articles used only the

Table 5.3 SAR Data in MSTAR Public Target Chips

Target	*Target Variations*	*Target Description*	*Data collection geometry*
T-72	3	T-72 tank	Elevations 15 and 17 degrees and azimuth 360 degrees
BMP-2	3	Infantry fighting vehicle	Elevations 15 and 17 degrees and azimuth 360 degrees
BTR-70	1	Armored personnel carrier	Elevations 15 and 17 degrees and azimuth 360 degrees

Table 5.4 SAR Data in MSTAR Public Mixed Targets CD1 and CD2

Target	*No. of Targets*	*Target Description*	*Data Geometry*
BTR-60	1	Armored personnel carrier	Elevations 15 and 17 degrees and azimuth 360 degrees
2S1	1	Self-propelled artillery	Elevations 15 and 17 degrees and azimuth 360 degrees
BRDM_2	1	Amphibious armored patrol car	Elevations 15 and 17 degrees and azimuth 360 degrees
T-62	1	Tank	Elevations 15 and 17 degrees and azimuth 360 degrees
ZIL_131	1	Army truck	Elevations 15 and 17 degrees and azimuth 360 degrees
ZSU_234	1	Antiaircraft self-propelled gun	Elevations 15 and 17 degrees and azimuth 360 degrees
D7	1	Bulldozer	Elevations 15 and 17 degrees and azimuth 360 degrees

Table 5.5 Number of Images of Each Target at Different Depression/Elevation Angles

Elevation/ Targets	*BMP2*	*BTR70*	*T72*	*BTR60*	*2S1*	*BRDM2*	*D7*	*T62*	*ZIL131*	*ZSU234*
17°	233	233	232	256	299	298	299	299	299	299
15°	587	196	582	195	274	274	274	273	274	274

public data set, while some of the reported results (mostly in the SPIE) database capture the strategy of the design and analysis. Given that 10 targets were made available, a large number of targets (32) were not made available.

Additionally, it is to be noted that the raw data collected to form the imagery was also available and released as the HRR radar data set. While it does not have the same appeal as the imagery, researchers were able to compare HRR 1D signals analysis with that of 2D SAR imagery. The HRR was used extensible for simultaneous tracking and identification (STID) [16], with other names provided as tracking and automatic target recognition (T-ATR), feature-aided tracing (FAT), or tracking and classification (TAC).

It has been 25 years since the MSTAR data collection, so sensor designs have improved, algorithms have been developed, and numerous results on real data collection. Thus, the entire data set is readily available to researchers. However, another approach was developed to determine the robustness of the approaches without the need to continue to conduct massive test plans, data analysis, and results reporting.

5.4.2 CVDome

The Civilian Vehicles Data Dome (CVDome) was developed to conduct ML research using synthetic RF data. Collecting measured RF data is expensive and time consuming. Synthetic RF data is a less expensive way of generating SAR target signatures in a controlled environment. The CVDome dataset contains simulated X-band scattering data for civilian vehicles (see Figure 5.8). By using computer-aided design (CAD) models of 10 civilian vehicles and high-frequency electromagnetic scattering code, AFRL produced fully polarized synthetic RF signatures of these targets. For each target, the synthetic RF data provides azimuth angles coverage of 360 degrees and elevation angles from 30 to 60 degrees. The data are in MATLAB format. Also included are back-projection algorithm code (MATLAB) to be used for image formation.

Once CVDome data is downloaded and unzipped, the database folders contain the code, a paper describing the data, and the data (Figure 5.9). Figures 5.9–5.11 provide screenshots on the data available. By opening the folder "Domes," there is a folder containing 10 targets (Figure 5.10). For each target, there are four MATLAB files containing phase history data. In the code folder, you will find the back-projection code and the image formation script to form the images (Figure 5.11).

Examples of the target phase histories are shown in Figure 5.12.

To improve the opportunity to research advanced ML and DL methods, the SAMPLE data set was released that combines both measured and synthetic data.

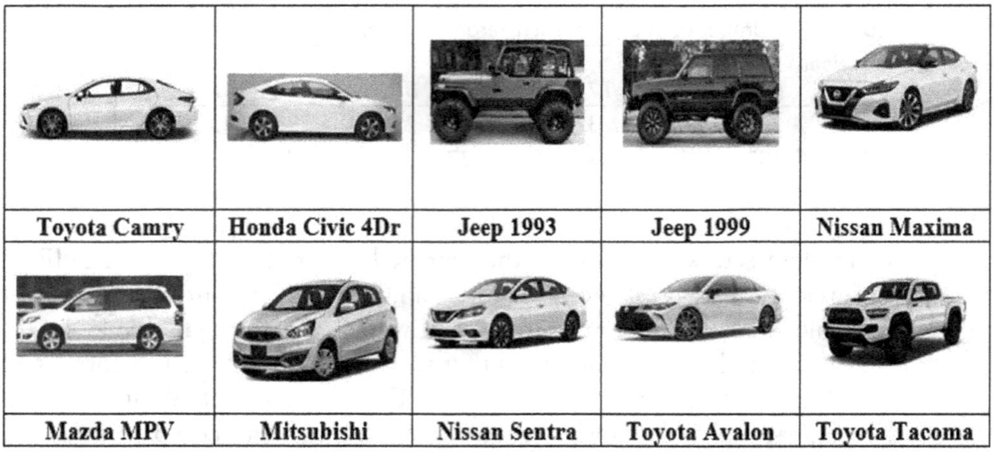

Figure 5.8 Civilian vehicles.

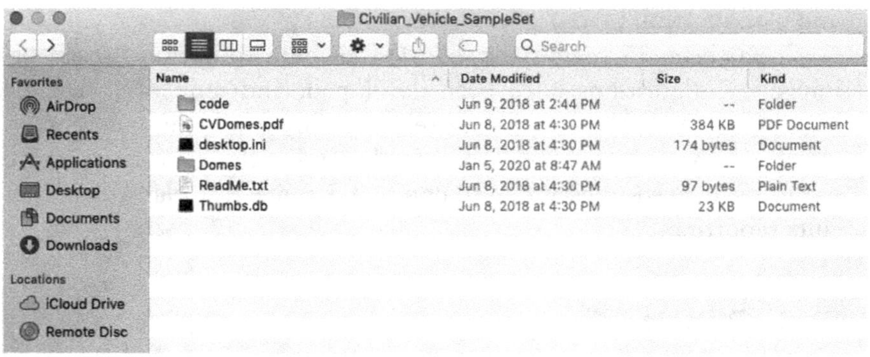

Figure 5.9 Civilian vehicles sample data.

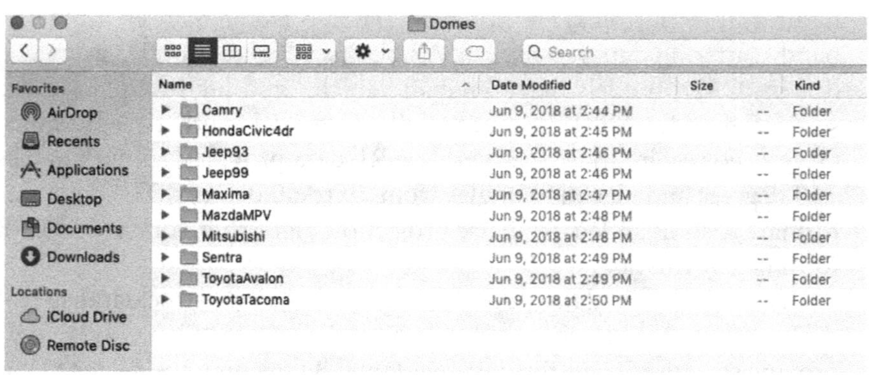

Figure 5.10 Civilian vehicles targets phase history data.

5.4.3 SAMPLE

The synthetic and measured paired labeled experiment (SAMPLE) dataset was developed by AFRL Sensors Directorate [46, 47] in an effort to enhance SAR database as well as developing advanced machine learning algorithms for cases of limited measured data. As mentioned earlier, RF/SAR data collection is expensive and time

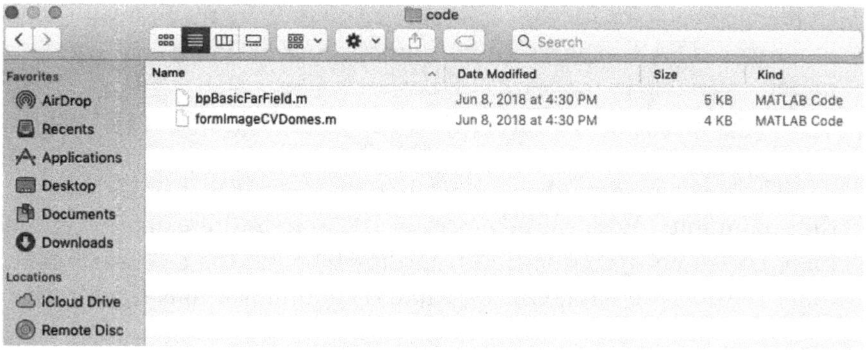

Figure 5.11 Civilian vehicles targets imaging code.

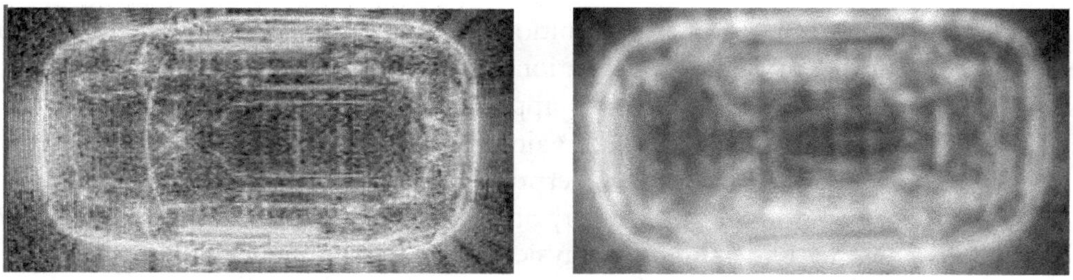

Figure 5.12 SAR image of a Jeep93 and a Toyota Camry.

consuming. Electromagnetic prediction code and CAD target models enable computation of synthetic RF target signatures. Although a supercomputer is typically needed for synthetic RF data generation, the SAMPLE process only requires a few hours of computation for high-fidelity target signatures. One prominent issue with synthetic RF target signature data that is too pristine compared to the measured target signature. In other words, the data distribution (pixel value) of synthetic RF targets and measured targets are different. Hence, one of the major objectives of the SAMPLE data set is finding algorithmic solutions to transform synthetic RF data to resemble measured RF data. In the ML literature, this is known as transfer learning (TL). Developing algorithms to fill in the gap between synthetic and measured RF data mismatch is one type of transfer learning. Other types of TL include matching data between video and radar imagery, infrared and video imagery, airborne and space sensors data; as well as between radar bandwidths, operating frequencies, and combinations among these variations.

The public released MSTAR SAR data set has been a valuable asset in the development of SAR ATR algorithms. The MSTAR data has been vital for achieving state-of-the-art target classification algorithm development. However, these data sets do not cover extended operating conditions (EOC) for operational scenarios. The EOC includes a large number of possible sensor parameters (e.g., frequency, bandwidths, types), target configurations, and environmental conditions. To cover the entire EOCs, SAR data collections will be extremely costly. Thus, synthetic data must be generated to augment measured data sets. As a result, AFRL produced SAMPLE, which consists of SAR imagery from the MSTAR data set and well-matched synthetic RF data. In addition to the data set, AFRL published four challenge problems for the researcher to investigate the best ways to classify targets in measured SAR imagery given synthetic SAR training imagery. Table 5.4 lists the targets available in the SAMPLE data set.

As a comparison, Figure 5.13 shows the 10 targets in measure and synthetic variations.

The SAR imagery data, as well as the 1D HRR, has shown great interest for the community. With the advances in DL from multimedia text and video imagery, there is also an open question regarding whether the DL methods would apply to other types of RF data. One example is communications signals data.

5.5 Communication Signals Data

Other than object/target recognition from RF imagery (i.e., SAR image classification), device or signal classification from RF sensing has become an active area of research. More specifically, the application of ML techniques in order to classify RF communication signals has gained attention from research community. An example of RF ML involves fingerprinting or identifying an electronic device (e.g., WiFi devices). In the past, fingerprinting approach was based on handcrafted methods, such as understanding physical layer, medium access control (MAC) layer, Internet Protocol (IP) addresses, and international mobile station equipment identity (IMEI) numbers [17]. However, these simple techniques can easily be spoofed. Location-based features, such as radio signal strength (RSS), angle of arrival (AoA), and channel state information (CSI) are susceptible to mobility and environmental changes [18]. Hence, research has focused on transmitter features that are inherent to a device's hardware. These features are unalterable and cannot be easily replicated by unauthorized users. It was found that manufacturing process of individual radios contributes to the unique signature of a given device. In particular, manufacturing imperfections are presented within the analog components (e.g.,

Table 5.6 List of Military Vehicles and Serial Numbers Used in a SAMPLE Data Set

Target Serial Number	*Target Name*	*Target Description*
B01	2S1	Self-propelled artillery
9563	BMP2	Infantry fighting vehicle
C71	BTR70	Armored personnel carrier
0AP00N	M1	Combat car/tank
MV02GX	M2	Combat tank/car (light)
T839	M35	Armored car
C245HAB	M548	Military cargo carrier
3336	M60	Military tank
812	T72	Military tank
D08	ZSU_234	Antiaircraft self-propelled gun

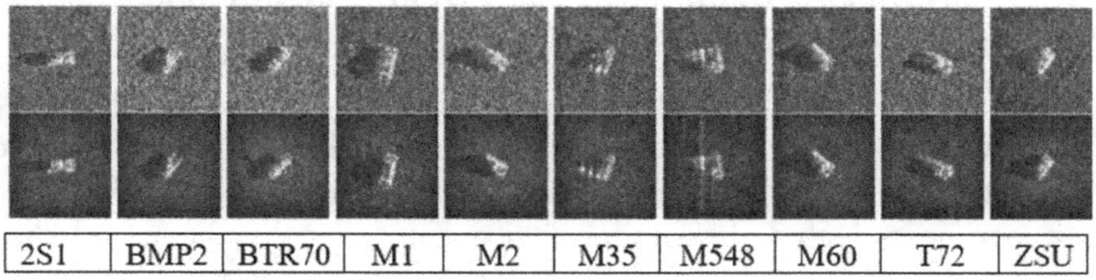

Figure 5.13 Measured (top row) and synthetic signatures (bottom row) of the vehicles used in SAMPLE data set. Visually, both synthetic and measured target's signatures look the same. However, when calculating pixel-wise comparison, these signatures differ [46].

digital-to-analog converters, band-pass filters, frequency mixers, and power amplifiers) of a device that differentiate radio devices, even if the manufacturer and make/model are identical. Another area of RF signal classification involves RF spectrum awareness. For example, one could be interested in finding signals of interest (SOI) in a certain frequency and bandwidth [19–21]. Spectrum awareness may involve identifying diverse signals of interest in a complicated background environment of anomalous signals or identifying ensemble signal activities.

5.5.1 RF Signal Data Library

The DARPA Radio Frequency Machine Learning Systems (RFMLS) program was launched in 2018 to incorporate ML solutions for RF spectrum understanding and exploitation. The goal of the RFMLS program was to develop the foundations for applying modern ML to the RF spectrum domain and develop practical applications in emerging spectrum problems. Traditionally, ML research has focused on problems in the domains of vision, hearing, and language. Progress in these areas has been significant in recent years, surpassing human abilities in some ways.

Researchers realized that there are opportunities to apply ML techniques for RF signal classification. Although the RF environment presents familiar ground for ML researchers, there are some important differences that make RFML unique. Compared to other domains, the RFML data rate is much higher. A high definition video stream at 60 Hz could be on the order of 3 Gbps (ignoring coding compression savings). A data set of 1 GHz of RF spectrum data has approximately 10 times that data rate and is not nearly as amenable to decimation in order to improve processing. Additionally, RF waveforms are typically captured and represented as complex numbers, underscoring the importance of both amplitude and phase of the signal. Although there has been interest recently in complex-valued neural networks, the technology for learning naturally in the complex plane is not fully developed and often relies on treating complex variables as two real numbers. The RF environment offers a more compelling reason for developing a native complex-valued approach. Other prominent differences between current ML applications and the RF domain exist and must be addressed by the RFMLS Program. The DARPA RFMLS program created several datasets for ML research. The following two RF signals data sets are available for ML research.

5.5.2 Northeastern University Data Set RF Fingerprinting

Northeastern University (NU) developed data sets for RF fingerprinting of bit-similar USRP X310 radios [22]. This data is publicly available and can be downloaded (www.genesys-lab.org/oracle) from ML research. They also published a paper on utilizing these data sets for ML-based WiFi device classification [23]. Many of the experiments include bit-similar devices, meaning they have same hardware, protocol, physical address, and MAC ID. NU published two data sets and both are available for research.

5.6 Challenge Problems with RF Data

Unlike video imagery, RF frequency imagery and signals have additional technical challenges. Raw RF data cannot be applied readily to feed the ML algorithms. For SAR data, in many cases after collection, image formation algorithms need to be applied. The SAR image formation requires knowledge of SAR signal processing and understanding the benefits and weaknesses of various imaging methods (e.g., range-Doppler, polar format processing, backprojection). Furthermore, whether to use a complex image or magnitude-only image may alter classification accuracy. There are computational costs involving utilization of complex imagery for classification. For RF communication signals, spectrogram imagery (by applying the fast Fourier transform) must be formed to feed signals into ML algorithms. Here, decisions must be made for proper utilization of both magnitude and phase information of signal contents. Finally, examples include combing video imagery of moving targets with the RF Doppler shifts for moving target recognition requiring data association for registration and positioning [24].

Another concern is sensor collection and management. With the large amount of data, one goal of ATR over various data is sensor management. Some elements of performance-driven sensor management include data collection, exploitation, and evaluation [25]. The key attributes are listed in Table 5.7. The first category is the time, space, and frequency coverage [26]. Typically, a ML approach would seek to balance these requirements to utilize the least amount of data as needed. The second category is the measures of performance (MOPs) of timeliness, accuracy, and availability [27]. The third category is measures of effectiveness (MOEs) of trusted, explainable, and autonomous. The final category is for measures of force effectives (MOFEs) of cost, usability, and extensibility [28]. While these categories have been well reported, the reason to repeat them here is that big data RF approaches that utilize ML methods need to consider the systems approach towards the design, development, and deployment. Future research and approaches will be reported on these subjects as the community continues to discuss how ML systems will be incorporated into MDC2 mult-modal information fusion systems.

With these desirable requirements and measures, the big data problems include reach goals in the form of challenge problems.

An example of a benchmark challenge problem provided to the ATR community has produced a set of challenge problems [29] such as SAR [30, 31] to address object recognition over sensor, target, and environment operating conditions. A

Table 5.7 List of Sensor Management Challenges

Persistent Coverage	Unblinking eye/omnipresent ear; revisit rate consistent with info/mission requirements
Spatial Coverage	Scalable level field of regard; rapid focus on interest area
Spectrum Coverage	Spectrum dominance and control exploitation of observable quantities
Access	Assured availability to cooperative/denied areas; survivability
Quality	Engagement Information with requisite accuracy
Timeliness	Required information at the right time
Trusted	Accurate sensing from uncorrupted data with attributable knowledge
Explainability	Efficient identification of relevant pieces of data
Autonomous	Robust, agile, and adaptable sensing abilities for sensors/networks
Cost	Cost-effective open architecture over manpower and fiscally constrained solution
Usability	Anticipatory observations/interactive engagements of adversary reactions/intents
Extensibility	Tailored performance optimized, performance-driven sensor and platform

scenario includes various kinematic target movements, possible sensor signals, target signatures, and terrain details. Figure 5.14 provides the general constructs of a challenge problem for ATR.

Scenarios provide data and support documentation for real world analysis either through analytical, simulated, or empirical results. However, as in software design, you can't control what you can't measure. One example of an open-source big data challenge is the wide-area motion imagery (WAMI) in the Columbus Large Image Format (CLIF) collection that includes baseline methods for image registration [32]. Big data from radar imagery can be combined with other sensor modalities, such as infrared [33], another multimodal video [34], and operating conditions solutions [35], in order to yield object tracking and identification [36] and event detection.

An ATR *challenge problem* includes:

- *Problem dfinition:* The scope and significance;
- *Data:* Applicable data for the defined problem;
  - Tools for reading and processing data
  - Suggestions on training and test sets
  - Characterization of the data
- *Goals:* Research questions and suggested experiments;
- *Metrics:* Guidance on reporting results;
- *Tools:* Baseline code and results that show reproducible minimum performance standards for the defined problem.

Figure 5.15 lists a few benefits of the challenge problem by providing goals, metrics, data, and tools. With the constraints of evaluation metrics, operation goals, computation resources, and data characterization for each situation; then

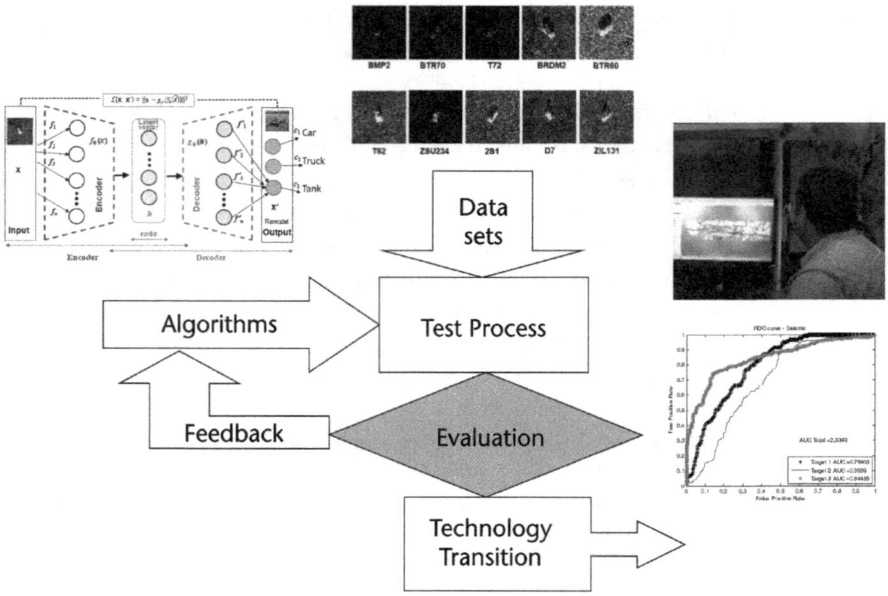

Figure 5.14 Challenge problem construct to support the evaluation process.

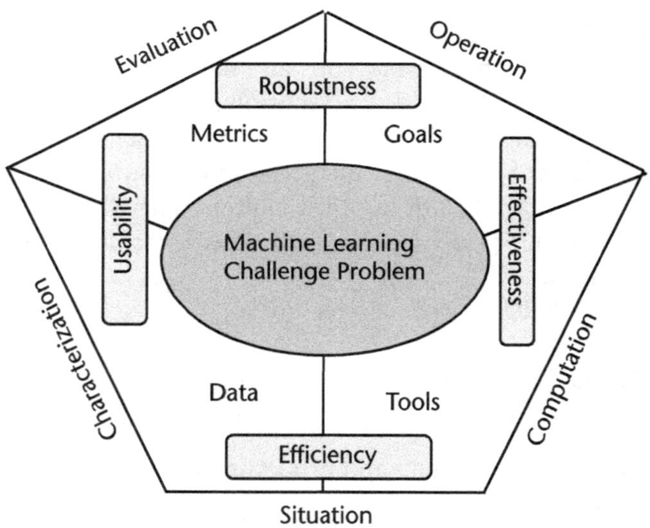

Figure 5.15 Elements of a ML challenge problem.

measures of effectiveness determine general performance over efficiency, effectiveness, usability, and robustness of ML systems for RF data.

As a final discussion, an evolution of comparative SAR data sets affords future techniques to be compared with the established techniques. Some challenge problems are shown in Table 5.8 [37].

Table 5.8 ATR Challenge Problems

Challenge Problem	*Challenge*				*Year*	*Reference*
	Reconstruction	*Detection*	*Tracking*	*Recognition*		
Standard SAR ATR evaluation experiments using the MSTAR public release data set		X	X	X	1998	[38]
Data Dome: Full k-space sampling for high-frequency radar research	X				2004	[39]
A challenge problem for detection of target in foliage		X			2006	[40]
Challenge problem for 2D/3D imaging of targets from volumetric data set in an urban environment	X				2007	[41]
A challenge problem for SAR-based GMTI in urban environments	X	X	X		2009	[42]
Civilian Vehicle Radar Data Domes	X			X	2010	[43]
A challenge problem for SAR change detection and data compression				X	2010	[44]
Wide-angle SAR data for target discrimination research				X	2012	[45]

5.7 Summary

This chapter began with the definition of big data and how RF data analytics includes velocity, volume, variety, veracity, and value. The RF data in motion (e.g., communications) and data at rest (e.g., images) require correspondence for ML techniques as the appropriate data in use (e.g., RF features). Typically, RF data size increases with the objective of a RF application including the number of targets, available sensors, and changing environmental conditions. To support such ML investigations with different sensor, environment, and target conditions, different SAR datasets were described for research. Additionally, RF communications data sets are available. These data sets lead to different challenge problems of which DL techniques can be explored. As new Automatic target recognition (ATR), exploitation (ATE), and learning (ATL) become available, the methods can leverage the existing data sets and methods for comparison.

References

[1] Blasch, E., R. Cruise, A. Aved, U. Majumder, and T. Rovito, "Methods of AI for Multimodal Sensing and Action for Complex Situations," *AI Magazine*, Vol. 40, No. 4, 2019, pp. 50–65.

[2] IBM's Definition of Big Data, https://www.slideshare.net/EdurekaIN/introduction-to-big-data-hadoop-i.

[3] Blasch, E., C. C. Insaurralde, P. C. Costa, Alta De Waal, and J. Pieter De Villiers, "Uncertainty Ontology for Veracity and Relevance," *Int'l. Conf. on Information Fusion*, July 2019.

[4] Zheng. Y., E. Blasch, and Z. Liu, *Multispectral Image Fusion and Colorization*, SPIE Press, 2018.

[5] Duda, R. O., P. E..Hart, and D. G. Stork, *Pattern Classification*, Second Edition, Wiley, 2000.

[6] Blasch, E., "Situation, Impact, and User Refinement," *Proc. of SPIE,* Vol. 5096, 2003.

[7] Blasch, E., C. Yang, and I. Kadar, "Summary of Tracking and Identification Methods," *Proc. of SPIE,* Vol. 9091, 2014.

[8] Blasch, E., and B. Pokines, "Analytical Science for Autonomy Evaluation," *IEEE National Aerospace and Electronics Conference* (NAECON), 2019.

[9] Blasch, E., S. Ravela, and A. Aved (eds.), *Handbook of Dynamic Data Driven Applications Systems*, Springer, 2018.

[10] Snidaro, L., J. Garcia Herrero, J. Llinas, and E. Blasch (eds.), *Context-Enhanced Information Fusion: Boosting Real-World Performance with Domain Knowledge*, Springer, 2016.

[11] Blasch, E., "Enhanced Air Operations Using JView for an Air-Ground Fused Situation Awareness UDOP," *AIAA/IEEE Digital Avionics Systems Conf.*, 2013.

[12] Xu, R., Y. Chen, et al., "BlendCAC: A Smart Contract Enabled Decentralized Capability-based Access Control Mechanism for IoT," *Computers*, May 2018.

[13] Snidaro, L., J. Garcia, et al., "Recent Trends in Context Exploitation for Information Fusion and AI," *AI Magazine*, July 2019.

[14] Blasch, E., J. Nagy, A. Aved, W. M. Pottenger, et al., "Context aided Video-to-Text Information Fusion," *Int'l.. Conf. on Information Fusion*, 2014.

[15] Xu, R., Y. Chen, et al., "BlendCAC: A BLockchain-ENabled Decentralized Capability-based Access Control for IoTs," *IEEE Blockchain*, 2018.

[16] Blasch E., and C. Yang, "Ten Methods to Fuse GMTI and HRRR Measurements for Joint Tracking and Identification," *Int'l Conf. on Info Fusion*, 2004.

[17] Majumder, U., and E. Blasch, "Machine Learning for Object Recognition from High Volume Radio Frequency Data," Workshop on Big Data Trends Shaping the Future of Photonics, SPIE 2017.

[18] Mukherjee, T., P. Kumar, D. Pati, E. Blasch, E. Pasiliao, and L. Xu, "LoSI: Large Scale Location Inference through FM Signal Integration and Estimation," *IEEE Big Data Mining and Analytics*, Vol. 2, No. 4, December 2019, pp. 319–348.

[19] Yang, C., T. Nguyen, and E. Blasch, "Mobile Positioning via Fusion Mixed Signals of Opportunity," *IEEE Aerospace and Electronic Systems Magazine,* Vol. 29, No. 4, April 2014, pp. 34–46,.

[20] Zulch,P., M. Distasio, T. Cushman, B. Wilson, B. Hart, and E. Blasch, "ESCAPE Data Collection for Multi-Modal Data Fusion Research," *IEEE Aerospace*, March 2019.

[21] Vakil, A., J. Liu, P. Zulch, E. Blasch, R. Ewing, and J. Li, "Feature Level Sensor Fusion for Passive RF and EO Information Integration," *IEEE Aerospace Conference*, 2020.

[22] Sankhe, K., M. Belgiovine, F. Zhou, S. Riyaz, S. Ioannidis, and K. R. Chowdhury, "ORACLE: Optimized Radio clAssification through Convolutional neuraL nEtworks," IEEE INFOCOM, May. 2019.

[23] Sankhe, K., M. Belgiovine, F. Zhou, et al., "No Radio Left Behind: Radio Fingerprinting Through Deep Learning of Physical-Layer Hardware Impairments," *IEEE Transactions on Cognitive Communications and Networking*, Vol. 6, No. 1, March 2020, pp. 165–178.

[24] Shen, D., E. Blasch, P. Zulch, et al., "A Joint Manifold Leaning-Based Framework for Heterogeneous Upstream Data Fusion," *Journal of Algorithms and Computational Technology*, Vol. 12, No. 4, 2018, pp. 311–332.

[25] Bryant, M., P. Johnson, B. M. Kent, M. Nowak, and D. Rogers, "Layered Sensing: Its definitions, Attributes, and Guiding Principles for AFRL Strategy Technology Development," June 2008, www.wpafb.af.mil/shared/media/document/AFD-080820-005.pdf.

[26] Blasch, E., C. Yang, I. Kadar, and G. Chen, "Net-Centric Layered Sensing issues in Distributed Target Tracking and Identification Performance Evaluation," *International Conference on Information Fusion*, 2008.

[27] Yang, C., I. Kadar, and E. Blasch, "Performance-Driven Resource Management in Layered Sensing," *International Conference on Information Fusion*, 2009.

[28] Blasch, E. P., E. Bosse, and D. A. Lambert, *High-Level Information Fusion Management and Systems Design*, Norwood, MA: Artech House, 2012.

[29] Arnold, G., T. Ross, L. Westerkamp, L. Carin, and R. Moses, "The ATR Center and ATRPedia," *Proc. of SPIE,* Vol. 6978, 2008.

[30] Mossing, J. C., and T. D. Ross, "An Evaluation of SAR ATR Algorithm Performance Sensitivity to MSTAR Extended Operating Conditions," *Proc. of SPIE,* Vol. 3370, 1998.

[31] Mishra, A. K., and B. Mulgrew, "Multipolar SAR ATR: Experiments with the GTRI Dataset," *IEEE Radar Conf.*, 2008.

[32] Mendoza-Schrock, O., J. A. Patrick, and E. Blasch, "Video Image Registration Evaluation for a Layered Sensing Environment," *Proc. IEEE Nat. Aerospace Electronics Conf (NAECON)*, 2009.

[33] Ling, H., L. Bai, E. Blasch, and X. Mei, "Robust Infrared Vehicle Tracking across Target Pose Change Using L1 Regularization," *Int. Conf. on Info. Fusion,* 2010.

[34] Wu, Y., E. Blasch, G. Chen, L. Bai, and H. Ling, "Multiple Source Data Fusion via Sparse Representation for Robust Visual Tracking," *Int. Conf. on Info. Fusion*, 2011.

[35] Mei, X., H. Ling, Y. Wu, E. P. Blasch, and L. Bai, "Minimum Error Bounded Efficient L1 Tracker with Occlusion Detection," *IEEE Computer Vision and Pattern Recognition*, 2011.

[36] Blasch, E. P., "Derivation of a Belief Filter for Simultaneous High Range Resolution Radar Tracking and Identification," Ph.D. Thesis, Wright State University, 1999.

[37] Gorham, B. D., E. G. Zelnio, L. A. Gorham, and F. D. Garber, "A Synopsis of Challenge Problems," *Proc. SPIE*, Vol. 8394, Algorithms for Synthetic Aperture Radar Imagery XIX, 2012.

[38] Ross, T., S. Worrell, V. Velten, J. Mossing, and M. Bryant, "Standard SAR ATR Evaluation Experiments Using the MSTAR Public Release Data Set," in *Proceedings of SPIE*, Vol. 3370, 1998.

[39] Naidu K., and L. Lin, "Data dome: Full k-Space Sampling Data for High-Frequency Radar Research," *Proceedings of SPIE*, Vol. 5427, 2004.

[40] Lundberg, M., L. Ulander, W. Pierson, and A. Gustavsson, "A Challenge Problem for Detection of Targets in Foliage," *Proceedings of SPIE*, Vol. 6237, 2006.

[41] Casteel, C. H., Jr., L. A. Gorham, M. J. Minardi, S. M. Scarborough, K. D. Naidu, and U. K. Majumder, "A Challenge Problem for 2D/3D Imaging of Targets from a Volumetric Data Set in an Urban Environment," in *Proceedings of SPIE*, Vol. 6568, 2007.

[42] Scarborough, S. M., C. H. Casteel, Jr., L. Gorham, et al., "A Challenge Problem for SAR-based GMTI in Urban Environments," *Proceedings of SPIE,* Vol. 7337, 2009.

[43] Dungan, K., C. Austin, J. Nehrbass, and L. Potter, "Civilian Vehicle Radar Data Domes," *Proceedings of SPIE*, Vol. 7699, 2010.

[44] Scarborough, S. M., L. Gorham, M. J. Minardi, et al., "A Challenge Problem for SAR Change Detection and Data Compression," *Proceedings of SPIE*, Vol. 7699, 2010.

[45] Dungan, K. E., J. N. Ash, J. W. Nehrbass, J. T. Parker, L. A. Gorham, and S. M. Scarborough, "Wide Angle SAR Data for Target Discrimination Research," *Proceedings of SPIE*, Vol. 8394, 2012.

[46] Lewis, Benjamin, et al., "A SAR dataset for ATR development: the Synthetic and Measured Paired Labeled Experiment (SAMPLE)," *Algorithms for Synthetic Aperture Radar Imagery XXVI*, Vol. 10987, International Society for Optics and Photonics, 2019.

[47] Scarnati, Theresa, and Benjamin Lewis, "A deep learning approach to the Synthetic and Measured Paired and Labeled Experiment (SAMPLE) challenge problem," *Algorithms for Synthetic Aperture Radar Imagery XXVI,* Vol. 10987, International Society for Optics and Photonics, 2019.

[48] Elizabeth R. Sudkamp, John W. Nehrbass, Eric Branch, Michael Levy, "Synthetic data accuracy sensitivity to CAD model accuracy using ATR-based metrics," *Proc. SPIE 10987, Algorithms for Synthetic Aperture Radar Imagery XXVI,* 2019

CHAPTER 6

Deep Learning for Single-Target Classification in SAR Imagery

6.1 Introduction

This chapter presents classification techniques of a single target (also referred to as an image chip [33]) in a SAR image. SAR single-image classification is similar to the Modified National Institute of Standards and Technology (MNIST) digit classification problem, wherein images are labeled for training and classification results are computed based on various machine learning (ML) methods. The classification problem is more challenging if the training images are not labeled. If the images are not labeled, then alternative techniques are used such as clustering help to organize targets into categories for semantic labeling as per a dictionary.

The MSTAR public released data set is widely used by the researchers for SAR-based object classification [1]. Given the reference to the various ML MSTAR papers throughout the book, this chapter will highlight another set of data known as CVDome [2]. The CVDome data set is a high-fidelity synthetic radar data set generated using detailed ray-tracing software methodologies.

The MSTAR data provides a useful scenario for research in SAR data collection, pattern recognition, and target tracking to support challenge problems with labeled data. Over the years, there have been goals to develop methods to simulate more numerous data sets to challenge the ATR methods across a variety of operating conditions [3] and scenarios [4] as well as software methods for developers to enhance variations [5]. The simulated data provides opportunities, but real-world conditions from collections can further explore developments in ML [6].

In general, a single-target classification algorithm involves preprocessing the SAR data, labeling the training data, designing a classification algorithm such as a DNN, training and validating the DNN model, testing the model with sequestered data, and finally analyzing the results with a confusion matrix [7]. Figure 6.1 summarizes the algorithmic steps for single-target classifications. Once the phase history data is collected, the first process is to form the imagery [8, 9].

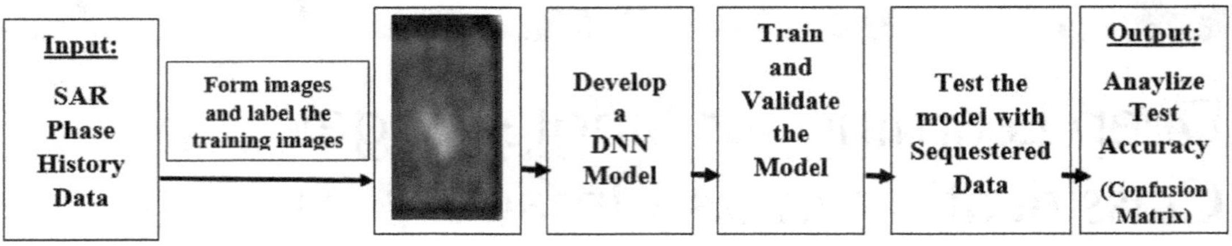

Figure 6.1 Algorithmic flow diagram involving single target classifications in SAR imagery.

6.1.1 Machine Learning SAR Image Classification

Most of the papers published on methods for SAR are focused on a comparative example of the MSTAR data set (~280), while a few (~25) focus on the extraction of HRR data (see Chapter 1). Another large section of the SAR literature focuses on polarimetric SAR (POLSAR), which can be used for scene and object recognition. Over 90% of the papers for single-object detection extract chips and perform single-target recognition. Some of the interesting papers conduct a performance analysis for the readers to appreciate a variety of techniques. Various approaches are presented throughout the text on the traditional approaches for the MSTAR analysis. As the introduction chapter highlighted, there was an initial analysis from 1998–2002 on ML methods for SAR. Then beginning in 2014, DL approaches were investigated for single-object SAR ATR. This review will highlight some of these techniques, although not all can be covered within the scope of this text.

ML approaches existed prior to the release of the MSTAR data, and numerous researchers adopted these ML methods to the MSTAR data. Marinelli et al. [10] conducted an experiment of the learning vector quantization (LVQ) algorithm applied to the MSTAR data set. To compare the methods, they divided the MSTAR data set of 360° into equal bin sizes of 20°, 40°, and 60° and compared the traditional ML method of k-means clustering to the LVQ using an ANN. With 100% probability of declaration (see Chapter 9), the results highlighted 96% probability of correct classification with the 20° binning with both k-means and LVQ. However, for the 60° binning, k-means had 87% and LVQ 92%. In 2000, Zhao et al. [11] presented three methods of SVM, quadratic mutual information (QMI), and principal component analysis with multiresolution (PCA-M) using an MLP. The highlighted results provide a baseline for the MSTAR analysis, as shown in Table 6.1. Hence, the SVM provides a useful classifier from these results.

Another development that became popular was that of *joint sparse representation* from multiple views. For single-object detection, one example is the simultaneous performance of a multivariate regression model and the recovery of sparse representations for the multiple views. In [12], the authors compare the kernel-based SVM with that of the sparse representation classifier (SRC).

For the three-target data set, Zhang et al. [13] presented optimistic results with 20 dimensions using the multispectral feature analysis (MSFA) achieving 99% with the SVM as compared to the PCA-SVM of 96% classification. From the 10-class MSTAR, the use of multiple looks was shown to enhance accuracy that

Table 6.1 Comparison on ML Approaches

Classifier	*Accuracy (%)*	*Rejection (%)*
Template	90.4	53.5
PCA-M	89.7	60
SVM	94.9	68.8
QMI	90.4	54.5

Table 6.2 Sparse Reconstruction

Single View	*Accuracy (%)*	*Multiple View*	*Accuracy (%)*
SVM	72.1	SVM	81.0
SRC	81.4	SRC	89.8
K-SVM	82.9	K-SVM	90.2
		J-SCR	93.2

From: Zhang et al. [14].

demonstrates the need to consider multiple views for single object target recognition. Hence, multilook fusion reduces uncertainty and enhances accuracy.

6.1.2 Deep Learning SAR Image Classification

With the emergence of deep learning (DL) in 2012 for visual image recognition, there was interest in utilizing the techniques for SAR image classification. Initial reasoning for this investigation was that previous approaches (e.g., the MSTAR program) for SAR image classification had shortcomings, including the requirement of a large database of target signatures (i.e., millions of images from various azimuth and elevation angles), poor classification performance for slight target variations, and slow computation. Hence a DARPA seedling program (MADLEARNING) was launched to determine the feasibility of DL for SAR image classification. However, since SAR imagery of targets is not literal, and thus may be challenging for visual recognition, the application of DL to classify SAR imagery was still in an exploratory phase. Interestingly, the MADLEARNING program demonstrated a promising capability of DL to classify SAR imagery. Hence, a DARPA Target Recognition and Adaption for Contested Environments (TRACE) program was funded to investigate DL methods more fully for the SAR image classification problem. Since then, researchers began to adopt the DL methods to the MSTAR data for object detection and classification.

Since 2015, various approaches have been presented, with one of the first by Morgan applying deep CNNs to the MSTAR data [15]. Other researchers also looked at variations of weight decay [16] and combinations of the CNN with the SVM [17]. Further studies on the MSTAR data compared the CNN with the convolution autoencoder [18]. Since the data set was small, efforts have been made to improve the robustness via augmented training samples [19] and semisupervised methods [20].

6.2 SAR Data Preprocessing for Classification

In many applications, SAR imagery is of interest due to its being an all-weather sensing modality and the invariance of target signatures with respect to distance from the sensor. In contrast, video imagery (i.e., electro-optical (EO) sensors) is favorable when the operating conditions of the sensor, environment, and target correspond to good illumination (e.g., sunlight), direct line of sight, and the absence of obscuration. The use of EO imagery may pose a challenge for many applications such as autonomous vehicles operating at night and in snow conditions (unless other sensors are integrated). Since SAR sensors can operate from long stand-off distances, then the data affords formation of high resolution, multipolarization images of large areas. A SAR system is capable of synthesizing a large aperture by periodically transmitting and receiving pulses from multiple locations along a trajectory and collecting the pulses reflected from the scene to form images (see Chapter 1). The information from the reflected pulses is saved as *phase history* (raw) data (or in-phase and quadrature, I/Q data) and is subsequently used to form human-interpretable images by applying any of a number of image formation algorithms.

Many of the current SAR-based object recognition (SAR-OR) systems use a fast Fourier transform (FFT) based algorithm to form the images from the phase history before analyzing the images for target detection and classification. For example, many efforts use images from the MSTAR public data set, which are formed using a FFT technique called the *polar format* (PFA) algorithm. One problem with the FFT approach, especially for size, weight, and power (SWaP) constrained systems, is that the image formation process is computationally expensive. In some cases, the image formation process requires more computation than a deep convolution neural network (DCNN) based classification algorithm.

The current focus of research in SAR-OR attempts to maximize the accuracy of the DCNN by using different DL algorithms models, *without* regard for computational cost. Researchers developing convolutional auto-encoders seek to maximize the accuracy of the classifier. Recent investigations seek network architectures and backpropagation algorithms to achieve gains in accuracy, effectiveness, and computational efficiency. For example, enhancements to SAR ML classification experiments would leverage both the in-phase and quadrature-phase data as output from a back-propagation algorithm for a complex input CNN. Likewise, to optimize the computational cost, object detection performance could utilize the range-Doppler domain (partially processed data).

SAR data offers many data domains for analysis such as: phase history, range-Doppler, polar format, and FFT back-projection [21–23]. Recently, the performance of a DCNN on SAR data sought ways to minimize the overall computational cost in a SAR object recognition system, while also trying to maintain the accuracy of the object classification [24] (See Figure 6.2). The four domains include:

- *Phase history* (PH, i.e., raw) data, where the phase history information from the aperture is directly represented as a grayscale plot of magnitudes;
- *Range-Doppler* (RD, i.e., partially processed) data, which is produced using a range-Doppler algorithm;

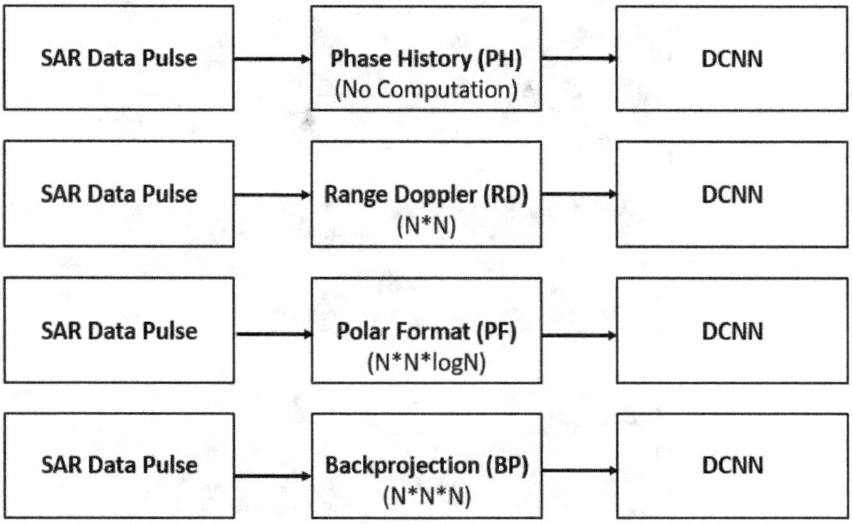

Figure 6.2 Depiction of four SAR image formation algorithms for object recognition approaches using (1) phase history (raw) data, (2) range-Doppler (partially-processed images) data, (3) polar format (fully formed images with artifacts) data, and (4) back-projection (the fully formed images) data.

- *Polar format* (PFA) algorithm to form a SAR image, which contains artifacts that do not comply with the assumption of only stationary point scattering centers on the surface of the ground plane;
- Backprojection (BP, baseline) algorithm is the fully formed images using a traditional FFT.

The BP approaches are often more computationally expensive than RD and PF approaches, but they offer an ability to directly incorporate elevation profile data available from digital terrain elevation data (DTED) sets. The BP domain is meant to represent a baseline approach to SAR object recognition for comparison purposes.

Each of the four approaches accepts phase history data for a single object image chip, forms an image (2D data plot) with some or no processing, and finally classifies the object with a DCNN-based algorithm.

6.3 SAR Data Sets

6.3.1 MSTAR SAR Data Set

The MSTAR data set was collected by the Sandia National Laboratory SAR sensor platform [25]. A subset of the overall data collected was then released to the public for ATR research, shown in Figure 6.3. The MSTAR public dataset contains several hundred SAR images of a variety of targets collected in varying conditions. The data collection aircraft had an X-band radar sensor configured for 1-foot resolution in spotlight mode to collect target data at 15-, 17-, 30-, and 45-degree depression angles. SAR image data was collected at varying azimuth angles in order to provide a 360-degree view of each given target. The images are 128 × 128 pixel chips,

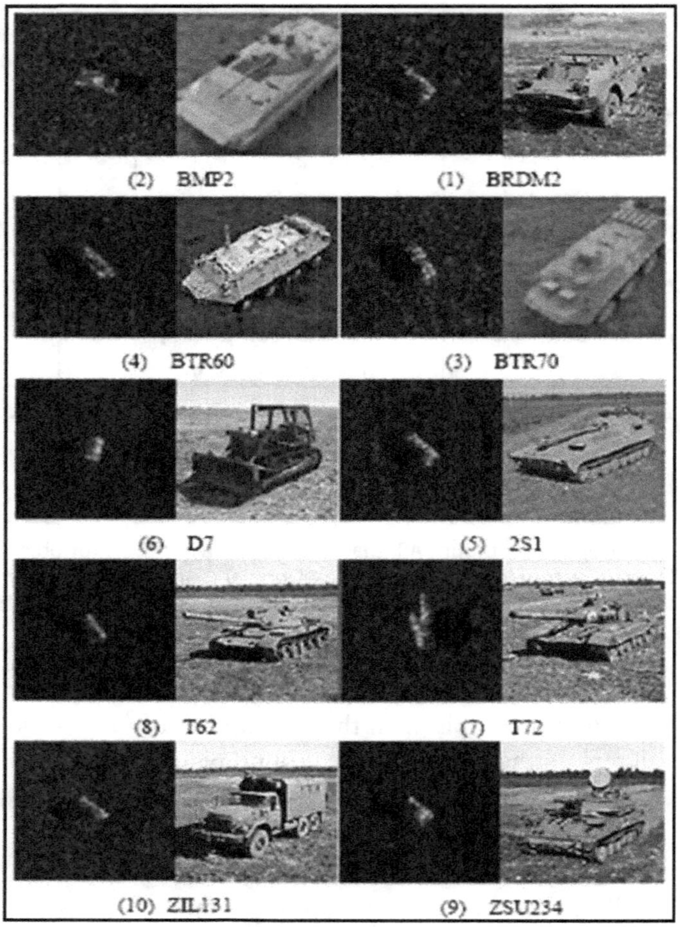

Figure 6.3 SAR imagery (next to electro-optical imagery) of various military vehicles.

each containing a target roughly centered within its background. Bulldozers, trucks, tanks, and antiaircraft vehicles are among the targets included in the data set.

For the MSTAR data, the targets were collected at different poses to facilitate matching and various perspectives for ATR as shown in Figure 6.4.

In addition to target data, Sandia collected clutter data of rural and urban background settings. The data set includes 100 full-scene images collected at a 15-degree depression angle. The purpose of these clutter images is to develop training and testing data to be used in multitarget detection tasks. Other radar data sources [26–29] and processing techniques are available for classification research.

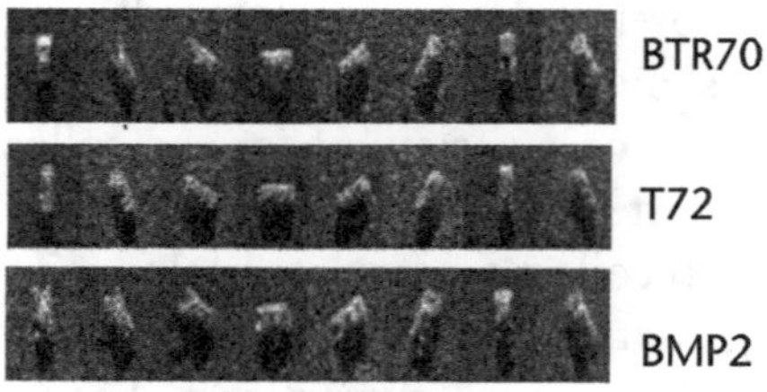

Figure 6.4 MSTAR imagery of different target poses.

6.3.2 CVDome SAR Data Set

The CVDome [30] is a synthetically generated SAR data set for ATR research (see Figure 6.5). The CVDome data set has been generated due to the limited availability of measured SAR phase history data. The scenes from which the data is collected involve single objects modeled with perfect electric conducting (PEC) surfaces, where the effects of the background clutter have been neglected. The resulting data is the simulated phase history information (raw pulses) for 1,024 ranges and 9,000 azimuth angles, from 0° to 359.96° at an azimuth sampling of 0.04°. The data for each object is available in multiple elevation angles such as 30, 40, 50, and 50°.

The CVDome experiment data uses a total of 10 models (classes) of objects. These targets include Camry, Honda Civic, Jeep93, Jeep99, Maxima, Mazda MPV, Mitsubishi, Sentra, Toyota Avalon, and Toyota Tacoma. The data, when loaded into MATLAB, is a 9000 × 1024 matrix of complex-valued single-target images for both the horizontal (*hh*) and vertical (*vv*) copolarizations. Each of the 9,000 rows corresponds to an azimuth sampling degree (i.e., corresponding to slow time), and each of the 1,024 columns represents a range bin (i.e., fast time). Figure 6.6 shows examples from the Jeep93.

For any given image, only 30° of azimuth samples are considered. To cover all azimuth angles for all objects, the starting azimuth angle is incremented from 0° to 330° in steps of 2°, so there are 166 submatrices per elevation per class. Figure 6.7 shows examples of image formation at 30°elevation angle, 50° integration angle, and 0° azimuth angle. From the data set, each of the submatrices is used as the phase history data, RD algorithm, PF algorithm, and BP algorithm.

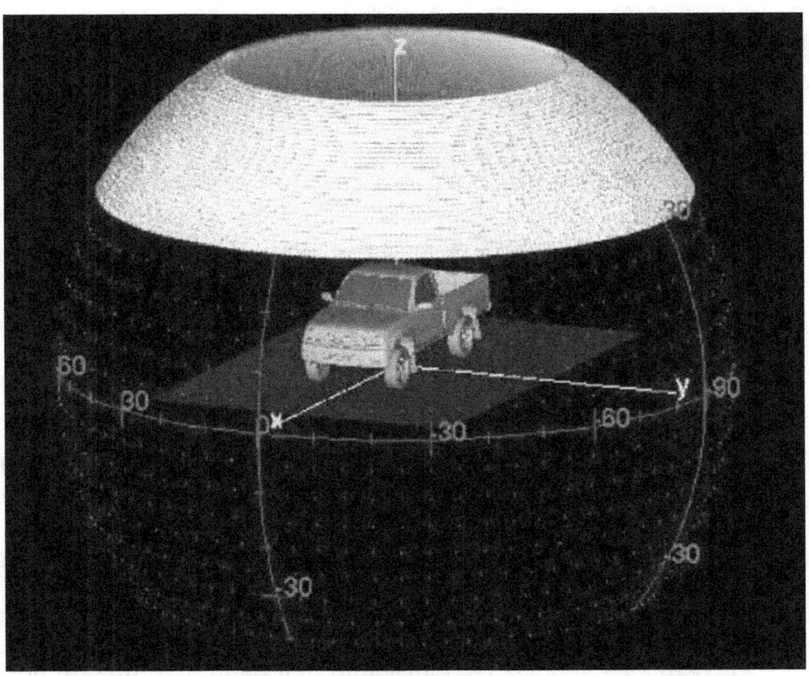

Figure 6.5 Visualization of elevation and azimuth angles used in computing simulated monostatic scattering. The viewing sphere is shown with a Tacoma CAD model.

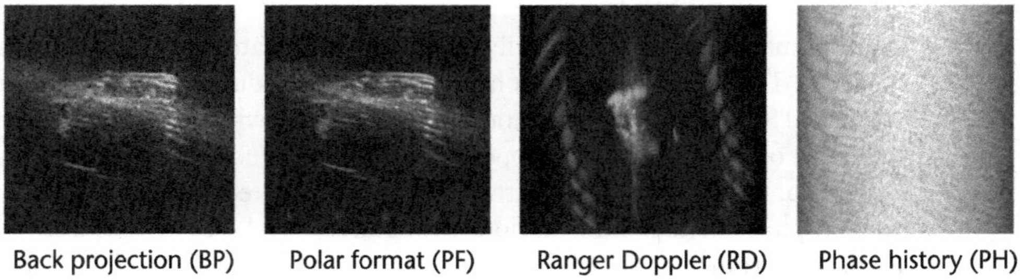

Figure 6.6 SAR images of Jeep93, at elevation 40, integration angle 30, starting at azimuth 90.

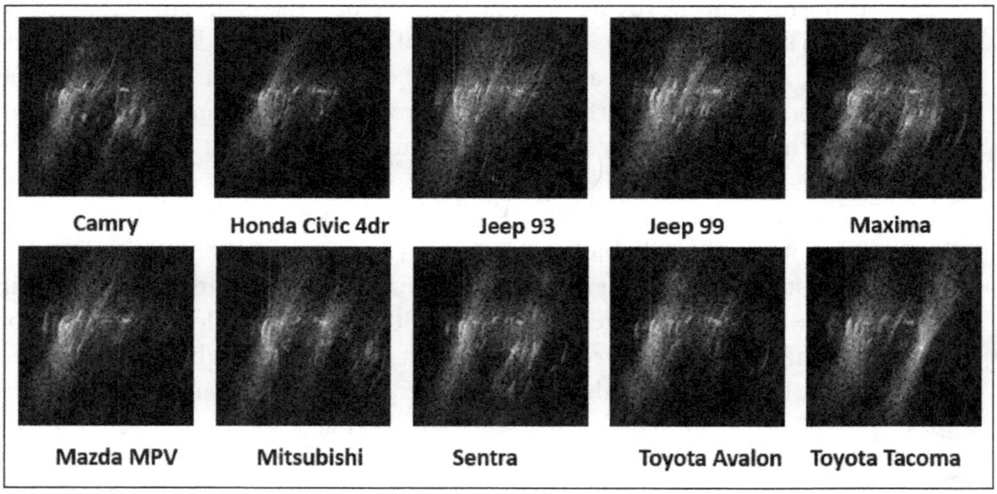

Figure 6.7 BP SAR images formed at elevation angle 30°, integration angle 50°, starting at azimuth 0°.

6.4 Deep CNN Learning

Once all the trained images have been generated and labeled, the next step is to build a classifier. The classifiers are in the form of DCNNs. DCNNs have shown to be very capable for the task of video image classification. The basic structure of the SAR DCNN can be adapted from the Cifar-10 example that comes with Caffe DNN model developed by U.C. Berkley [31].

6.4.1 DNN Model Design

The DCNN can be made deeper and wider based on the selection of various parameters. The idea behind making the network deeper is that in general, a greater number of convolutional layers in a DCNN yields a more complex and global network for better extracting and learning features. For the phase history data, where the features are not localized, it is important to have a network that can learn these complex and global features. It has also been shown that a deeper network can provide better accuracy (to a point), and that depth is a more important factor than the convolution filter size. It is important to note that the larger the network, the more memory is required, which has proven to be a computational performance bottleneck.

Based on a series of experiments, a network with five convolutional layers provides suitable results for all tests. The DCNN design is shown in Figure 6.8. Ultimately, the final network consists of five convolutional layers with a kernel size of 5, a stride of 1, and a pad of 2, with a DCNN characterized by 5 ReLU layers and five pooling layers each having a kernel size of 3, a stride of 2, and a pad 0. After the fifth pooling layer, the data in the network passes to a fully connected (FC) layer and then finally to a second FC layer that has one output for each class in the data set. The final stage of the network is to output the final classification probabilities by using a softmax function. The initial weights of the convolutional layers were randomly initialized with Gaussian weight fillers.

The backpropagation algorithm used to update the weights during training is a stochastic gradient decent algorithm as implemented in the Caffe solver file [32]. The parameters of the solver are shown in Table 6.3.

6.4.2 Experimentation: Training and Verification

Using the preprocessed data sets supports the verification of software techniques for experimental laboratory settings. When the systems have been verified with the collected data, field tests for operational scenarios with live equipment and users validate the robustness of the methods.

After all the images (i.e., 128 × 128-pixel lossless JPEGs) are generated, the next step is to create the training, verification, and test data sets for use with the DCNNs. For each image-domain (i.e., PH, RD, PFA, BP), four lists are compiled, one for each type of data set. In first experiment scenario, the images are formed from azimuth angles (0°-360°) and elevation angles of 30°, 40°, 50°, and 60°. The

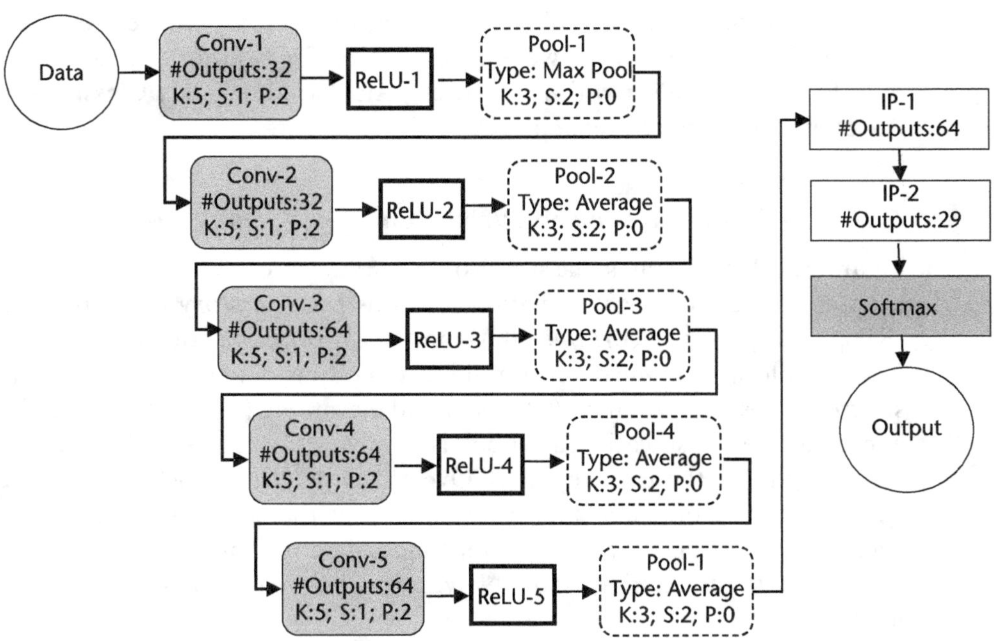

Figure 6.8 A typical DCNN model architecture.

Table 6.3 Backpropagation Solver Parameters from the Caffe Solver File

Solver Parameters	
test_itr	100
test_interval	500
base_lr	0.001
Momentum	0.9
weight_decay	0.004
lr_policy	"fixed"
solver mode	GPU

experiment randomly picks 80% of the images for training and verification and 20% for testing. In a second experiment scenario, images are formed from azimuth angles (0°–360°) and take elevation angles 30°, 40°, and 60° for training and verification and elevation angle 50° for testing data.

Next, the system creates lightning memory–mapped databases (LMDBs) using Caffe's *convert_imageset* program. The input of the program is the set of text files, and the output is the set of LMDB directories (train, verification, and test). Then, the image mean is computed for the each of the training sets using Caffe's *compute_image_mean* program. The last step in the procedure is to train each of the classifiers three separate times from scratch and to record the results of the training and testing for each. The accuracy of the networks is measured in two ways. During training, top-1 and top-5 accuracies are measured on a validation set using Caffe's *train* program. Once the networks have trained for 100,000 iterations each, the top-1 and top-5 accuracies of the deployed model are then measured on the test data.

Among CNNs of choice, one may pick any of the widely used models. For the experiments with the CVDome data set, five different models are used. These are ResNet18 (Residual Network), MobileNetV2 (good for mobile applications), ShuffleNet (MobileNet with low power), AlexNet (ImageNet Classification), and AlexNet-slim (a simple version of AlexNet).

6.4.3 Evaluation: Testing and Validation

Figure 6.9 demonstrates classification accuracy across various DNN models for the different SAR image formation methods *phase history* (PH), *range-Doppler* (RD), *polar format algorithm* (PFA), and *backprojection* (BP, baseline) algorithm. These results are obtained when training and testing data (imagery) were randomly selected across all azimuths (0°–360°) and elevations (30°, 40°, 50°, and 60°). As expected, the classification accuracy is almost 99% across all DNN models and SAR imaging methods (except PH). Overall, MobileNetv2 yields the best accuracy.

Figure 6.10 presents similar results except that instead of random data for training and testing, specific elevation angles data were used for training and other elevation (50°) for testing. MobilNetv2 provided best accuracy across all imaging methods.

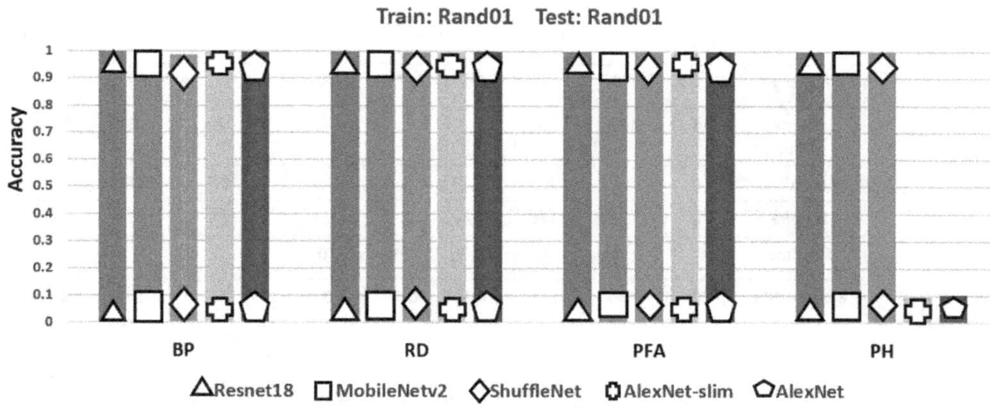

Figure 6.9 Classification accuracy across DNN models and SAR imaging methods for random train and test data.

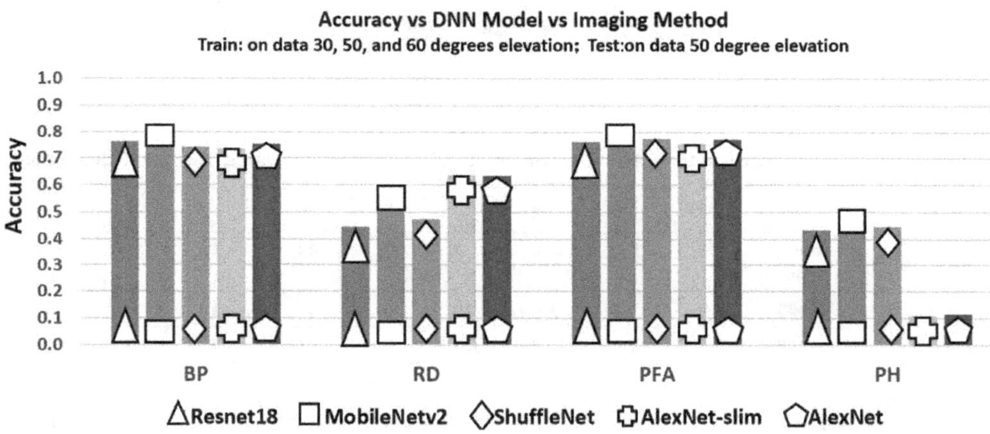

Figure 6.10 Classification accuracy across DNN models and imaging for training on elevation angle 30°, 40°, and 60° and testing on elevation angle 50° data.

6.4.4 Confusion Matrix Analysis

In this section, confusion matrices provide a convenient means for examining target classification accuracy. Notice that the experiment revealed that MobilNetv2 provides the best overall classification accuracy across all imaging methods. Thus, it is important to examine classification accuracy across the different imaging methods and for testing with respect to various elevation angles and the selection of random data. Figure 6.11 highlights the classification accuracy of MobilNetv2 on images formed using BP.

Since backprojection provides the most accurate SAR imagery, Figure 6.11 shows that testing accuracy is the same for elevation angles 40° and 50°. However, for PF imaging in Figure 6.12, notice that the accuracy decreases for testing at the elevation angle 40° data set. This result follows from the fact that certain elevation angles are important for training. Further, imaging artifacts for PF also play a role for reduced classification accuracy. Finally, Figure 6.13 and Figure 6.14 present decreased classification accuracy, as these imaging methods do not seem to extract the salient features of the targets.

Confusion matrix for train on 30, 40, 60 degrees elevation, test on 50 degree elevation

True Label \ Predicted Label	Camry	Honda Civic	Jeep93	Jeep99	Maxima	MazdaMPV	Mitsubishi	Sentra	Toyota Avalon	Toyota Tacoma
Camry	1038	6	1	0	212	0	0	26	9	8
Honda Civic	4	1017	0	8	8	112	7	43	58	43
Jeep93	0	6	906	81	2	83	28	57	121	16
Jeep99	0	15	176	972	0	23	41	10	39	24
Maxima	65	0	1	0	1105	0	0	89	1	39
MazdaMPV	12	69	111	0	10	1033	22	9	33	1
Mitsubishi	4	0	0	0	39	0	1234	16	7	0
Sentra	14	0	0	0	45	0	32	1195	14	0
Toyota Avalon	8	36	6	0	5	40	219	5	926	55
Toyota Tacoma	3	3	0	0	3	0	0	4	37	1250

(a)

Confusion matrix for train on 30, 50, 60 degrees elevation, test on 40 degree elevation

True Label \ Predicted Label	Camry	Honda Civic	Jeep93	Jeep99	Maxima	MazdaMPV	Mitsubishi	Sentra	Toyota Avalon	Toyota Tacoma
Camry	949	0	0	0	263	0	36	0	52	0
Honda Civic	0	1125	0	3	0	42	1	0	129	0
Jeep93	0	9	1236	47	0	1	1	0	6	0
Jeep99	0	0	79	1142	0	77	0	0	2	0
Maxima	242	0	0	0	1033	4	0	21	0	0
MazdaMPV	0	54	13	15	0	1052	0	0	166	0
Mitsubishi	0	0	0	6	0	134	1068	0	92	0
Sentra	84	0	1	0	429	2	6	777	1	0
Toyota Avalon	0	56	18	49	0	27	71	1	1078	0
Toyota Tacoma	0	0	0	2	2	25	8	2	0	1261

(b)

Figure 6.11 Classification accuracy of MobilNetv2 on BP formed images. (a) Training on 30°, 40°, and 60° elevation data and testing on 50° elevation data; classification accuracy was recorded approximately 82%. (b) Training on 30°, 50°, and 60° elevation data and testing on 40° elevation data; classification accuracy was recorded approximately 82%. (c) Training on mixed (30°, 40°, 50°, and 60°) elevation data and testing on mixed elevation data; classification accuracy was recorded approximately 99.8%.

Confusion matrix for train and test on mixed (30, 40, 50, 60 degrees) elevation

True Label	Camry	Honda Civic	Jeep93	Jeep99	Maxima	MazdaMPV	Mitsubishi	Sentra	Toyota Avalon	Toyota Tacoma
Camry	1016	0	0	0	1	0	36	0	52	0
Honda Civic	0	1027	0	0	0	0	1	0	129	0
Jeep93	0	0	1064	0	0	0	1	0	6	0
Jeep99	0	0	0	1014	0	0	0	0	2	0
Maxima	0	0	0	0	1045	0	0	21	0	0
MazdaMPV	0	0	0	0	0	1051	0	0	166	0
Mitsubishi	0	0	0	0	0	0	1039	0	92	0
Sentra	0	0	0	0	0	0	6	1022	1	0
Toyota Avalon	0	0	0	0	0	0	71	1	1062	0
Toyota Tacoma	0	0	0	0	0	0	8	2	0	1057

Predicted Label

(c)

Figure 6.11 (continued)

A comparison of the Figures 6.12–6.14 reveals interesting information relevant for the processing of radar imagery data, which is formed from the complex-valued I/Q data channels.The raw PH data chips are input into the DCNN processing. The actual PH images appear as random noise to humans, regardless of whether one examines the data as two I and Q images or as magnitude and phase images. The random-noise property follows from the general holographic property of complex-valued SAR imagery that is transduced in the collection of the I/Q radar signal returns. The remarkable result of Figure 6.14 is that DCNN architectures can provide a nontrivial level of classification capability, whereas any human examining the same input PH image data chips is likely to have absolutely no classification capability. In effect, it seems that silicon-based ANNs are actually more flexible than biological neural networks. In order to obtain these results, the machine-based ANN must effectively learn the 2D Fourier transform, or at least some approximation thereof, which defines the relationship between the raw PH data chips and the simple RD target images. It is challenging for any user to accomplish the same feat, even if presented with complex-valued data as pairs of I and Q images or as magnitude and phase images. Specifically, the human visual system is not designed to classify the SAR PH data, regardless of how much training data is available.

Validation of the MobilNetv2 on images formed using BP and PF using subject matter experts and the human-machine interface demonstrated subjective approval of the techniques for application use. It is worthwhile to note that the data in this experiment is ideal, wherein the input is comprised of *single object* image chips

Confusion matrix for train on 30, 40, 60 degrees elevation, test on 50 degree elevation

True Label \ Predicted Label	Camry	Honda Civic	Jeep93	Jeep99	Maxima	MazdaMPV	Mitsubishi	Sentra	Toyota Avalon	Toyota Tacoma
Camry	706	0	0	0	474	0	0	118	0	2
Honda Civic	0	942	8	4	3	159	17	59	82	26
Jeep93	2	0	1163	6	9	0	3	24	45	48
Jeep99	3	0	77	1162	0	29	6	1	0	22
Maxima	111	0	0	0	880	0	18	268	0	23
MazdaMPV	9	3	9	36	73	1125	4	13	0	28
Mitsubishi	0	0	0	0	13	0	1200	80	5	2
Sentra	9	0	0	0	95	0	23	1171	1	1
Toyota Avalon	0	7	0	0	72	8	29	66	1111	7
Toyota Tacoma	0	0	0	0	0	0	0	2	0	1298

(a)

Confusion matrix for train on 30, 50, 60 degrees elevation, test on 40 degree elevation

True Label \ Predicted Label	Camry	Honda Civic	Jeep93	Jeep99	Maxima	MazdaMPV	Mitsubishi	Sentra	Toyota Avalon	Toyota Tacoma
Camry	988	0	0	0	280	0	14	1	16	0
Honda Civic	0	1091	13	6	0	2	4	0	184	0
Jeep93	0	5	1032	227	0	14	12	0	10	0
Jeep99	0	0	14	1260	0	23	0	2	1	0
Maxima	156	0	3	0	1091	11	21	4	14	0
MazdaMPV	0	36	14	18	0	1023	0	0	209	0
Mitsubishi	0	9	38	49	4	86	987	0	127	0
Sentra	84	11	0	0	391	27	10	732	45	0
Toyota Avalon	0	105	0	12	0	7	16	0	1157	3
Toyota Tacoma	1	0	0	0	28	15	0	1	1	1254

(b)

Figure 6.12 Classification accuracy of MobilNetv2 on polar format (PFA) formed images. (a) Training on 30°, 40°, and 60° elevation data and testing on 50° elevation data; classification accuracy was recorded approximately 83%. (b) Training on 30°, 50°, and 60° elevation data and testing on 40° elevation data; classification accuracy was recorded approximately 82%. (c) Training on mixed (30°, 40°, 50°, and 60°) elevation data and testing on mixed elevation data; classification accuracy was recorded almost 100%.

Confusion matrix for train and test on mixed (30, 40, 50, 60 degrees) elevation

True Label	Camry	Honda Civic	Jeep93	Jeep99	Maxima	MazdaMPV	Mitsubishi	Sentra	Toyota Avalon	Toyota Tacoma
Camry	1051	0	0	0	1	0	0	0	0	0
Honda Civic	0	1069	0	0	0	0	1	0	0	0
Jeep93	0	0	1045	0	0	0	1	0	0	0
Jeep99	0	0	0	1041	0	0	0	0	2	0
Maxima	0	0	0	0	1018	0	0	0	0	0
MazdaMPV	0	0	0	0	0	1062	0	0	0	0
Mitsubishi	0	0	0	0	0	0	1017	0	0	0
Sentra	0	0	0	0	0	0	0	1015	1	0
Toyota Avalon	0	0	0	0	0	0	0	1	1051	0
Toyota Tacoma	0	0	0	0	0	0	0	0	0	1030

Predicted Label

(c)

Figure 6.12 (continued)

of relatively small size and the objects are modeled without background clutter. In an operational object recognition (OR) system, the image sizes would be much larger, contain background clutter, and involve multiple objects in a single frame. Therefore, the accurate extraction of the RD and PH image chips would be a great challenge in the detection stage. Also, it is acknowledged that for many systems, the processing in the image formation stage is not a concern. For SWaP-constrained OR systems, operating with RD and/or PH images may provide useful computation and energy savings with tolerable accuracy loss.

Figures 6.12–6.14 offer insight after further thoughts and investigations. In particular, the simplest form of input into DCNN architectures is that of the raw PH data of the I/Q radar data, which yields some nontrivial correct classification capability over that of a user. It seems that a sufficiently deep DCNN architecture can learn some basic approximation of the 2D Fourier transform. However, if developers, as managers of DCNN processing, select the preprocess raw PH data via 2D Fourier transforms to yield crude RD target image chips, then the resultant DCNN processing yields improved results. If more accurate preprocessing of the PH or BP image formation are available to the DCNN input data, then even more improved classification results. As an aside, it is expected that PFA and BP yield similar results, since the primary advantage of BP over PFA applies when wavefront curvature effects are important over the span of a target, which is typically unimportant for the mobile targets considered herein.

Confusion matrix for train on 30, 40, 60 degrees elevation, test on 50 degree elevation

True Label	Camry	Honda Civic	Jeep93	Jeep99	Maxima	MazdaMPV	Mitsubishi	Sentra	Toyota Avalon	Toyota Tacoma
Camry	861	1	9	95	278	0	0	55	0	1
Honda Civic	6	833	13	29	0	187	82	0	143	7
Jeep93	0	0	881	85	1	256	51	0	2	24
Jeep99	0	0	240	972	0	55	0	0	0	33
Maxima	111	0	0	52	1105	0	0	30	0	2
MazdaMPV	0	41	25	8	3	1179	0	0	25	19
Mitsubishi	3	50	0	0	0	31	1103	0	113	0
Sentra	83	8	10	60	257	19	1	831	23	8
Toyota Avalon	0	124	29	0	20	168	91	0	836	32
Toyota Tacoma	1	0	0	9	27	4	0	141	0	1118

Predicted Label

(a)

Confusion matrix for train on 30, 50, 60 degrees elevation, test on 40 degree elevation

True Label	Camry	Honda Civic	Jeep93	Jeep99	Maxima	MazdaMPV	Mitsubishi	Sentra	Toyota Avalon	Toyota Tacoma
Camry	775	6	94	27	234	20	132	6	6	0
Honda Civic	0	283	105	112	22	193	540	0	37	0
Jeep93	0	2	909	187	13	141	11	21	1	0
Jeep99	0	3	297	892	13	34	53	4	4	0
Maxima	156	60	25	11	936	49	24	34	12	3
MazdaMPV	0	3	137	177	15	674	265	4	25	0
Mitsubishi	0	111	30	82	0	52	990	26	4	0
Sentra	322	0	109	0	324	18	165	348	14	0
Toyota Avalon	16	15	135	60	15	251	344	9	455	0
Toyota Tacoma	2	0	12	31	382	0	55	37	0	781

Predicted Label

(b)

Figure 6.13 Classification accuracy of MobilNetv2 on RD formed images. (a) Training on 30°, 40°, and 60° elevation data and testing on 50° elevation data; classification accuracy was recorded approximately 75%. (b) Training on 30°, 50°, and 60° elevation data and testing on 40 elevation data; classification accuracy was recorded approximately 54%. (c) Training on mixed (30°, 40°, 50°, and 60°) elevation data and testing on mixed elevation data; classification accuracy was recorded almost 100%.

Confusion matrix for train and test on mixed (30, 40, 50, 60 degrees) elevation

True Label	Camry	Honda Civic	Jeep93	Jeep99	Maxima	MazdaMPV	Mitsubishi	Sentra	Toyota Avalon	Toyota Tacoma
Camry	1039	0	0	0	1	0	0	0	0	0
Honda Civic	0	1007	0	0	0	0	1	1	0	0
Jeep93	0	0	1064	0	0	0	1	0	0	0
Jeep99	0	0	0	1014	0	0	0	0	0	0
Maxima	0	0	0	0	1033	0	0	0	0	0
MazdaMPV	0	0	0	0	0	1072	0	0	0	0
Mitsubishi	1	0	0	0	0	0	1034	0	0	0
Sentra	0	0	0	1	0	0	0	1048	1	0
Toyota Avalon	0	0	0	0	0	1	0	1	1020	0
Toyota Tacoma	0	0	0	0	0	0	0	0	1	1068

Predicted Label

(c)

Figure 6.13 (continued)

6.5 Summary

In this chapter, the algorithmic steps were presented involving the classification of single targets in SAR imagery. DCNN performance can benefit from using available well-thought-out preprocessing, rather than asking the DCNN to learn everything. We highlighted that it can be quite advantageous to select the type of preprocessing or any type of prior knowledge to incorporate into the overall analysis. For SAR image classification, results improve when using the vast knowledge of SAR image formation developed over the past decades. Thus, DCNN algorithms for single-target classification should not be used to replace previous ML methodologies when such methods have been developed based on detailed and accurate signal modeling and processing. Instead, the goal is to determine the best means of combining the benefits of both approaches. In particular, ML and DCNN approaches are well poised for portions of the overall problem space for which image analysts have validated weak or incomplete knowledge of the signal models and the machine utilizes physically-verified processing algorithms.

Confusion matrix for train on 30, 40, 60 degrees elevation, test on 50 degree elevation

True Label \ Predicted Label	Camry	Honda Civic	Jeep93	Jeep99	Maxima	MazdaMPV	Mitsubishi	Sentra	Toyota Avalon	Toyota Tacoma
Camry	634	83	163	119	100	9	137	50	8	1
Honda Civic	90	536	128	45	16	203	35	79	156	16
Jeep93	13	6	695	254	3	37	2	1	17	276
Jeep99	62	18	204	642	16	27	104	4	2	225
Maxima	171	17	160	115	562	34	115	46	69	15
MazdaMPV	28	190	6	122	0	481	67	75	164	171
Mitsubishi	49	61	0	42	37	82	864	55	16	15
Sentra	248	39	83	64	86	79	117	481	22	80
Toyota Avalon	199	86	88	9	167	65	115	9	484	131
Toyota Tacoma	5	12	39	216	17	62	7	20	5	884

Predicted Label

(a)

Confusion matrix for train on 30, 50, 60 degrees elevation, test on 40 degree elevation

True Label \ Predicted Label	Camry	Honda Civic	Jeep93	Jeep99	Maxima	MazdaMPV	Mitsubishi	Sentra	Toyota Avalon	Toyota Tacoma
Camry	890	5	0	103	25	53	3	91	0	134
Honda Civic	237	573	82	36	129	120	23	4	96	4
Jeep93	56	99	560	203	95	115	21	3	18	134
Jeep99	94	37	81	838	44	92	20	45	31	22
Maxima	85	114	190	47	661	6	72	62	59	8
MazdaMPV	33	153	20	60	5	462	286	83	73	129
Mitsubishi	41	177	72	15	7	17	862	77	16	20
Sentra	304	58	17	161	288	2	43	333	3	95
Toyota Avalon	18	41	72	55	87	2	123	219	655	32
Toyota Tacoma	5	0	7	133	9	24	127	0	5	994

Predicted Label

(b)

Figure 6.14 Classification accuracy of MobilNetv2 on phase history (PH) formed images. (a) Training on 30°, 40°, and 60° elevation data and testing on 50° elevation data; classification accuracy was recorded approximately 48%. (b) Training on 30°, 50°, and 60° elevation data and testing on 40° elevation data; classification accuracy was recorded approximately 52%. (c) Training on mixed (30°, 40°, 50°, and 60°) elevation data and testing on mixed elevation data; classification accuracy was recorded almost 100%.

Confusion matrix for train and test on mixed (30, 40, 50, 60 degrees) elevation

True Label	Camry	Honda Civic	Jeep93	Jeep99	Maxima	MazdaMPV	Mitsubishi	Sentra	Toyota Avalon	Toyota Tacoma
Camry	1017	0	0	0	1	0	0	0	0	0
Honda Civic	0	1012	0	0	0	1	1		1	0
Jeep93	0	0	1104	0	0	0	1	0	0	0
Jeep99	0	0	0	1053	0	0	0	0	0	0
Maxima	0	1	0	0	1021	0	0	0	0	0
MazdaMPV	0	0	0	0	0	1058	0	0	0	0
Mitsubishi	1	0	0	0	0	0	1053	0	0	0
Sentra	0	0	0	1	0	0	0	1030	1	0
Toyota Avalon	0	0	1	0	0	1	0	0	1048	0
Toyota Tacoma	0	1	0	0	1	0	0	0	0	1036

Predicted Label

(c)

Figure 6.14 (continued)

References

[1] Majumder, U., and E. Blasch, "Machine Learning for Object Recognition from High Volume Radio Frequency Data," *Workshop on Big Data Trends Shaping the Future of Photonics*, SPIE DCS, 2017.

[2] Dungan, K., C. Austin, J. Nehrbass, and C. Potter, "Civilian Vehicle Radar Data Domes," in *Algorithms for Synthetic Aperture Radar Imagery XVII* (E. G. Zelnio, F. D. Garber eds.), *Proc. of SPIE,* Vol. 7699, 2010.

[3] Majumder, U., M. Minardi, E. Blasch, et al., "Radar Signals Dismount Data Production," *Proc. of SPIE,* Vol. 6237, 2006.

[4] Blasch, E., U. Majumder, and M. Minardi, "Radar Signals Dismount Tracking for Urban Operations," *Proc. of SPIE,* Vol. 6235, 2006.

[5] Gorham, L. A., and L. J. Moore; "SAR Image Formation Toolbox for MATLAB," *Proc. SPIE,* Vol. 7699, Algorithms for Synthetic Aperture Radar Imagery XVII, 2010.

[6] Blasch, E., U. K. Majumder, T. Rovito, P. Zulch, and V. Velten, "Automatic Machine Learning for Target Recognition," *Proc. SPIE,* Vol. 10988, 2019.

[7] Liu, M., J. Shi, Z. Li, C. Li, J. Zhu, and S. Liu, "Towards Better Analysis of Deep Convolutional Neural Networks," *IEEE Transactions on Visualization and Computer Graphics*, Vol. 23, No. 1, 2017, pp. 91–100.

[8] Burns, B. L., and J. T. Cordaro, "A SAR Image-Formation Algorithm that Compensates for the Spatially-Variant Effects of Antenna Motion," *Proc SPIE,* Vol. 2230, 1994.

[9] Richards, M. A., *Fundamentals of Radar Signal Processing*, New York: McGraw-Hill, 2005.

[10] Marinelli, M. P., L. M. Kaplan, and N. M. Nasrabadi, "SAR ATR Using Learning Vector Quantization," *Proc. SPI,E* Vol. 3647, Applications of Artificial Neural Networks in Image Processing IV, 1999.

[11] Zhao, Q., J. C. Principe, V. L. Brennan, D. Xu, and Z. Wang, "Synthetic Aperture Radar Automatic Target Recognition with Three Strategies of Learning and Representation," *Opt. Eng.*, Vol. 39, No. 5, 2000, pp. 1230–1244.

[12] Zhang, H., N. Nasrabadi, T. Huang, and Y. Zhang. "Joint Sparse Representation Based Automatic Target Recognition in SAR Images," *Proc. SPIE,* Vol. 8051, Algorithms for Synthetic Aperture Radar Imagery XVIII, May 2011.

[13] Zhang, X., L. Jiao, S. Zhou, N. Zhou, and J. Feng, "Adaptive Multiparameter Spectral Feature Analysis for Synthetic Aperture Radar Target Recognition," *Optical Engineering,* Vol. 51, No. 8, Aug. 2012.

[14] Zhang, H., N. M. Nasrabadi, Y. Zhang, and T. S. Huang, "Multi-View Automatic Target Recognition Using Joint Sparse Representation," *IEEE Transactions on Aerospace and Electronic Systems*, Vol. 48, No. 3, 2012.

[15] Morgan, D. A. E., "Deep Convolutional Neural Networks for ATR from SAR Imagery," *Proc. SPIE,* Vol. 9475, Algorithms for Synthetic Aperture Radar Imagery XXII, 2015.

[16] Wilmanski, M., C. Kreucher, and J. Lauer; "Modern Approaches in Deep Learning for SAR ATR," *Proc. SPIE,* Vol. 9843, Algorithms for Synthetic Aperture Radar Imagery XXIII, 2016.

[17] Wagner, S. A., "SAR ATR by a Combination of Convolutional Neural Network and Support Vector Machines," *IEEE Transactions on Aerospace and Electronic Systems*, Vol. 52, No. 6, 2016, pp. 2861–2872.

[18] El Housseini, A., A. Toumi, and A. Khenchaf, "Deep Learning for Target Recognition from SAR Images," *2017 Seminar on Detection Systems Architectures and Technologies (DAT)*, 2017.

[19] Yan, Y., "Convolutional Neural Networks Based on Augmented Training Samples for synthetic Aperture Radar Target Recognition," *Journal of Electronic Imaging,* Vol. 27, No. 2, 2018, pp. 023024-1–023024-10.

[20] Gao, F., W. Shi, J. Wang, A. Hussain, and H. Zhou, "A Semi-Supervised Synthetic Aperture Radar (SAR) Image Recognition Algorithm Based on an Attention Mechanism and Bias-Variance Decomposition," *IEEE Access*, Vol. 7, 2019, pp. 108617–108632.

[21] Jakowatz, Jr., C. V., and N. Doren, "Comparison of Polar Formatting And Back-Projection Algorithms for Spotlight-Mode SAR Image Formation," *Proc. SPIE,* Vol. 6237, Algorithms for Synthetic Aperture Radar Imagery XIII, 2006.

[22] Park, J., P. T. P. Tang, M. Smelyanskiy, D. Kim, and T. Benson, "Efficient Backprojection-Based Synthetic Aperture Radar Computation With Many-Core Processors," in *Proceedings of the International Conference on High Performance Computing, Networking, Storage and Analysis*, 2012.

[23] Kelly, S. I., M. Yaghoobi, and M. E. Davies, "Sparsity-Based Autofocus for Under-Sampled Synthetic Aperture Radar," *IEEE Transactions on Aerospace and Electronic Systems*, Vol. 50, No. 2, 2014, pp. 972–986.

[24] Majumder, U., E. Christiansen, Q. Wu, N. Inkawhich, and E. Blasch. "High-Performance Computing for Automatic Target Recognition in Synthetic Aperture Radar Imagery," *Proc. SPIE, Cyber Sensing*, Vol. 10185, 2017.

[25] MSTAR Public Targets, https://www.sdms.afrl.af.mil/index.php?collection=mstar&page=targets.

[26] Lu, L., R. Wang, and L. Wugao. "A Target Detection Method in Range-Doppler Domain from SAR Echo Data," *Object Recognition Supported by User Interaction for Service Robots,* 2002, pp. 91–94.

[27] Brito, A. E., S.H. Chan, and S.D. Cabrera, "SAR Image Superresolution via 2-D Adaptive Extrapolation," *Multidimensional Systems and Signal Processing,* Vol. 14, Nos. 1–3, 2003, pp. 83–104.

[28] Jakowatz, Jr., C. V., D. E. Wahl, D. A. Yocky, B. K. Bray, W. J. Bow, Jr., and J. A. Richards, "Comparison of Algorithms for Use in Real-Time Spotlight-Mode SAR Image Formation," *Proc. SPIE,* Vol. 5427, Algorithms for Synthetic Aperture Radar Imagery XI, 2004.

[29] Wilmanski, M., C. Kreucher, and A. Hero, "Complex Input Convolutional Neural Networks for Wide angle SAR ATR," *2016 IEEE Global Conference on Signal and Information Processing (GlobalSIP),* 2016, pp. 1037–1041.

[30] Naidu, K., and L. Lin, "Data Dome: Full k-Space Sampling Data for High-Frequency Radar Research," *Proc. SPIE,* Vol. 5427, Algorithms for Synthetic Aperture Radar Imagery XI, 2004.

[31] Jia, Y., "Convolution Layer," [Online], http://caffe.berkeleyvision.org/tutorial/layers/convolution.html.

[32] Jia, Y., E. Shelhamer, J. Donahue, et al., "Caffe: Convolutional Architecture for Fast Feature Embedding." in *Proceedings of the 22nd ACM International Conference on Multimedia* (ACM), 2014, pp. 675–678.

[33] Lewis, Benjamin, Theresa Scarnati, and Ernest Parke, "Investigating the saliency of SAR image chips," *Algorithms for Synthetic Aperture Radar Imagery XXVII,* Vol. 11393, International Society for Optics and Photonics, 2020.

CHAPTER 7

Deep Learning for Multiple Target Classification in SAR Imagery

7.1 Introduction

This chapter describes multiple target classifications for SAR imagery. For deep learning-based image classification, the algorithm implementation begins by obtaining a collection of simple image chips (i.e., a single target is the image) for training, validation, and testing. For example, MNIST data sets are a collection of imagery of 10 digits (0, 1, 2, …., 9). Many researchers use the MNIST datasets for the training and validation of deep learning (DL) models and for evaluating testing performance. In previous chapters, we presented DL for SAR image classification based on a single input image chip. Most of the DL-based SAR image classification examples found in the literature involve such single target classification. However, in many applications, SAR imagery contains multiple targets (some known and others unknown) and clutter. Hence, this chapter addresses issues of target classification involving the detection of all targets and then the classification of all detected targets. Furthermore, the classification task must include the declaration of known or unknown targets (or clutter) and their corresponding locations within the image.

Groups of targets can be defined over many domains, including that for automotive cars [1] and various targets in SAR imagery [2]. In fact, initial MSTAR publications had detection capabilities involving a group of targets [3]. Such a group can include stationary targets, stationary movers, or dynamic targets. The stationary cluster is characterized by scenarios in which mobile vehicles are in fixed positions. Similarly, stationary movers have no bulk translational motion but can include various moving parts, such as moving turrets, engine vibration, and opening doors. In contrast, there is nonzero bulk translational motion for moving targets, and such motion can occur for a group, as with a cluster, convoy, or caravan. Velten et al. [4]. investigated the target invariances for distinguishing between various stationary targets. Blasch et al. [5, 6], have explored the moving target analysis of a target groups, as shown in Figure 7.1.

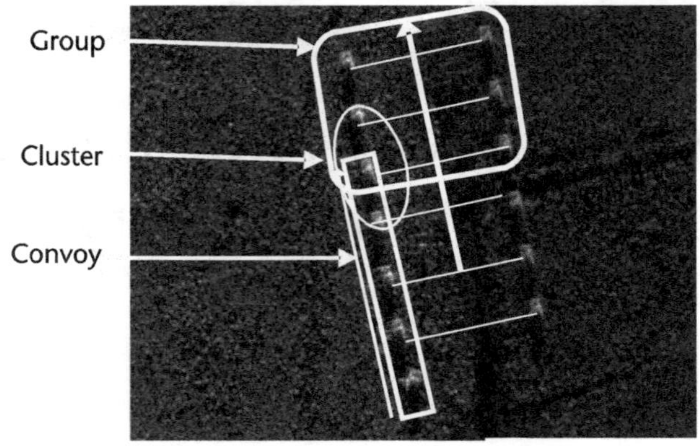

Figure 7.1 Multitarget group concept.

In attempting to discriminate between potential targets, features aid in making these choices involving classification selection [7]. With the use of features, various approaches applied to the MSTAR data for multitarget analysis include improving track attributes [8], tracking through occlusions [9], and maintaining track [10]. With the RF analysis of target signatures, additional results include target identification assessment [11, 12], user refinement [13], and generalized group tracking [14].

These MSTAR experiments have provided opportunities to test multitarget assessment from HRR profiles and SAR images [15]. Attempts to coordinate multitarget assessment with this data have included information theory [16], belief theory [17], and estimation [18]—all with a focus on the target pose [19] and separating various targets within groups. The value of the pose in the SAR/HRR multitarget assessment results in performance models [20] and validation of multilook SAR ATR [21]. Sensor-management and decision-level fusion using RF data have demonstrated various approaches for mitigating the effects of clutter [22], determining sensor-to-target sensing [23], and optimizing platform routes [24].

By using the MSTAR data, group target detection has been focused on the profiles of the imagery in support of various group configurations. Techniques based upon the Gaussian ML classifier [25], feature likelihood classifier by Douglas et al. [26], and scene statistics [27] have been used to determine the group of targets in a scene. Subsequent research with the advances in ML and DL has provided additional interest for group target classification with the region-based CNN [28], CAD-model comparisons [29], and generative adversarial networks [30].

7.2 Challenges with Multiple-Target Classification

Single-image classification does not require a target detection step. Thus, technical issues (e.g., algorithm implementations, overall systems design) associated with single-target classification are less important than multiple-target classification. Detection algorithms have had an impact on classification and on overall AI/ML systems performance. An optimum detection algorithm with an associated optimized threshold value may be able to detect all the possible targets within a given scene.

The various sensors, parameters, and environments (e.g., clutter, noise) may dictate the optimum selected detection threshold. Furthermore, a higher threshold value may detect unwanted objects and clutter as possible targets, thus increasing the false alarm rate (FAR). On the other hand, a lower threshold value may reject possible targets and therefore bring serious issues pertaining to the omission of critical targets in the detection process. For the case in which images contain excess noise or clutter, it may be necessary to include a preprocessing step (e.g., compressive sensing techniques) in order to clean up the images, thus improving detection performance. The following sections overview the most widely used detection algorithms:

- Constant False Alarm Rate (CFAR) detector;
- Region-based CNN(R-CNN);
- You only look once (YOLO);
- Multiscale CNN (MS-CNN).

7.2.1 Constant False Alarm Rate Detector

The constant false alarm rate detector was designed to maintain a constant probability of false alarm (Pfa). The algorithm uses an alpha (α) parameter to set the PFA. For a given image, an $N \times N$ window is shifted across the image. At each location (x,y) of the image, the average of the pixels in the target region (μ_T) and the average of the pixels in the background region (μ_B) are computed and compared, as shown in Figure 7.2. Typically, the pixels in the guard region are ignored.

Consider an image where background region (B) is (60 × 60) pixels, guard region (G) is (30 × 30) pixels, and target region (T) is (10 × 10) pixels. In this case, CFAR algorithm takes the input as (B, G, T) = (60, 30, 10). Further, consider alpha (α) parameter is set to be 0.6. Then, CFAR method outputs a target is detected at location (x,y) and the resulting (10×10) pixels as:

$$\text{Target}(x,y) = \begin{cases} 1, & \mu_T * \alpha > \mu_B \\ 0, & \text{otherwise} \end{cases}$$

Figure 7.2 CFAR concept. T is the target region, G is the guard region, and B is the background region.

7.2.2 Region-Based Convolutional Neural Networks (R-CNN)

CFAR-based detection involves finding many possible targets in the image and then cropping out each of these targets as individual image chips. Then, it requires recording the actual image frame from which the target was extracted and the corresponding pixel location within the image. This process, although relatively simple and easily understood, requires extraneous computation and bookkeeping (e.g., labeling.). Hence, researchers have developed algorithms that will avoid the shortcomings of the CFAR method.

R-CNN: Target detection with the R-CNN is based on a selective search (SS) process (Girshick, et.al [35]) that extracts 2,000 regions (of possible targets) in the input image, instead of an unlimited number of possible targets. These candidate regions were referred to as *region proposals*. The SS is essentially a segmentation algorithm that starts with small segments in the image and then combines similar segments to make a large and meaningful object in the image. The chief issue with this naïve region proposal approach for target detection is that it's computationally expensive to be implemented for real-time object detection and classification. The R-CNN has been applied to multitarget ship detection from SAR [31].

Fast R-CNN: To overcome these computational issues, another version of R-CNN was developed, referred to as "Fast R-CNN" Girscick, et.al. [36]. In Fast R-CNN, instead of feeding 2,000 region proposals to the CNNs, the entire image frame is fed to the CNN to generate a convolutional feature map. The convolution operation is performed only once per image, and a feature map is generated from it. After creating the convolutional feature map, a selective search-based region proposal process is applied. This approach reduces the number of unwanted regions, as these are not actual features of targets.

Faster R-CNN: R-CNN and Fast R-CNN are based on a selective search method and both apply region-proposal processing in order to extract (i.e., detect) possible targets in the image frame. One major issue with the selective search technique is that run-time is slow, and thus, performance of the overall network is affected. To mitigate this issue, another approach for an object detection algorithm avoids the use of the selective search algorithm. This approach is named faster R-CNN. Like the fast R-CNN algorithm, the entire image frame is fed into the convolutional network, which provides a convolutional feature map. Now, instead of using the selective search algorithm on the convolutional feature map to generate the region proposals, a separate network is used to predict the region proposals. The predicted region proposals are then reshaped using a region-of-interest pooling layer. These regions are then used to classify the objects in the image frame, such as the example of the image ship detection [32].

7.2.3 You Only Look Once

YOLO is an object detection algorithm that finds objects by analyzing a much larger region in the image instead of a local region. The algorithm first splits the image into grids. It then combines multiple grids to find possible targets. It can add more grids if no targets are found. A single convolutional network predicts the bounding boxes and the class probabilities for these boxes. The bounding boxes having the class probability above a threshold value are selected and used to locate the object within the image.

YOLO is much efficient than the Faster R-CNN algorithm. The limitation of YOLO algorithm is that it is not optimum to find small objects within an image frame.

7.2.4 R-CNN Implementation

This section presents the challenges associated with target detection. In particular, the results of faster R-CNN implementation on the xView data set (Figure 7.3), which is one of the largest and most diverse publicly available object-detection datasets are analyzed. The xView data set accelerates progress in four computer vision frontiers: (1) reduces minimum resolution for detection, (2) improves learning efficiency, (3) enables discovery of more object classes, and (4) improves detection of fine-grained classes. Though this data is from video imagery (i.e., electro-optical sensor [33]), the analysis is representative of that which can be applied for SAR exploitation. The statistics of the dataset are provided in Table 7.1, and the characteristics are given in Table 7.2.

- Highlights of xView data set:
  - Collected using DigitalGlobe WV3 imaging satellite

Figure 7.3 A sample of xView imagery with many objects. (http://xviewdataset.org.)

Table 7.1 xView Data Statistics

Image size	*~3,000 × 3,000 pixel (px)*
# images	846
# classes	60
# objects/image	Min = 1 Avg. = 711 Max = 7618
Ground sample distance (GSD)	0.3 m

Table 7.2 xView Data Challenges and Advantages

Challenges	*Advantages*
Inconsistent labeling	Large data set
Large class imbalance	High-resolution imagery
Fine-grained classes	Consistent perspective
Very Small objects	Little object obscuration
Densely packed objects	

- Over 1 million objects total
- 60 object classes
- Covers 1,400 square kilometer at 0.3m GSD resolution

- Organizers of xView Data set:
  - Defense Innovation Unit Experimental (DIUx)
  - National Geospatial-Intelligence Agency (NGA)

xView Data Preprocessing

- Split training set manually:
  - Training: 678 (80%)
  - Validation: 168 (20%)
- Rotate the imagery: {10°, 90°, 180°, 270°}
- Flips: horizontal and vertical
- Chip full-frame images (20% overlap)

An example of the xView imagery (Figure 7.4) is similar to the scenario and chip extraction process as applied to the MSTAR data set. Machine and deep learning examples have been applied to the data set to demonstrate the opportunities for multitarget separation.

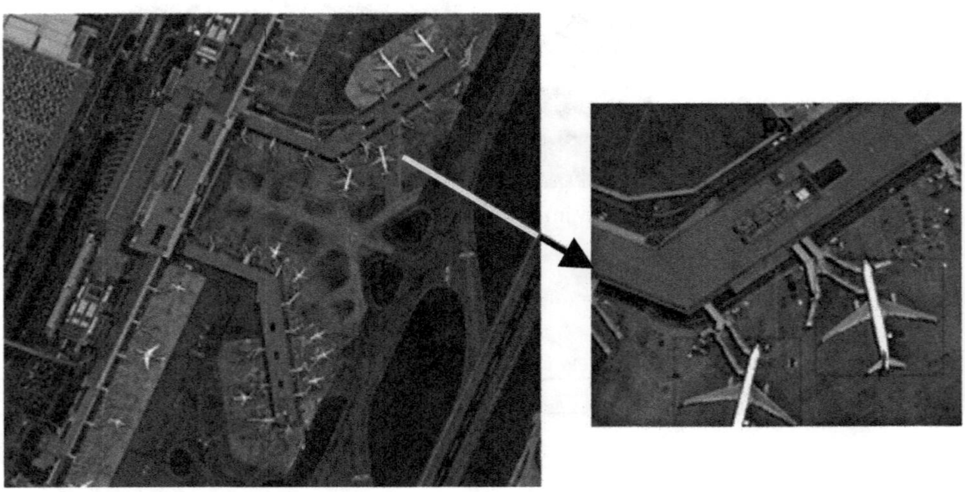

Full frame image: 3195×3216 px   Chipped image: 600×600 px

Figure 7.4 xView image frame and chipped image.

Faster R-CNN Algorithm Implementation Steps

One can start with faster R-CNN implementation by using available code:

- Get the PyTorch implementation w/ CUDA optimizations:
- Over 70 tunable flags and hyperparameters.
- Key hyperparameters are chip size, anchor scale, anchor stride, and region-of-interest (RoI) size.
- Anchor boxes serve as the initial guesses for detection. Considering anchors of various sizes is critical for detecting objects of different sizes.

An example of the Faster R-CNN is provided in Figure 7.5 for an image scene.

Algorithmic Steps

Step 1: Calculate intermediate feature map using pretrained CNN.

Step 2: Define a set of anchor boxes.

Step 3: Feed all anchors to the region proposal network (RPN) in order to generate the relevant outputs.

- Probability that the box contains foreground object
- Bounding box regression deltas/errors

Step 4: Force the top N most confident RoIs into same size feature map using ROI pooling

Step 5: Use R-CNN to classify each of these RoI-pooled feature maps and to predict bounding box regression deltas/errors.

7.3 Multiple-Target Classification

Multiple-target classification involves four major steps: (1) preprocessing, (2) detection, (3) classification, and (4) results and analysis (i.e., performance evaluations of the classification algorithms against extended operating conditions). We provide

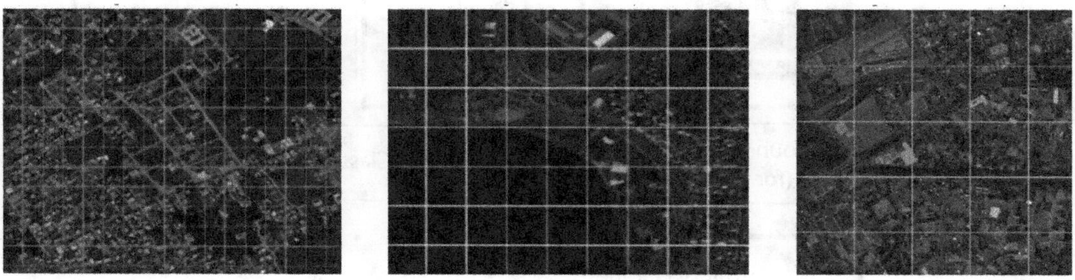

Figure 7.5 Faster R-CNN chip size experiments, considering four chip sizes: 200 × 200 (left), 400 × 400 (middle), 600 × 600 (right). Large chip size improves targets feature and classification accuracy at the cost large memory requirement.

details on these steps below (although the details step 4 will be developed more fully in Chapter 9). Figure 7.6 summarizes the algorithmic steps involving multiple target classification.

7.3.1 Preprocessing

Preprocessing involves signal processing of the radar or communication pulses. It may include phase corrections, motion compensation, and image processing or noise reduction. For SAR imagery, often noise and clutter reduction is necessary to separate real targets from the clutter and unwanted targets. By reducing clutter and noise with a discrete wavelet transform, unwanted targets are avoided in the classification task, and thus reduce the false alarm rate. In many cases, preprocessing for noise reduction may not be necessary. However, these concepts are described herein to guide the reader for cases in which it is necessary. Another reason for including this step is to mitigate noise-induced adversarial issues.

7.3.2 Two-Dimensional Discrete Wavelet Transforms for Noise Reduction

Wavelet transforms are traditionally used to produce a time-frequency representation of the input signal. They are a valuable tool for multiresolution analysis and subband coding, as opposed to the frequency analysis attained with the Fourier

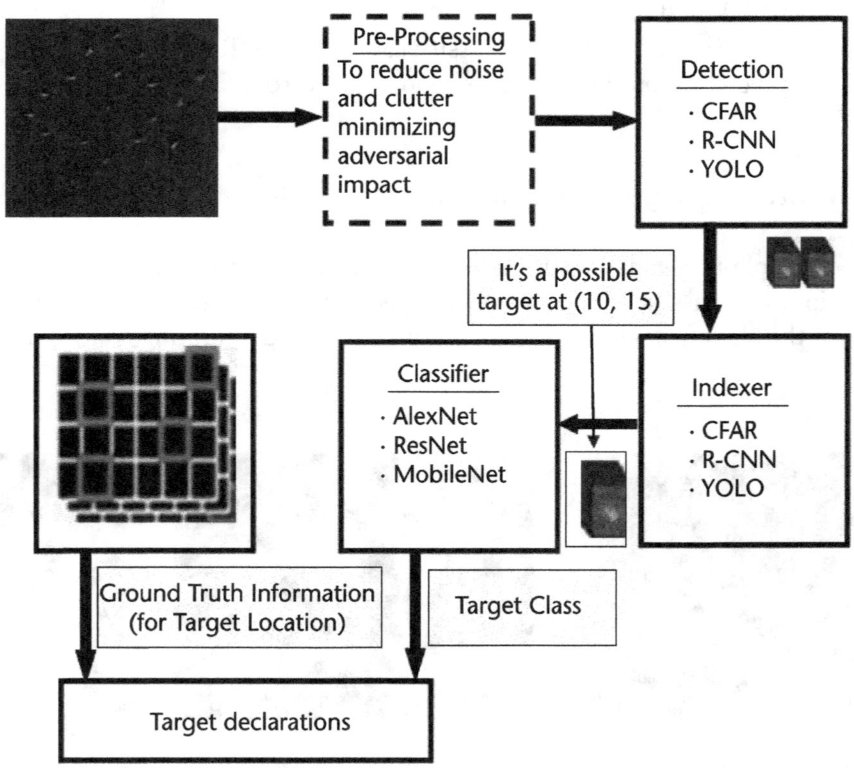

Figure 7.6 Multiple-target classification flow diagram.

transform. In the image domain, a two-dimensional discrete wavelet transforms (2D-DWT) compares time and spatial positioning (Figure 7.7). At each scale (level), the image is decomposed into frequency bins using a method of successive half-band high-pass filtering (HPF) and low-pass filtering (LPF) operations. There are several sets of common wavelet functions including: Haar, Symlets, Daubechies, and Coiflets—many of which support image fusion [34].

For example, at scale 1, the original image is first filtered with a half-band HPF and a half-band LPF across the rows of the image. The results of both filters are then filtered again with a LPF and a HPF along the columns of the image. As a result, there are four sets of wavelet coefficients produced: HPF to HPF (HH), HPF to LPF (HL), LPF to HPF (LH), and LPF to LPF (LL) (i.e., the approximation). At the next scale (scale 2), the LL coefficients (also called the approximation) from scale 1 are input into the filters, and the same processing is repeated.

Wavelets are intriguing for use with SAR imagery due to the high-frequency speckle noise that is characteristic of SAR imagery. It is expected that the objects would exist in the lower frequencies of the image, so their signatures would not get filtered out by the high-pass filters of the first few scales. Figure 7.8 shows scale 1 and scale 2 approximations (i.e., LL subimages). At scale 1, the approximation contains frequencies (in radians) 0 to $\pi/2$. At scale 2, the approximation contains frequencies 0 to $\pi/4$.

Since the objects in the images are relatively small, a two-dimensional stationary wavelet transform is applied that does not subsample at each scale. During the analysis, the results maintain the dimensionality of the original image, as shown in Figure 7.9.

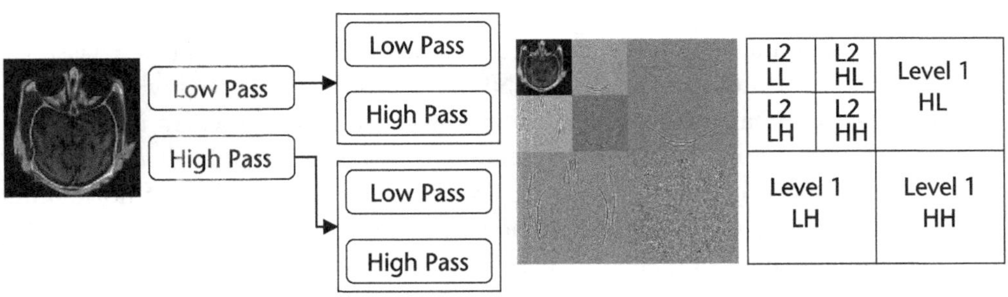

Figure 7.7 Discrete waveform transform.

(a) (b) (c)

Figure 7.8 (a) Original SAR image, (b) the four subimages at scale 1, and (c) the four subimages at scale 2.

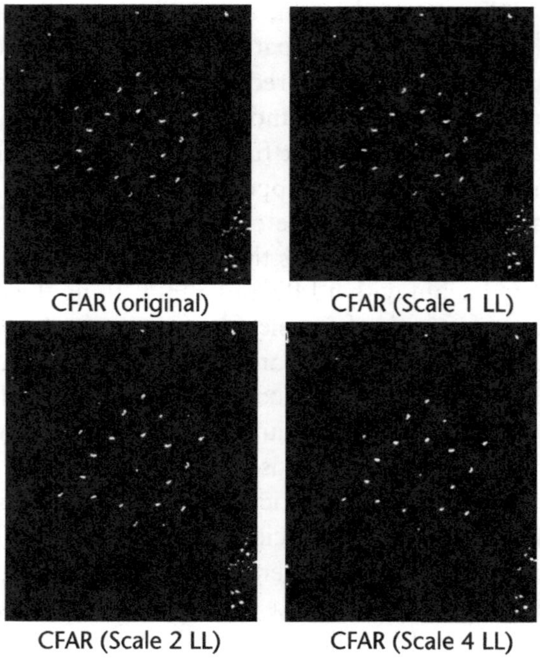

Figure 7.9 SAR Imagery 2D-wavelet transform. Scale 4 LL looks the best to eliminate unwanted target-like clutter.

7.3.3 Noisy SAR Imagery Preprocessing by L1-Norm Minimization

Among various parameters, large-scene object detection and classification accuracy depend on image quality. In general, DNNs are trained to achieve a desired recognition accuracy for a set of targets. However, DNNs become tuned to the training data used and may not generalize to new unseen data artifacts. Classification accuracy of a previously trained DNN is significantly reduced when classification is run on an image that has been altered with additive noise. An example of a target with image noise is shown in Figure 7.10. In recently published paper [37] it was found that preprocessing can reduce the deleterious impact of noise-induced low classification accuracy. The L1-norm minimization approach consists of applying compressive sensing inspired preprocessing techniques to noisy images which substantially improves the classification accuracy, as shown in Figure 7.11 and Table 7.3.

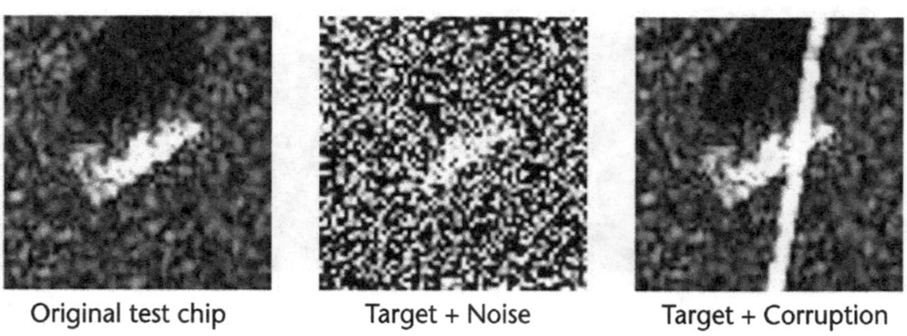

Figure 7.10 SAR image with noise and corruption.

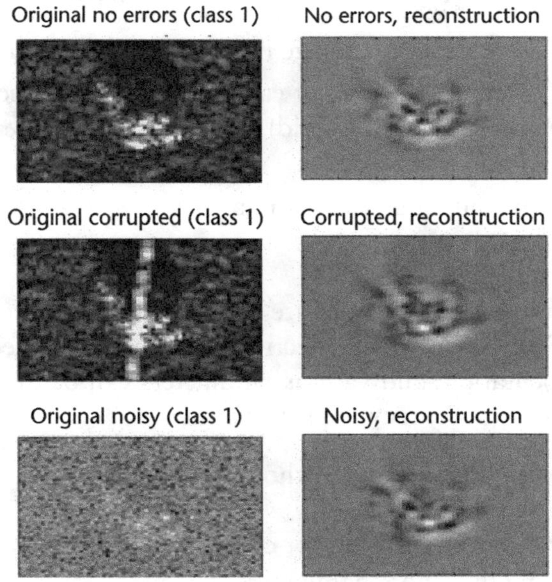

Figure 7.11 SAR image (a) before and (b) after preprocessing (L1norm minimization).

Table 7.3 SAR Image Classification Accuracy without and with Preprocessing

	Without Preprocessing (Pcc)	*With Preprocessing (Pcc)*
Standard SAR Imagery	86%	94%
Corrupted SAR Imagery	32%	82%
Noisy SAR Imagery	15%	90%

Preprocessing substantially increases classification accuracy for noisy and corrupted SAR data sets.

7.3.4 Wavelet-Based Preprocessing and Target Detection

Wavelet-based SAR imagery noise reduction and target detection processing is summarized as follows:

Step 1: Run 2D-WT with db-4 coefficients on the original SAR imagery at 2 scales to obtain the scale 1 LL (s1.LL) and scale 2 LL (s2.LL) images.

Step 2: Perform CFAR detection independently on s1.LL and s2.LL, which results in two binary images.

Step 3: Perform an AND operation on the binary images to create one composite binary image.

- CFAR(s1.LL) && CFAR(s2.LL) = composite binary image.
- If CFAR detects something in the same position across both images, it is worth investigating further.

Step 4: Find areas of white in this composite binary image using the 'bwboundaries' method from the Image Processing Toolbox in MATLAB.

Step 5: Compute the area of each boundary returned. If the boundary is large enough (per a fixed threshold), then compute the center pixel and chip 128 × 128px around that location.

Step 6 Save all chips as JPG images in a *special directory* and next move to classification.

An example is shown in Figure 7.12 with both the group detection and a target.

The CFAR-based target detection algorithm presented here is a very quick and fast implementation, and various parameters can be tuned for detection accuracy improvement for accuracy improvements.

1. Use other LL images. Although scale-4 appears interesting, other scales can be used as well.
2. Try different methods for creating composite images.
3. Tuning the CFAR parameters:
   α, B, G, and T
   Changing area threshold when chipping
4. For computation speed improvements, CFAR detection algorithms can be parallelized.

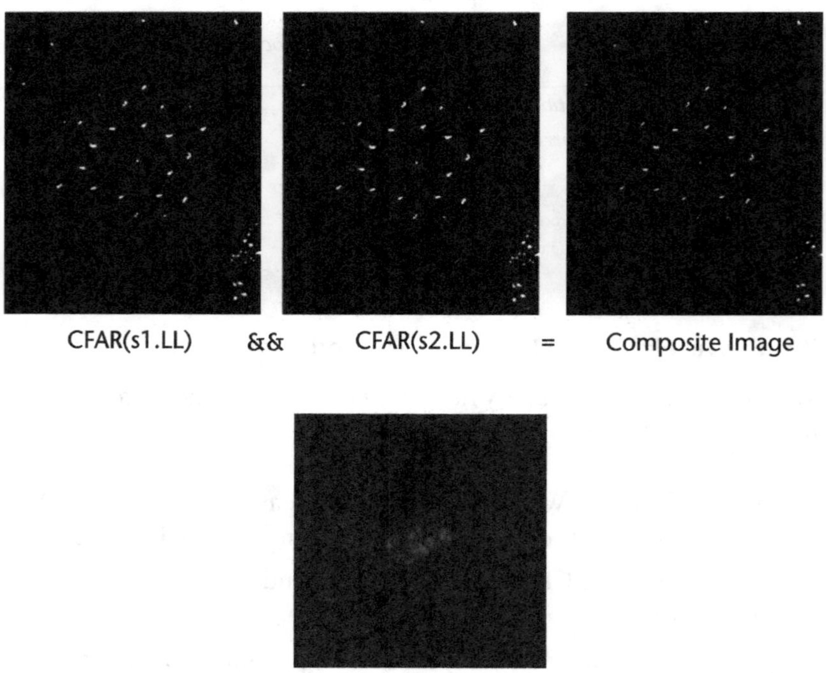

Figure 7.12 SAR Image (top row) after 2D wavelet transform. The image (composite image) to the right shows (visually) reduction of clutter-like targets. One of the target chips (bottom row) after detection.

7.4 Target Classification

Results of the detection stage is a directory containing the 128 × 128 (or other desired size) pixel image chips. For example, chip_330 × 827.jpg represents location (330,827) of the center pixel of the detected object in the image. The first step of the classification is the development of a DCNN model. Although one may choose to develop their own DCNN model, it is often the case that engineers apply available DL models from the GITHUB or online sources. In terms of software, researchers often use PyTorch (python code) or TensorFlow. Here, the DCNN was implemented using PyTorch. Figure 7.13 shows detail of the DCNN architecture, and the following are the key parameters for the model:

1. DCNN
   - Five convolution/ReLU/pooling layers;
   - Two fully connected layers;
   - Trained for 100,000 iterations;
   - Learning rate of 0.01;
   - 5 × 5 and 3 × 3 convolution filters with stride of 1 and 2.
2. Data
   - 25 classes, including clutter class;
   - Number of training/validation/test images: 191,788/63,929/63,930;
   - Top-1/top-5 test accuracy after training for 100k iterations is 99.6%/100%.

Once the classifier is initialized, then it performs one forward pass (prediction) for each chip in the detection directory. The results of the classification processing are written to a text file of the format: "chip_name \t prediction \n", where

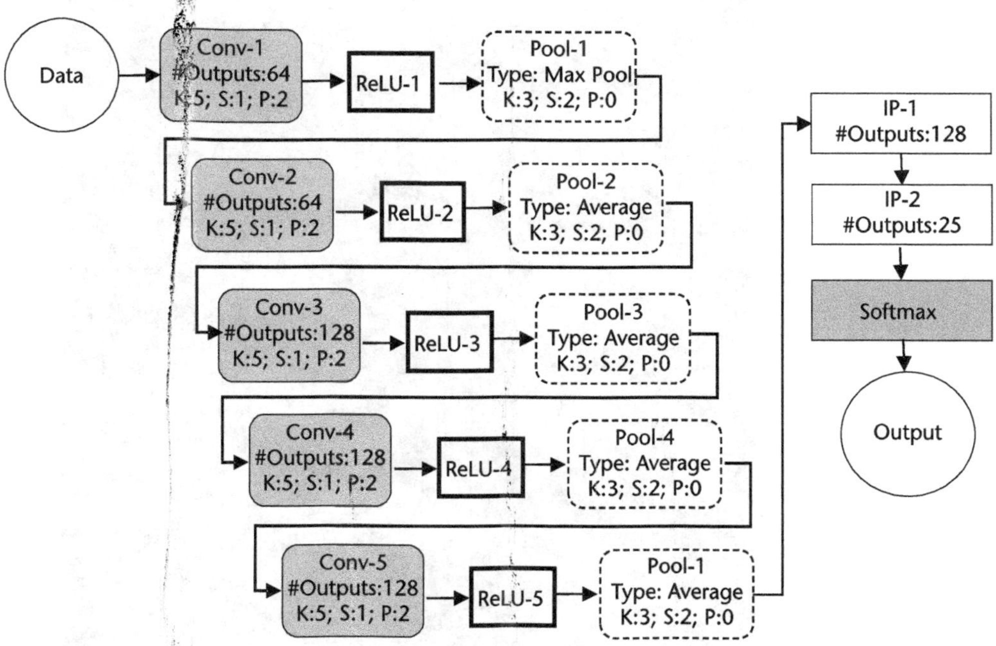

Figure 7.13 DCNN model for multiple-target classification.

"prediction" is the target level/number. The following are typical classification results stored in a text file for analysis.

7.5 Multiple-Target Classification: Results and Analysis

Once detection and classification tasks have been performed, the results can be analyzed for the performance of the DCNN model. For a given SAR image (with multiple targets), the first analysis involves the determination whether the CFAR processing was able to detect all possible targets. The second task involves an analysis, among all of the detections, of whether the DCNN was able to classify known targets as the real targets and declaring valid geolocation estimates. The detected targets can be shape-coded for visual representation that is conducive for human analysis. For example, all the detected objects can be circles. Second, all the correctly detected and classified objects/targets can be rectangles. Third, all the clutter regions that were detected and correctly classified as clutter can be triangles. Fourth, all the clutter that was misclassified as object/target can be pentagons. Finally, all the objects/targets that were detected but misclassified can be diamonds. The following is a summary of object detection and classification per SAR image frame, with the example in Figure 7.14 and Figure 7.15.

- SAR Image Frame # 1

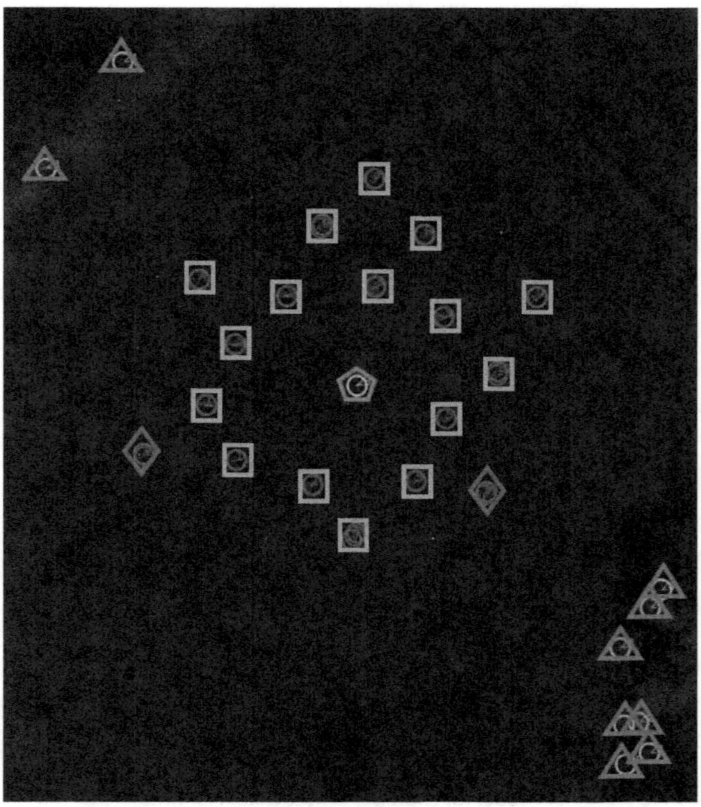

Figure 7.14 Analyzing multiple-target detection and classification.

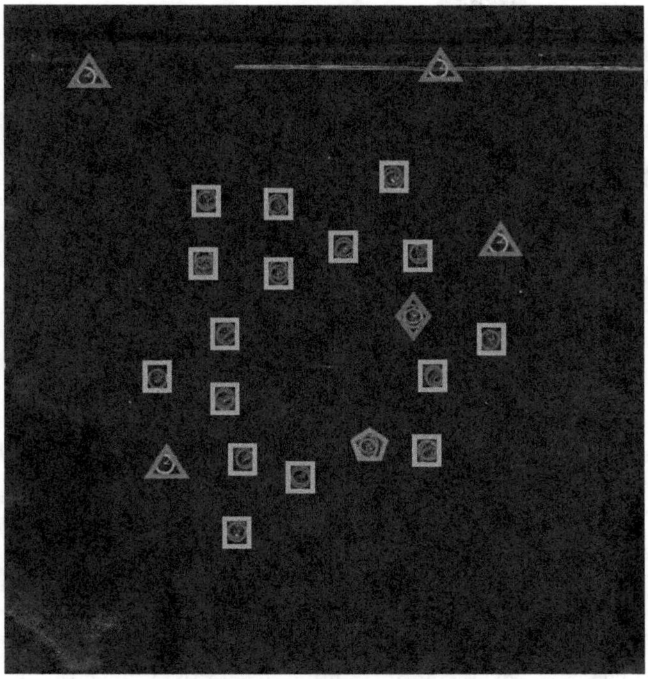

Figure 7.15 Analyzing multiple-target detection and classification.

- Total Detections = 29
  - 100% of objects were detected
- Target detection and classification:
  - Circle = detected objects
  - Rectangle = correctly detected and classified objects
  - Triangle = clutter that was detected but correctly classified as clutter
  - Pentagon = clutter that was misclassified as an object/target
  - Diamond = object that was detected but misclassified
- SAR image frame #2
- Total detections = 24
  - 100% of objects were detected
- Target detection and classification:
  - Circle = detected objects
  - Rectangle = correctly detected and classified objects
  - Triangle = clutter that was detected but correctly classified as clutter
  - Pentagon = clutter that was misclassified as an object/target
  - Diamond = object that was detected but misclassified

7.6 Summary

This chapter discussed algorithms available for multiple-target detection and classification from SAR imagery. Detection algorithms implementations are challenging for cases in which targets are of different types (i.e., heterogeneous) and sizes. However, homogeneous target detection is much easier as shown using the xView data set. An example of end-to-end multiple-target detection and classification processing using SAR imagery provides a general implementation pipeline..

References

[1] Bartsch, A., F. Fitzek, and R. H. Rasshofer, "Pedestrian Recognition Using Automotive Radar Sensors," *Adv. Radio Sci.*, Vol. 10, 2012, pp. 45–55.

[2] Wissinger, J., R. B. Washburn, N. S. Friedland, et al., "Search Algorithms for Model-Based SAR ATR," *Proc. SPIE,* Vol. 2757, Algorithms for Synthetic Aperture Radar Imagery III, 1996.

[3] Novak, L. M., G. J. Owirka, and W. S. Brower, "An Efficient Multitarget SAR ATR Algorithm," *Proc. 30th IEEE Asilomar Conf. Signals, Syst. Computer.*, Vol. 1. 1998, pp. 3–13, [Online], http://dx.doi.org/10.1109/ACSSC.1998.750815.

[4] Velten, V. J., "Synthetic Aperature Radar (SAR) Geometric Invariants: Experimental Results," *Proc. SPIE,* Vol. 3721, Algorithms for Synthetic Aperture Radar Imagery VI, 1999.

[5] Blasch, E. P., and L. Hong, "Sensor Fusion Cognition Using Belief Filtering for Tracking And Identification," *Proc. SPIE,* Vol. 3719, Sensor Fusion: Architectures, Algorithms, and Applications III, 1999.

[6] Blasch, E. P., and L. Hong, "Set Theory Correlation Free Algorithm for HRRR Target Tracking," *Nat. Symposium on Sensor and Data Fusion,* JHU MD, May 1999.

[7] Blasch, E. P., J. J. Westerkamp, J. R. Layne, L. Hong, F. D. Garber, and A. Shaw, "Identifying Moving HRR Signatures with an ATR Belief Filter," *Proc. SPIE,* Vol. 4053, April 2000.

[8] Blasch, E. P., T. Connare, and F. Scarpino, "Improving Track Attributes with Group Information Feedback," *Int. Conf. on Info Fusion - Fusion 2001*, Aug 2001.

[9] Blasch E. P., and T. Connare, "Group Tracking of Occluded Targets," *Proc. SPIE,* Vol. 4365, April 2001.

[10] Blasch E.P., and T. Connare, "Improving Track Maintenance Through Group Tracking," *Proc of the Workshop on Estimation, Tracking, and Fusion; A Tribute to Yaakov Bar Shalom*, Monterey, CA, May 2001, pp. 360 –371.

[11] Connare, T., E. Blasch, J. Schmitz, F. Salvatore, and F. Scarpino, "Group IMM Tracking Utilizing Track and Identification Fusion," *Proc. of the Workshop on Estimation, Tracking, and Fusion; A Tribute to Yaakov Bar Shalom,* Monterey, CA, 2001, pp. 205–220.

[12] Blasch, E. P., and T. Connare, "Feature-Aided JBPDAF Group Tracking and Classification Using an IFFN Sensor," *Proc. SPIE,* Vol. 4728, April 2002.

[13] Blasch, E. P., and S. B. Plano, "JDL Level 5 Fusion Model: User Refinement Issues and Applications in Group Tracking," *Proc. SPIE,* Vol. 4729, April 2002.

[14] Blasch, E., "Group Tracking, Using Fusion of Simultaneous Track and ID Information", in NATO ASI: *Data Fusion for Situation Monitoring, Incident Detection, Alert and Response Management*, 2004.

[15] Blasch, E. P., "Fusion of HRR and SAR Information for Automatic Target Recognition and Classification," *Int. Conf. on Info Fusion - Fusion 99,* Las Vegas, 1999, pp. 1221–1227.

[16] Blasch, E. P., "Information-Theory-Based Feature-Aided Tracking and Identification Algorithm for Tracking Moving and Stationary Targets through High-Turn Maneuvres Using Fusion of SAR and HRR Information", *Proc. of SPIE,* Vol. 4727, April 2002.

[17] Blasch, E., and C Yang, "Ten Methods to Fuse GMTI and HRRR Measurements for Joint Tracking and Identification," *Int'l Conf. on Info Fusion*, 2004.

[18] Yang, C., E. Blasch, W. Garber, and R. Mitchell, "A Net Track Solution to Pose-Angular Tracking of Maneuvering Targets in Clutter with HRR Radar," *Radar Signal Processing, Asilomar Conf. on Signals, Systems and Computers,* Nov. 4, 2007.

[19] Yang, C., and E. Blasch, "Pose-Angular Tracking of Maneuvering Targets with High Range Resolution Radar (HRR)," *Int'l Conf. on Info Fusion,* 2008.

[20] Cong, S., L. Hong, and E. Blasch, "Performance Model for Joint Tracking and ATR with HRR Radar," *Proc. of SPIE,* Vol. 6970, April 2008.

[21] Kahler, B., and E. Blasch, "Robust Multi-Look HRR ATR Investigation through Decision-Level Fusion Evaluation," *Int'l Conf. on Info Fusion,* 2008.

[22] Kahler, B., and E. Blasch, "Decision-Level Fusion Performance Improvement from Enhanced HRR Radar Clutter Suppression," *J. of. Advances in Information Fusion,* Vol. 6, No. 2, 2011, pp. 101–118.

[23] Yang, C., L. Kaplan, and E. Blasch, "Performance Measures of Covariance and Information Matrices in Resource Management for Target State Estimation," *IEEE Transactions on Aerospace and Electronic Systems*, Vol. 48, No. 3, 2012, pp. 2594–2613.

[24] Yang, C., L. M. Kaplan, E. Blasch, M. Bakich, "Optimal Placement of Heterogeneous Sensors for Targets with Gaussian Priors," *IEEE Transactions on Aerospace and Electronics Systems*, Vol. 49, No. 3, 2013, pp. 1637–1653.

[25] Liu, Z., P. He, and Z. Bai, "A New Algorithm for Recognizing Vehicle Group Targets in High-Resolution SAR Images," *Proc. SPIE,* Vol. 5286, *Third International Symposium on Multispectral Image Processing and Pattern Recognition*, 2003.

[26] Douglas, J., M. Burke, and G. J. Ettinger, "High-Resolution SAR ATR Performance Analysis," *Proc. SPIE,* Vol. 5427, Algorithms for Synthetic Aperture Radar Imagery XI, 2004.

[27] Blasch, E., M. Hensel, and J. L. "Jackson, Scene Radar Clutter Modeling for Change Detection," *Algorithms for Synthetic Aperture Radar Imagery XI,* Vol. 5427, 2004, pp. 283–292.

[28] Long, Y., X. Jiang, X. Liu, and Y. Zhang, "SAR ATR with Rotated Region Based on Convolution Neural Network," *IEEE International Geoscience and Remote Sensing Symposium (IGARSS),* 2018.

[29] Sudkamp, E. R., J. W. Nehrbass, E. Branch, and M. Levy, "Synthetic Data Accuracy Sensitivity to CAD Model Accuracy Using ATR-Based Metrics," *Proc. SPIE, Algorithms for Synthetic Aperture Radar Imagery XXVI*, Vol. 10987, 2019

[30] Zheng, C., X. Jiang, and X. Liu, "Multi-Discriminator Generative Adversarial Network for Semi-Supervised SAR Target Recognition," *IEEE Radar Conference (RadarConf)*, 2019.

[31] Lie, Y., M.- H. Zhang, P. Xu, and Z.- W. Guo, "SAR Ship Detection Using Sea-Land Segmentation-Based Convolutional Neural Network," *International Workshop on Remote Sensing with Intelligent Processing (RSIP),* 2017.

[32] Wang, R., F. Xu, J. Pei, C. Wang, Y. Huang, J Yang, and J Wu, "An Improved Faster R-CNN Based on MSER Decision Criterion for SAR Image Ship Detection in Harbor," *IEEE International Geoscience and Remote Sensing Symposium*, 2019.

[33] Savakis, A., N. Nagananda, J. P. Kerekes, et al., "Change Detection in Satellite Imagery with Region Proposal Networks," *Defense Systems Information Analysis Center (DSIAC) Journal*, Vol. 6, No. 4, 2019, pp. 23–28.

[34] Zheng. Y., E. Blasch, and Z. Liu, *Multispectral Image Fusion and Colorization*, SPIE Press, 2018.

[35] R. Girshick, J. Donahue, T. Darrell and J. Malik, "Rich Feature Hierarchies for Accurate Object Detection and Semantic Segmentation," *2014 IEEE Conference on Computer Vision and Pattern Recognition,* Columbus, OH, 2014.

[36] R. Girshick, "Fast R-CNN," *2015 IEEE International Conference on Computer Vision (ICCV),* Santiago, 2015.

[36] R. Girshick, "Fast R-CNN," *2015 IEEE International Conference on Computer Vision (ICCV),* Santiago, 2015.

CHAPTER 8

RF Signal Classification

8.1 Introduction

RF communication systems are present in many aspects of our daily life. For example, mobile telecommunication systems (smartphones) are built on wireless RF communications. In the home, radios and televisions are connected to a wireless access point (WAP) via RF signals such as WiFi 802.11. In fact, modern homes are built on the concept of IoT devices wherein all the electronics (home security systems, smartphones, smart watches, radios, televisions, remote car starters, personal assistant devices) can be connected to WiFi and Bluetooth devices. *Cognition*, the ability to gain understanding through sensing, enables context awareness. Hence, the connected devices introduce the concept of cognitive RF systems where RF systems can sense the environments autonomously and help users effortlessly make decisions. In an *industrial setting*, a cognitive RF system can be thought of as a network of RF identification (RFID) transponders (both transmitter and receiver). In a *surveillance system*, multiband cognitive radar can adapt due to environmental changes. Finally, cognitive communications are evident with software-defined radio (SDR), network (SDN), and infrastructure (SDI).

ML algorithms evolved from classifying images [1], speech [2], behaviors [3], and text [4] as well as providing recommendations among different choices. ML applications are evident in medical diagnosis, financial markets, IoT [5], and cybersecurity [6]. Also, in a *multidomain scenario* using information from satellites and broadcast signals, cognition through ML builds patterns of tracks and behaviors [7]. Most recently, ML has been considered for classifying radio RF communication signals and self-driving cars. The focus of this chapter is how ML can be used to classify RF communication signals. The chapter discusses how ML can be used for developing a cognitive RF system [8, 9] such as enhancements in software signal processing, analysis of received signals analysis, and hardware emitter classification.

A RF cognitive system can sense the environment and adjust parameters (frequency, bandwidth, code, etc.) based on functionality. Cognitive RF approaches utilize ML methods. Invoking the popular observe, orient, decide, act (OODA) loop, a cognitive radios/RF approach also creates plans by learning from the collections (Figure 8.1). For example, a cognitive RF system can use *antenna beamforming*

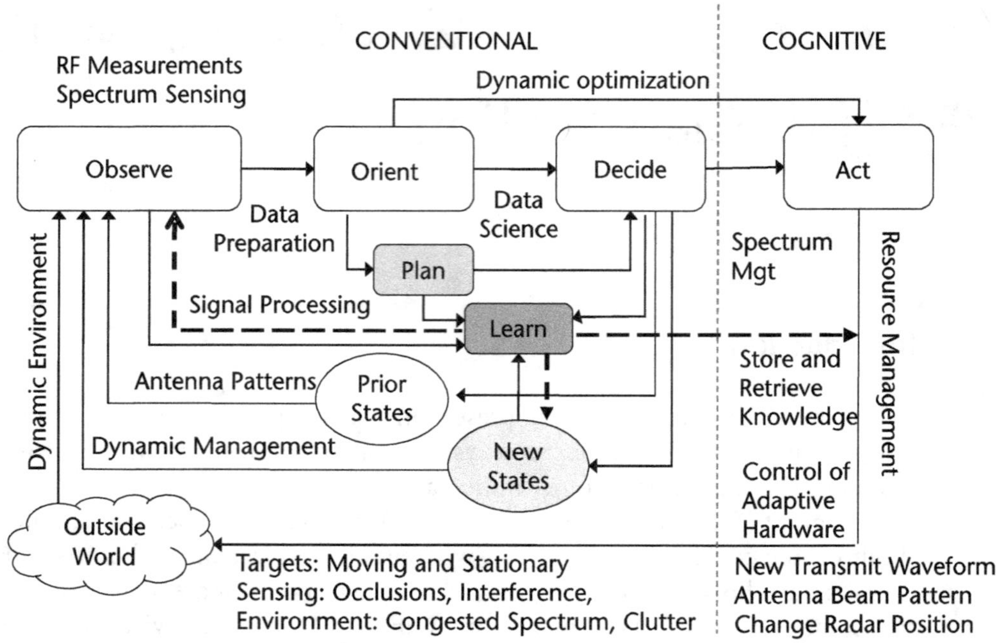

Figure 8.1 Cognitive RF design.

to send signals to a particular spatial area to avoid interference or improve transmission. In fact, the IEEE Wi-Fi 802.11ac standard used in many modern routers employs beamforming to improve reception. To maximize probability of target detection in airborne antenna radar, linear filters (e.g., Wiener) and nonlinear filters are used for adaptive space-time adaptive processing (STAP) to reduce noise and clutter. Machine learning supports parallel, distributed, and adaptive processing and noise reduction [10]. Multipath fading using ML supports analysis in complex scenarios [11]. Recent results using deep learning have demonstrated clutter suppression to improve target detection over STAP methods [12].

With the prevalence of big data from IoT systems such as phones, appliances, drones, traffic lights, security systems, and environmental sensors, there now is a need to apply DL to RF signals. ML provides dimension reduction, data classification, and pattern recognition. Deep learning (DL) applied to the RF spectrum could understand the composition of the radio frequency spectrum—discerning the kinds of signals occupying it, differentiating important signals from the background, and identifying those that don't follow the rules. Hence, DL would enable robust RF performance in complex scenarios.

According to the DARPA RFMLS program [13], there is power in utilizing ML techniques to enhance RF radar, communication, and control technology. Examples include the ability to:

- Discern subtle but inevitable differences in the RF signals from what otherwise are identical [14] and to distinguish these from signals intended to spoof or hack into these devices [15];

- Understand and trust what is happening in the Internet of Things and to stand up an RF forensics capability to identify unique and peculiar signals amongs the mix of signals [16];
- Recognize a specific transmitter based on the unique RF fingerprint naturally imparted by hardware imperfections within that transmitter [17];
- Provide spectrum awareness from narrow to wide bandwidths among software-defined radios, spectrum sharing, and other signals for communication and navigation [18];
- Modify transmit waveforms to enhance robust sensing and communication [19].

Combining deep learning (cognition) with RF signals analysis has benefits of:

- *Feature learning:* Identify and characterize RF signals in various settings. An example is detecting signal strength variations to determine the presence/absence of jamming.
- *Salient attention:* Direct attention to what is potentially important in the RF spectrum in which it is operating. An example is determining a signal of interest of a communications signal in a frequency band usually devoted to radar signals.
- *Sensor configuration:* Tune receiver locations and receptivity to signals and features the systems deem to be most effective at accomplishing the task at hand. An example is automatically placing the antennas and power selections to maximize target detection.
- *Waveform/signal synthesis:* Synthesize waveforms tailored to specific RF devices they emanate to give software defined radios (SDR) for improved identify verification. An example is in a crowded and adversarial spectrum, where there is the opportunity to generate and discriminate the particular digital waveforms that increase performance.

With the many options to improve RF signals analysis for sensing, communication, and navigation, the next sections review these concepts toward deep learning applications. Section 8.2 reviews communications signals with details on analog and digital modulation, shift keying for digital communications, and wireless technology. Section 8.3 describes three types of deep learning methods for RF signals, including (1) channel analysis, (2) I/Q imbalance, and (3) emitter analysis. Section 8.4 provides a discussion and Section 8.5 offers conclusions.

8.2 RF Communications Systems

The objective of a RF communication system is to move data or messages from one location (source) to another location (destination). The source and destination are connected by a communication channel. A communication channel can be a physical medium (such as cable connecting two computers) or a wireless channel (such as WiFi). In order to send data, the source uses a transmitter (Tx); and similarly, in order to receive the data, the destination employs a receiver (Rx) to process and understand the data. Figure 8.2 is a simple block diagram of a digital spread spectrum (SS) RF communication system. SS RF communications are important as the narrowband signal is spread in the frequency domain to enhance a wider bandwidth, among communications benefits such as (a) secure transmission (e.g., utilize frequency hopping), (b) resilient detection (e.g., mitigate adversarial jamming), (c) robust performance (e.g., operate in the presence of interference and noise), (d) limited power (e.g., in satellite downlinks), and (e) increased usage (e.g., enabling multiple-access).

All digital communication systems have a block diagram similar to Figure 8.2; however, they differ from each other primarily in spreading approaches and modulation schemes employed for data transmission and reception in RF systems. Modulation, along with bandwidth and frequency technologies, has an impact on data rate, bit error rate, signal security, signal interference, and overall systems complexity. To increase the channel capacity, RF communications systems utilize diversity over (a) *space:* multiple (ground, air, space) mobile of base stations, (b) *frequency:* more than one carrier frequency is used, (c) *time:* information is sent with time spacings, and (d) *polarization:* orthogonal polarization of signal. The next sections overview analog and digital signals modulation fundamentals that are utilized in contemporary DL RF approaches.

8.2.1 RF Signals Analysis

For RF wireless transmission of a signal as an electromagnetic wave, the system is allocated a certain bandwidth around a carrier frequency as shown in Figure 8.3. The RF systems operate within a passband that spans some particular bandwidth (B) that is centered about some particular passband frequency (f_c). Thus, an analog stream of signals (or digital stream of information bits) is upconverted to a passband frequency within the allocated bandwidth. *Passband signals* are determined by the shift of the baseband signal, and once that spectrum (B) is occupied, no other entity can share the channel for the purpose of communications. Hence, everyday RF communication uses such as FM radio, WiFi, cellular like 4G, 5G, and Bluetooth are all passband signals that execute transmission within their respective

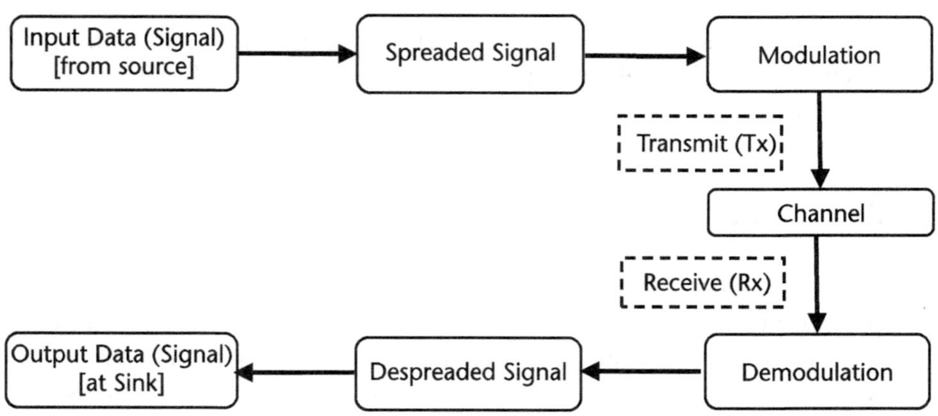

Figure 8.2 Block diagram of a spread spectrum communication system.

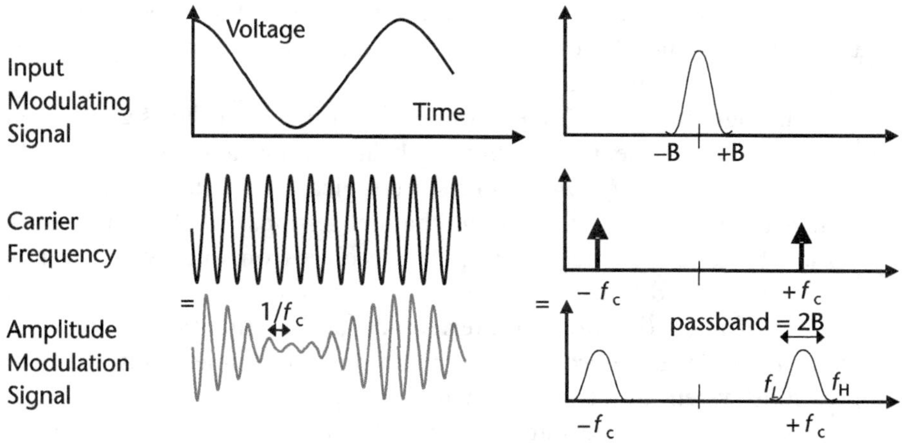

Figure 8.3 Passband signal.

frequency bands sharing the same physical medium [20]. To shift the spectrum to a designated carrier frequency is accomplished by multiplying, or mixing, it with a sinusoidal waveform as show in Figure 8.3. Note that:

- In time domain, a pulse amplitude modulation waveform can be multiplied with a carrier sinusoid at frequency f_c;
- Time domain multiplication between two signals is frequency domain convolution between their spectra;
- The spectrum of a pure sinusoid at any frequency is composed of two impulses at that frequency;
- The convolution of a signal with a unit impulse results in the same signal *parked* at its allocated slot.

The resulting passband waveform propagates through the RF environment, which is typically called the *channel* for the purposes of RF communications analysis. The communications receiver transduces the RF waveform at the receiver antenna front end, and the resulting signal waveform is typically not identical to that which was transmitted from the sender.

One difference between the transmitted signal waveform and that which enters the receiver front end is that the *multipath and fading effects* can be present in the intervening channel. Often, it is possible to assume that the majority of these effects depend linearly on the transmitted signal, so that the communications channel can be modeled according to an impulse response function $h(t)$ of time. The response function gives the received waveform due to a transmitted signal which is an impulse function that is nonzero only at $t = 0$ and is zero for $t \neq 0$. Thus, if a general passband signal waveform $x(t)$ is transmitted through a channel that is modeled by the impulse response $h(t)$ that characterizes the multipath and fading channel effects, then the signal waveform at the front end of the receiver is given by:

$$y(t) = h(t) * x(t) \tag{8.1}$$

That is, the physical effects of multipath and fading depend on the actual signal that was transmitted according to a one-dimensional convolution (denoted by $*$) in time with the impulse response $h(t)$ of the channel.

The signal at the receiver front end can include other signals emanating from the local environment, including both intentional and unintentional interference sources. These *interference sources* are typically modeled as an additive contribution relative to the convolutional channel response of (8.1). In addition, thermal noise effects can also be present in the receiver front end, which results from noise that is generated within the receiver electronics components themselves. These sources of noise (intentional interference, and unintentional interference) are collectively modeled according to the function $n(t)$ of time. Thus, the overall communications system model is assumed to comply with the standard linear form used for analysis of received signals in complicated multipath, fading, and nonuniform noise environments (e.g., [21–22]):

$$y(t) = h(t) * x(t) + n(t) \tag{8.2}$$

Analogously, in the frequency domain:

$$Y(f) = H(f)F(f) + N(f) \tag{8.3}$$

And for digital systems:

$$\mathbf{y}_c(n) = \mathbf{H}\mathbf{x}(n) + \mathbf{n}_b(n) \tag{8.4}$$

where

$\mathbf{H} = [\mathbf{h}(\theta_0)\ , \ldots., \mathbf{h}(\theta_{L-1})]$ is a $D \times L$ array steering matrix
$\mathbf{x}(n) = [\mathbf{r}_1(0), \ldots, \mathbf{r}_L(N-1)]$ is $L \times N$ array signal matrix
$\mathbf{n}_b(n) = [\mathbf{n}(0), \ldots, \mathbf{n}(N-1)]$ is a complex Gaussian noise matrix $N(0, \mathbf{I}\sigma^2)$

with θ as the direction of arrival angle. The baseband sinusoidal signal $s(t)$ is convenient for understanding the fundamentals of communications modulation and demodulation and is expressed as:

$$s(t) = A(t)\cos(\theta(t)) \tag{8.5}$$

where $A(t)$ is the time-varying amplitude and $\theta(t)$ is the time-varying angle. For the case in which the phase function has the form $\theta(t) = \omega_c(t) + \alpha(t)$, (8.5) can be written as:

$$s(t) = A(t)\cos\left[\omega_c(t) + \alpha(t)\right] \tag{8.6}$$

where $\omega_c(t)$ is the carrier frequency and $\alpha(t)$ is the phase modulation. Often, $A(t)$ is called the envelope of the signal $s(t)$. At a fixed, t, then the sine wave is represented as:

$$s(t) = A_c \cos[2\pi f_c t + \phi] \tag{8.7}$$

where the angle θ is represented as the frequency (f) and phase (ϕ).

8.2.2 RF Analog Signals Modulation

Some of the common modulation techniques are amplitude modulation (AM), frequency modulation (FM), phase modulation (PM), and pulse modulation (PuM). For convenience, Figure 8.4 presents both the analog and digital modulation methods.

In amplitude modulation, the phase of the signal $\alpha(t)$ is kept constant while the envelope can be changed in proportion to the modulating signal $f(t)$. Hence, an amplitude modulated signal can be written as:

$$s_{am}(t) = f(t)\cos(\omega_c(t)) \tag{8.8}$$

Within AM, there are techniques called AM with double-sideband suppressed carrier (DSB-SC) and AM with double-sideband large carrier (DSB-LC). DSB signals can be converted to single-sideband (SSB) to use a particular band efficiently. Vestigial-sideband (VSB) is a modulation scheme that is used to mitigate some issues with SSB or DSB modulation. In frequency modulation, the amplitude $A(t)$ is kept constant while the frequency $\omega_c(t)$ can be varied in proportion to the modulating signal $f(t)$. For the case of FM modulation, there are narrowband FM and wideband FM signals. In phase modulation, the phase is varied in proportion to the modulating signal $f(t)$. In pulse-amplitude modulation (PAM), the amplitude is constant within pulses and is changed in proportion to the sample values of the modulating signal. Among various types of pulse modulations, pulse position

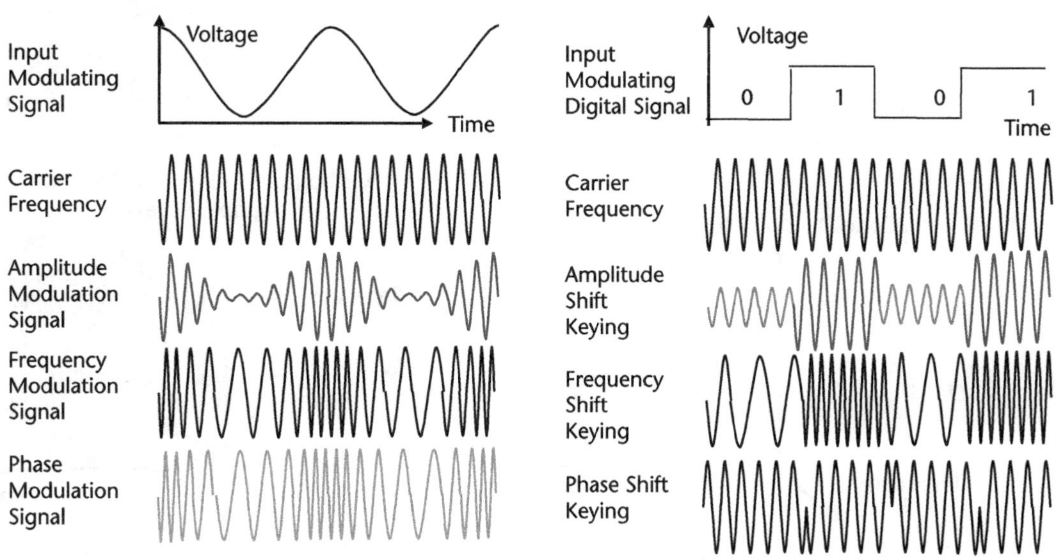

Figure 8.4 Signal modulation and shift keying over amplitude, frequency, and phase.

modulation (PPM), pulse code modulation (PCM), and pulse width modulation (PWM) are also commonly used in communication systems.

8.2.3 RF Digital Signals Modulation

For digital communication, if there are different *spreading sequences*, then the system modulates the carrier signal with digital data (input) and then performs data transmission. The different dimensions over which modulation is performed to separate signals in frequency (FDMA), time (TDMA), and code division multiple access (CDMA) [23], as well as wireless standards orthogonal frequency division multiple access (OFDMA) - as shown in Figure 8.5. Current techniques include OFDMA [24] and wideband CDMA (WCDMA).

Correspondingly, the modulation of digital data can be decomposed into keys using shifting approaches. Figure 8.4 illustrates the key assignment of 1s and 0s based on the shifts in the signal amplitude and phase. The common approaches include amplitude-shift keying (ASK), frequency-shift keying (FSK), phase-shift keying (PSK), and their combinations. Several other important modulations are quadrature amplitude modulation (QAM), quaternary PSK (QPSK), M-ary orthogonal FSK, and amplitude phase keying (APK). Each of these modulation techniques have advantages and disadvantages in terms of bit error rate and bandwidth efficiency. Among the variations of PSK, binary PSK (BPSK) and differential PSK (DPSK) are most commonly used. For example, in direct sequence spread spectrum (DSSS), BPSK modulation is used. The DSSS system allows CDMA that reduces

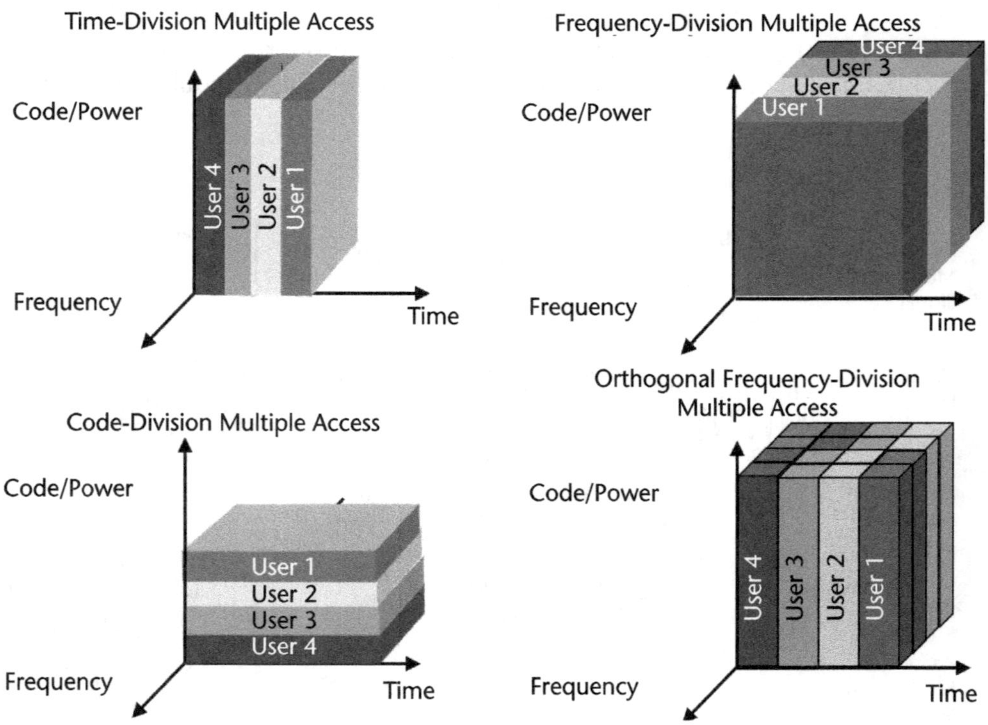

Figure 8.5 Partition-based multiple access strategies, including division amongst users based on particular slots in time, frequency, and spreading code.

interference among users, and hence is an important technique for satellite [25, 26] and avionics [27] communication systems. The CDMA transmitter architecture is shown in Figure 8.6. To transmit a digital message over a wireless spectrum, there is a need for baseband digital upconversion of the digital data achieved with shift keying, which is presented in the next section.

8.2.4 RF Shift Keying

Many common communications systems apply PSK, in which different phases (ϕ) within the in-phase (I) and quadrature (Q) contributions of the modulation and demodulation processes are used to send some number of symbols across a communications channel. Note, from Figure 8.7 that Q and I signals have the same frequency but are different in phase by 90°. PSK can be used with any number of phase values with a corresponding number of possible symbols. The only requirement is that the sender and receiver both agree on the selected mapping. For example, binary phase shift keying applies the 2 phase values of {0°, 180°}. Quadrature phase shift keying uses the 4 phases {0°, 90°, 180°, 270°}, as demonstrated in Figure 8.7.

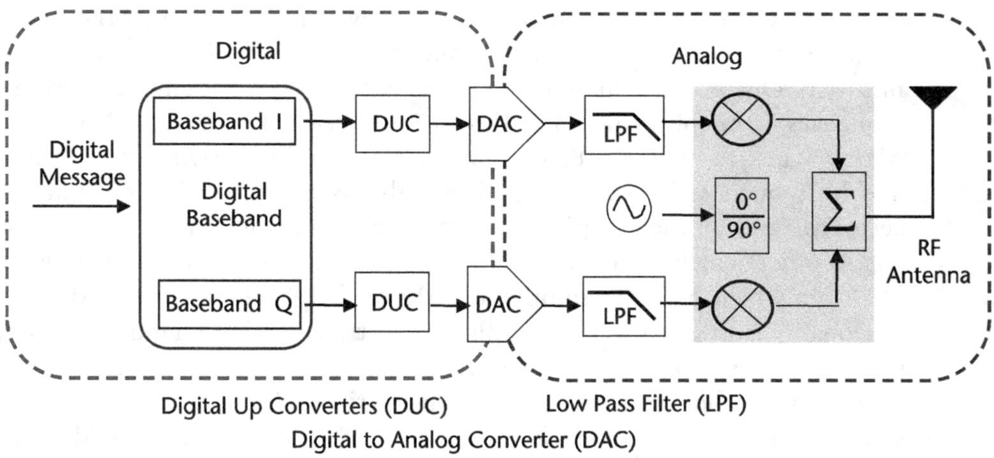

Figure 8.6 Interim Standard 95 (IS-95) transceiver for cellular CDMA.

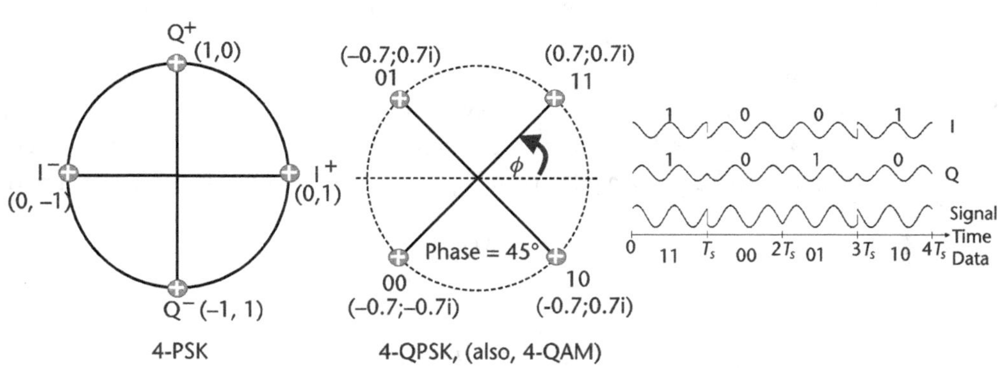

Figure 8.7 4-Phase shift keying and quadrature amplitude modulation.

The notation N-PSK conveys the general transmission of N symbols via an equal number of possible phase values. For example, 8-PSK uses 8 different phase values $\alpha(t) = 0°, 45°, 90°, 135°, 180°, 225°, 270°, 315°$, in sending a single symbol from the transmitter to the receiver to achieve 8 different data symbols. For 8-PSK modulation, each symbol conveys 3 bits of information, corresponding to the $2^3 = 8$ different phase values. Specifically, it is possible to define the following bit sequences corresponding to each of phase values: {0° ⇒ 100}, {45° ⇒ 001}, {90° ⇒ 010}, {135° ⇒ 011}, {180° ⇒ 100}, {225° ⇒ 101}, {270° ⇒ 110}, and {315° ⇒ 111}. The particular choice of mapping between a given 8-PSK symbol and the corresponding sequence of 3 bits is not unique. That is, other mappings can be selected arbitrarily.

Another form of modulation/demodulation is based on FSK, wherein a communications system can employ some number of frequency slots for sending symbols from one node to another. For example, suppose that a communications system which spans the frequencies between 30 and 40 MHz is allocated into frequency slots with the following minimum and maximum frequencies (in MHz): {30 → 32}, {33 → 35}, {36 → 38}, and {39 → 41}. Each of the four selected frequency subbands in this notional example are separated by a 1-MHz guard band in order to mitigate the effects of interference between adjacent channels. The corresponding center frequencies of each of these bands is given by {31, 34, 37, 40} MHz. Each of the center frequencies can be selected to map to a set of $2^2 = 4$ different sequences of 2 bits each. One possible mapping between center frequencies and corresponding bit sequences is the following: {31 MHz ⇒ 00}, {34 MHz ⇒ 01},{37 MHz ⇒ 10}, {40 MHz ⇒ 11}. As with the PSK example, the mapping between frequency symbols and bit sequences is not unique—it is only necessary for the transmitter and receiver to agree on the mapping prior to the transmission of any frequency symbols.

Another widely used mode of modulation/demodulation scheme is based on QAM, in which both the amplitude $A(t)$ and the phase $\alpha(t)$ are modulated simultaneously in order to send symbols over a transmission channel. QAM yields a specific set of symbols on the constellation diagram that is comprised of the *I value* of a given symbol, as determined by the horizontal axis, and the *Q value* of the symbol plotted along the vertical axis. Thus, for QAM modulation/demodulation, there are some number of symbols on the I-Q constellation diagram that form a rectangular grid of symbols that are equally spaced along both the horizontal and vertical directions. Common implementations of QAM include QAM-16 (with a rectangular grid of 4 × 4 symbols, shown in Figure 8.8), QAM-64 (8 × 8 symbols), and QAM-256 (16 × 16 symbols).

To implement QAM, IQ modulators generate RF signals from baseband information for digital communication systems. The I/Q data resulting from sampling of a signal allows all the amplitude, frequency and phase information contained in the signal to be preserved. From Eq (8.7), the I/Q signals are determined from:

$$I(t) = A(t)\cos(\phi(t)) \tag{8.9}$$

$$Q(t) = A(t)\sin(\phi(t)) \tag{8.10}$$

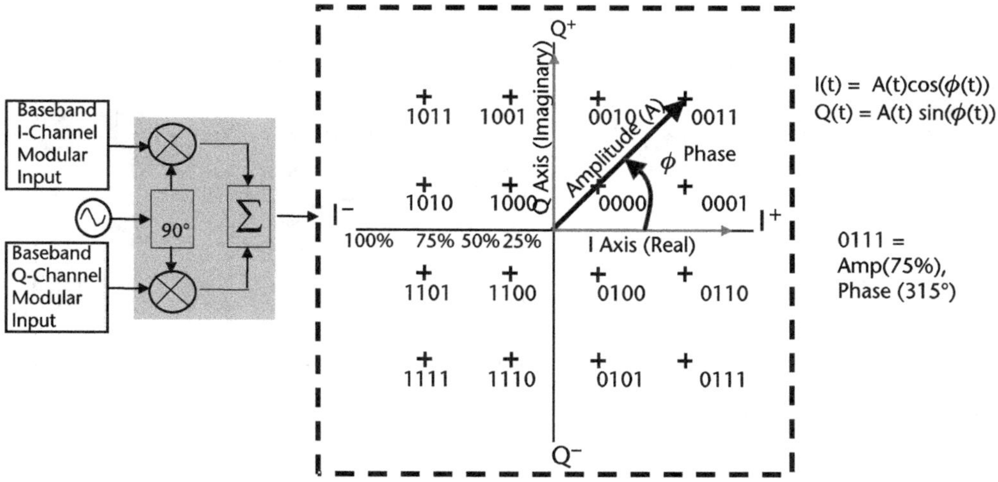

Figure 8.8 16-Quadrature amplitude modulation.

To accurately analyze or demodulate the signal in different ways, the two double-balanced mixers combine the I/Q signals. By targeting specific points in amplitude and phase, high order modulation is created. For example, from Figure 8.8, with 16-QAM, each symbol can convey four bits of information [28]. Virtually any type of RF modulation can be generated with IQ modulation, within the center frequency, bandwidth, and accuracy capabilities of the modulator device. Hence, the I/Q data produced from the modulator provides information for deep learning applications.

8.2.5 RF WiFi

One type of RF signal that is becoming ubiquitous in urban environments is WiFi 802.11, due to its prevalent use in wireless networks that use the standard Transport Control Protocol /Internet Protocol (TCP/IP) suite of the open internet. Most implementations of WiFi apply a strategy for sharing spectrum among a large number of users based on OFDM (see Figure 8.5). The OFDM multiplexing scheme applies multiple orthogonal subcarrier signals that have overlapping spectral properties. For example, the ODFM transmitter (Figure 8.9) employs QAM modulation to support the digital conversion of signals for RF transmission. Likewise, the receiver of a given subcarrier signal can be extracted with minimal interference from the other orthogonal subcarrier signals. OFDM is well-suited for environments with strong fading and multipath due to complicated channel conditions wherein many users can be packed into the RF channel. Then, each subcarrier can be modulated using PSK or QAM in order to carry the actual information symbols over the channel.

Another issue mentioned herein is that most modern communications systems favor FDMA, as with OFDMA for WiFi signals, over that of CDMA which has been applied in some earlier generations of communications systems, shown in Figure 8.10 [29]. The use of FDMA for wireless 4G follows a trend that is expected to continue in order meet the requirements to improve susceptibility to interference

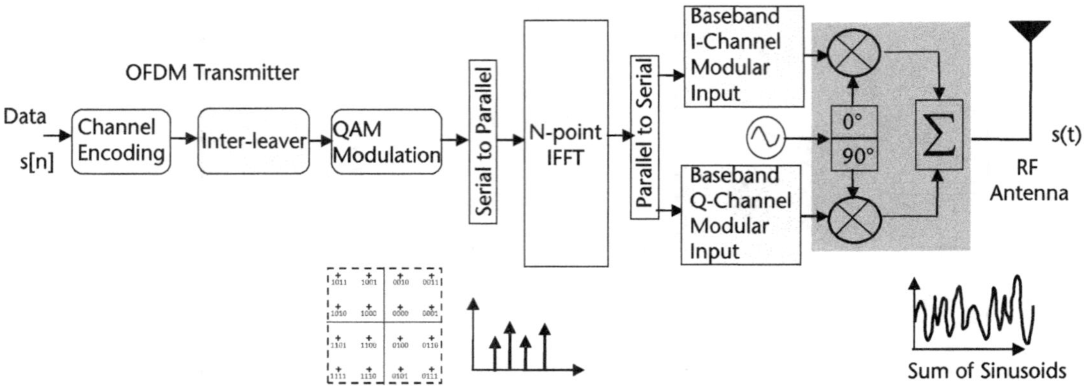

Figure 8.9 OFDM transmitter.

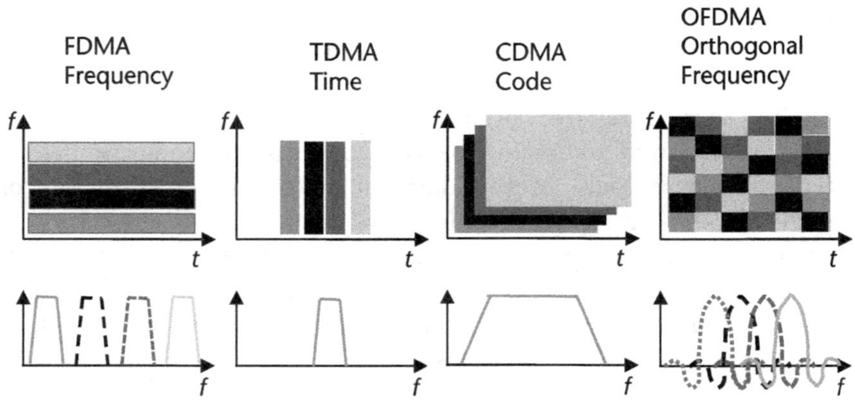

Figure 8.10 Multiple access methods.

and to pack many different users in a very dense and contested communications environment.

Figure 8.10 describes the three common types of partitioning of a channel among a number of users. However, WiFi802.11 does not use channel partitioning. Instead, it uses a shared access protocol strategy referred to as *Carrier Sense Multiple Access with Collision Avoidance* (CSMA/CA). When multiple users try to utilize a common RF communications channel, collisions can occur between two different transmitters that are attempting to transmit their RF symbols simultaneously. For such scenarios, it is likely that the current demodulation of the intended symbols will lead to high error rates so that the overall attempted communications fail. Thus, it is necessary for the network to be able to deal with such multiple users with CSMA/CA.

Typically, in CSMA/CA request to send (RTS) and clear to send (CTS) messages are transmitted in order to reduce the probability of collisions occurring within a contentious RF transmission environment, as shown in Figure 8.11 [30]. CSMA/CA is necessary in order to deal with the reality that mobile phone antennas are small and transmit with weak effective radiative power (ERP), whereas the base stations and wireless access points (WAP) typically have higher ERP. Thus, two different mobile phones in a particular scenario may be able to detect and communicate with

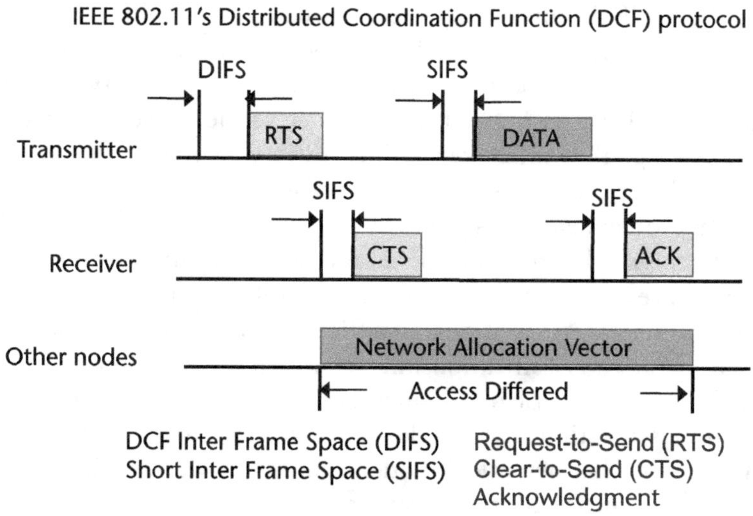

Figure 8.11 Carrier Sense Multiple Access with Collision Avoidance Protocol.

the WAP, but they may not be able to detect each other. When two devices cannot detect each other, it leads to the hidden node problem wherein the individual mobile phones are unaware of transmissions of the other and thus are required to rely upon WAP coordination via RTS and CTS message in order to maintain proper communications in a contentious environment. Given the approaches for signal transmission for communication, then the utility of ML is in signal detection.

8.2.6 RF Signal Detection

The first step in the overall processing of the receiver front end is typically to downconvert the input signal $y(t) = h(t) * x(t) + n(t)$ to some lower frequency, either at an intermediate frequency or to signal baseband. For the convenience of discussion, assume that all signals entering the receiver antenna are downconverted to baseband before the ML algorithms are applied. The basic reason for choosing the downconverted signals is to reduce the size of the input data sets and correspondingly the requirements applied to the input and storage of signal data sets. That is, the Nyquist theorem requires that the signal must be sampled at a frequency that is no lower than twice that of the maximum frequency of the underlying signal in order to accurately reproduce the original analog signal. A signal at baseband typically requires far fewer total samples of the signal waveform than the corresponding waveform at its original passband frequency, often by a factor of 10 or more. However, ML algorithms could be applied to intermediate or passband frequencies, as in the case of SDRs, wherein the full passband signal can be sampled.

Multiple methods can be applied in order to detect and estimate desired signals for such nonideal collection conditions. For example, mobile phones and other communications systems can use rake receivers and adaptive frequency allocation, as is applied with 4G FDMA [31]. The ability of the receiver to obtain accurate estimates of the transmitted symbols depends primarily on the signal-to-interference-plus-noise ratio (SINR). Thus, straightforward methods exist to demodulate signals in order to generate accurate estimates of the transmitted signals, provided that the SINR is sufficiently high. However, the ability to distinguish one emitter from another, in the absence of cooperative information, is generally quite a difficult problem. Traditional ML approaches use the SVM, AdaBoost, and least squares SVM (LS-SVM) that were applied to the phase data ϕ as the signal weights [32]. The signal-to-interference (SIR) ratio was used in the ML analysis to update the RAKE receivers, as shown in Figure 8.12.

For problem elements that are relatively well understood and straightforward, as with the modulation and demodulation of communications symbols, standard methods are applied that do not use or invoke ML methods. However, for problem elements that are exceptionally difficult and are poorly understood, as with the noncooperative classification and identification of communications transmitters, then ML techniques can enhance performance. That is, for parts of the overall signals classification that are described accurately through the use of a particular physics-based mathematical model, then the understanding should be invoked to predict or estimate those particular elements of the problem for which accurate models exist. In contrast, for problem elements that are fundamentally an enigma or for which the understanding is more art than science, then ML should be applied. Therefore, a reasonable approach to solving the problem of identifying a transmitter's manufacturer or identification is to apply ML methods. To determine the transmitter function, an equivalent knowledge of the receiver is needed, as shown in Figure 8.13.

The signal observed is a function of the real-world physical transmitting and receiving communications systems, as opposed to the idealized communications devices invoked in theoretical analyses. Specifically, theoretic analyses often assume that the transmitted signal waveform exhibits perfect quality with respect to amplitude $A(t)$, frequency $\omega_c(t)$, and phase $\alpha(t)$. That is, symbols are transmitted

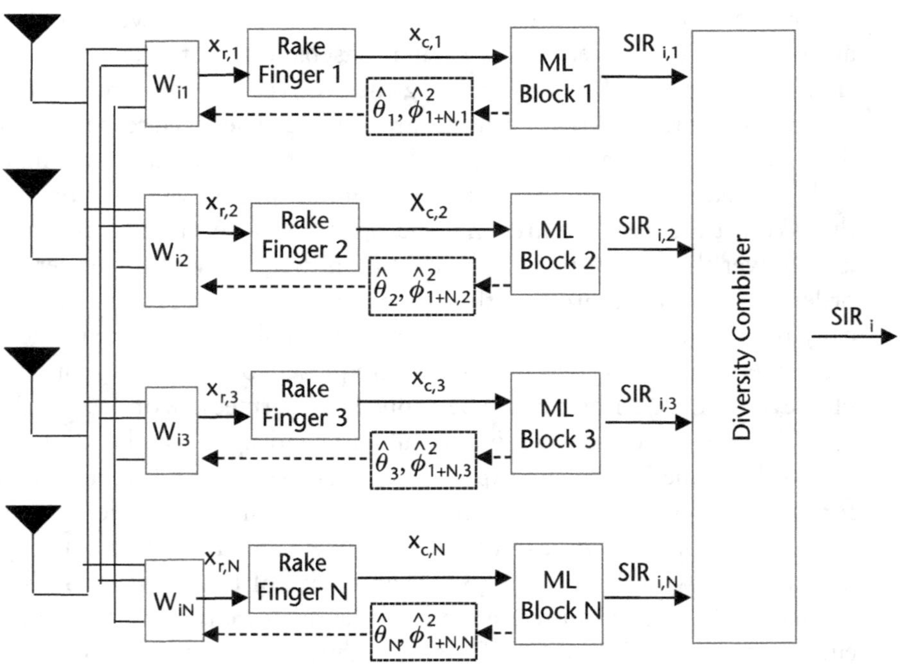

Figure 8.12 CDMA power control and direction of arrival estimation using ML.

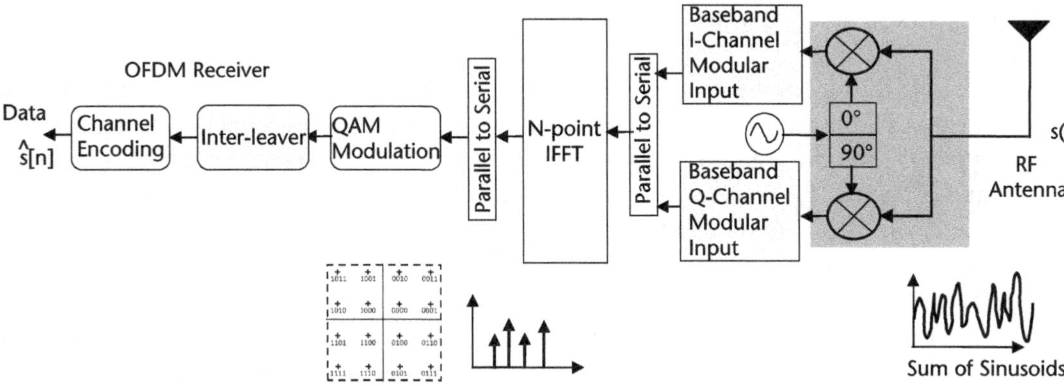

Figure 8.13 OFDM receiver.

according to the theoretic analytic equations, with infinitely sharp temporal transitions of $\{A(t), \omega_c(t), \alpha(t)\}$, according to the modulation type. For example, with theoretically idealized amplitude modulation, the amplitude $A(t)$ changes instantaneously in time, from one amplitude value to a different value corresponding to the transition from one symbol to the next.

In contrast, real physical transmitting devices exhibit small deviations in the signal amplitude $A(t)$, frequency $\omega_c(t)$, and phase $\alpha(t)$ away from the functional forms that are expected ideally. That is, for physically realistic transmitters with actual circuit elements used to generate the transmitted signal, the amplitude does not change instantaneously, but primarily occurs over some small, albeit nonzero, interval in time. In addition, the rapid symbol transition can occur with a relatively unique shape of the signal that is dependent on the actual circuit elements selected in the fabrication process. Of course, each manufacturer is free to design a unique configuration of circuit elements that generates the required transmission signals and receives them at the other end of the channel. These circuit differences can yield small differences in the resulting communications waveforms.

With the imprecision of *timing transitions* due to hardware components, the same basic concepts apply to frequency modulation, phase modulation, or any combinations of these modulation approaches. That is, the frequency cannot change instantaneously from one symbol to another but instead occurs over a small but nonzero interval in time. Furthermore, the shape of the signal for the generation of one symbol transitioning to another can be different when comparing two different manufacturers. Thus, a potentially unique signal shape can provide a type of signature for the classification and identification of RF emitters. These same concepts apply to *phase transitions* as well, because the signal phase cannot change instantaneously in time from one symbol to another. In addition, these issues apply to all generic modulations that can involve the encoding of information via the intended variations in signal amplitude $A(t)$, frequency $\omega_c(t)$, and phase $\alpha(t)$, as is desired to occur with the convenient theoretic ideals of signal generation and reception. The unintended component is the residual signal component that is not present in the theoretical models. These signal contributions are difficult to understand via physical models, and thus are a good candidate to benefit from ML methods. Thus, these unintended signal variations are effectively the small subtle signal components that are being sought for exploitation via ML.

8.3 DL-Based RF Signal Classification

The operating conditions for RF classification include the sensor, environment, and target. The largest variations between two different communications signals are typically the differences in the patterns of symbol sequences actually transmitted and received. However, it is the weaker unintended variations of the signal that are really being sought for exploitation by ML architectures. Hence, the *sensing* of RF emissions leads to researchers seeking DL methods to disambiguate signal transmissions.

To develop machine learning based classification algorithms there is a need for relevant data over different *environmental* conditions. In fact, to ensure that classification results are robust and reliable, ML algorithms must be trained with data from a variety of complex scenarios. Hence, several experiments must be designed, and data must be collected that conform to real operational environments. To augment cases in which complex scenarios are not available for data collection, *adversarial machine learning* affords simulation to stress the performance of ML algorithms in diverse environments.

The target of classification is typically the transmitter. ML methods have been applied to classify the model number of a given transmitter fabricated by a particular manufacturer. If the manufacturing process leads to uncontrolled variations of a transmitter product, even for different actual transmitters that all correspond to the same model number and the same manufacturer, the subtle differences can be learned over within-class variations.

8.3.1 Deep Learning for Communications

Examples of DL for communications are relatively recent. In 2017, Farsad et al. [33] sought to emulate a communication channel estimation using DL without any knowledge of the channel model. The strategy included a clustering analysis, along with RNNs and a LSTM to determine the data sequence and, although the communication included chemical transmissions, it was a reasonable assessment. In 2018, Gui et al. [34] looked at nonorthogonal multiple access (NOMA) for increasing system capacity and spectral efficiency. LSTM was used to determine the channel matrix composed of OFDM subcarriers employing BPSK. Results highlighted that the LSTM outperformed a physics-based hard choice of signal detection with Gaussian noise and a Rayleigh fading channel. Another access approach was the DL-aided sparse code multiple access (D-SCMA) [35]. The DNN-based autoencoder used on a simulated ODFM data outperformed a traditional message passing algorithm.

8.3.2 Deep Learning for I/Q systems

One common example of DL is to detect the transmitter, which is a function of the hardware components that perform the I/Q modulation, each with unique signatures for the devices. Zha et al. [36] constructed (1) single-shot multibox detector (SSD) networks for signal detection and (2) multi-input CNNs for modulation recognition for cognitive radio detection. In the research, they assess different (BPSK, QPSK, QPSK. and PSK) phase methods. Hence the data set included BPSK, QPSK, orthogonal QPSK (OQPSK), 8PSK, 16QAM, of 16APSK, 32APSK, and 64QAM. The study compared multi-input CNNs, cumulants, SVM, CNN for IQ waveforms, and CNN for the constellation. The time-frequency spectrum supports signal detection and the vector analysis supports the modulation analysis, demonstrating the ability of ML to discern the signals.

Sankhe et al. [37] present a CNN to determine the signals and tested at an indoor and outdoor facility for a covert channel. They called their method impairment shift keying (ISK) to account for the introduced controlled modifications to the radio transmitter hardware, which distorts regular standards-compliant waveforms, such as WiFi, with only a 1% increase in bit error rate. Using the I/Q channels, they were able to control the throughput. In another effort, ORACLE (Optimized Radio clAssification through Convolutional neuraL nEtworks) demonstrated the use of channel estimation for feedback-driven transmitter-side modifications at the receiver to increase CNN classifier performance [38]. The implementation included testing a CNN over collected IQ samples from 16, high-end X310 universal software radio peripheral (USRP) SDRs with a B210 radio as a receiver. Results indicate accuracy around 99% while the impairment from multipath fading and reflection offer additional experiments.

Northeastern University (NU) developed data sets for RF fingerprinting of Bit-similar USRP X310 radios. Bit-similar conveys that they have same hardware, protocol, physical address, and MAC identification. The RF fingerprinting data set is publicly available (www.genesys-lab.org/oracle). A paper using the data provides a baseline for ML-based WiFi device classification [39].

Example: RFAL: Radio Frequency Adversarial Learning (RFAL)

From Chapter 4, current deep learning methods utilize the CNN, RNN, LSTM, and GANs. Bringing these advances together for RF signals analysis, RFAL demonstrates adversarial learning for RF transmitter identification and classification [12, 43].

In order to learn the discriminating fingerprints (features) of similar transmitters, Roy et al. [43] used through USRP radios of the same kind, namely B210 from Ettus Research. The overall setup for signal generation and reception is shown in Figure 8.14. The B210s were programmed to transmit random data on 904 MHz using QPSK modulation with the configuration parameters of Table 8.1. A GNU-Radio was used for signal processing and data transmission. The flow graph is presented in Figure 8.15. The modulated signal was transmitted through the USRP sink block. For the receiver, a RTL-SDR (https://www.rtl-sdr.com/) captured over-the-air raw I/Q data and wrote it onto a file.

To determined robustness and statistically significant, multiple experimental runs were performed for each configuration, including:

- GAN, to distinguish adversarial transmitters from trusted ones;
- CNN, with 2D convolutions to exploit the correlation in collected signal data of the trusted transmitters for trusted transmitter identification;
- DNN, to classify the trusted transmitters;

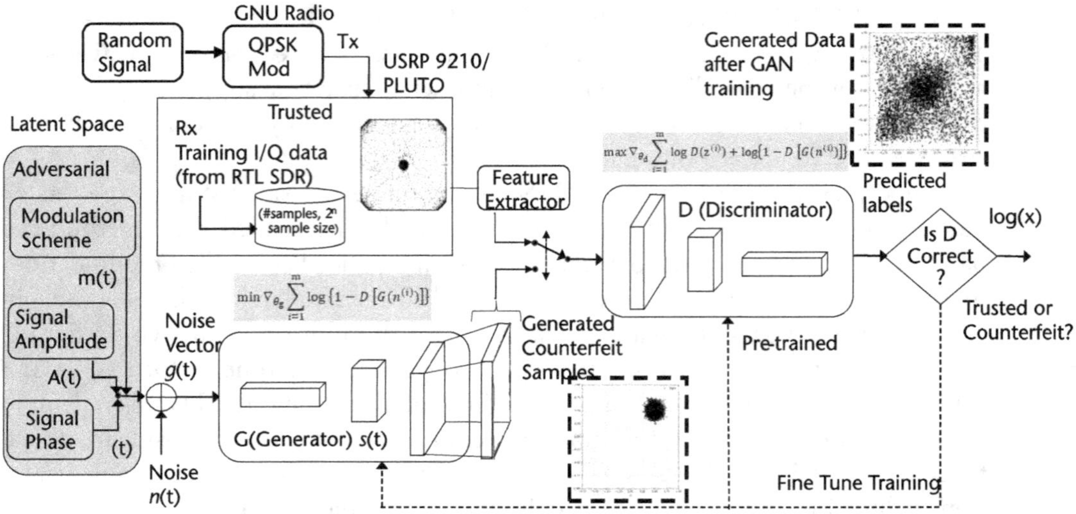

Figure 8.14 RFAL GAN architecture.

Table 8.1 Transmission Configuration Parameters

Parameters	*Values*
Transmitter gain	45 dB
Transmitter frequency	904 MHz (ISM)
Bandwidth	200 KHz
Sample size	1024
Samples/transmitter	40,000
# Transmitters	2 to 8

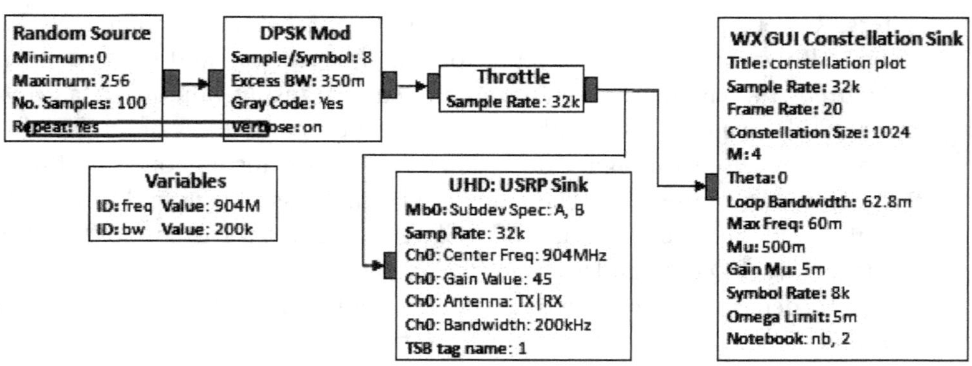

Figure 8.15 GNU radio graph for data collection for USRPs.

- RNN, with both LSTM and gated recurrent units (GRU) cells to improve the accuracy of trusted transmitter classification by exploiting the temporal aspect of the signal data.

Once the GAN training converges, the generator has learned the data distribution of the I/Q values from the known transmitters, and hence it starts to generate counterfeit data that succinctly captures the actual distribution of the I/Q values. Figure 8.16 and Table 8.2 present the final results after 2,000 samples.

8.3.3 Deep Learning for RF-EO Fusion Systems

With the goal of developing solutions in data fusion, Experiments, Scenarios, Concept of operations, And Prototype Engineering (ESCAPE) dataset was created. The dataset contains multimodal signatures using real and simulated simultaneous infrared (IR), full motion video (FMV), and passive radio frequency (P-RF) data collected from moving emitters [45]. Manifold learning approaches were applied to the dataset [40, 41], which reduced the data complexity to determine the associated features between the RF emitters and the change detections in the images.

Previous results demonstrated the benefit of a CNN to detect the human occupancy of an enclosed indoor space using the raw I/Q data of passive RF signals. For the ESCAPE data set, however, the I/Q data in its raw format was ineffective for the purposes of vehicle detection. The I/Q data was processed to generate a *2D histogram,* which is an estimation of the probability density function of the P-RF data. The ML methods include information fusion over the RF and EO data as shown in Figure 8.18 of the fusion of EO and RF neural network (FERNN) architecture.

For the stand-alone RF neural network, the classification accuracy could rise to 100% accuracy within four epochs for differentiation based on unique histograms formed by the different transmission waveforms. In order to balance training and testing data for the standalone RF and fusion networks, the simulations used for testing were limited to ones whose transmission waveforms are as different as

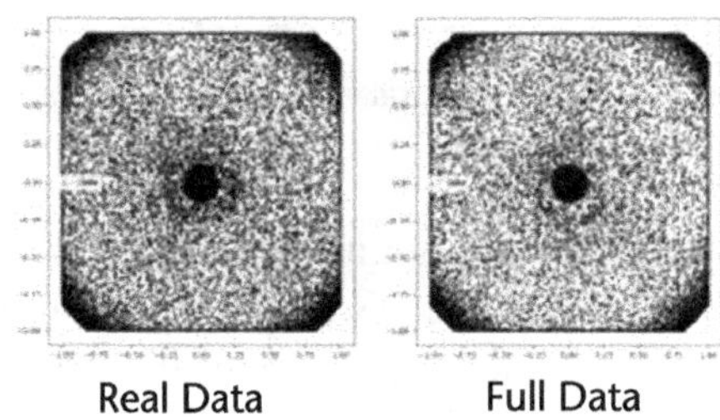

Figure 8.16 GAN results from generator (2,000 samples).

Table 8.2 GAN Results from Generator (2,000 Samples)

	Accuracy (%)				
Transmitters	*4*	*8 (30-dB SNR)*	*8 (20-dB SNR)*	*8 (10-dB SNR)*	*8 (0-dB SNR)*
CNN	89.07	81.59	81.3	78.64	51.53
DNN	96.49	94.60	94.60	92.24	85.12
RNN-LSTM	97.40	95.78			
RNN-GRU	97.85	97.06	97.02	95.64	92.3

Figure 8.17 ESCAPE EO frame and corresponding P-RF histogram (simulation 9, two vehicles, NADIR RF receiver from 2G, 3G).

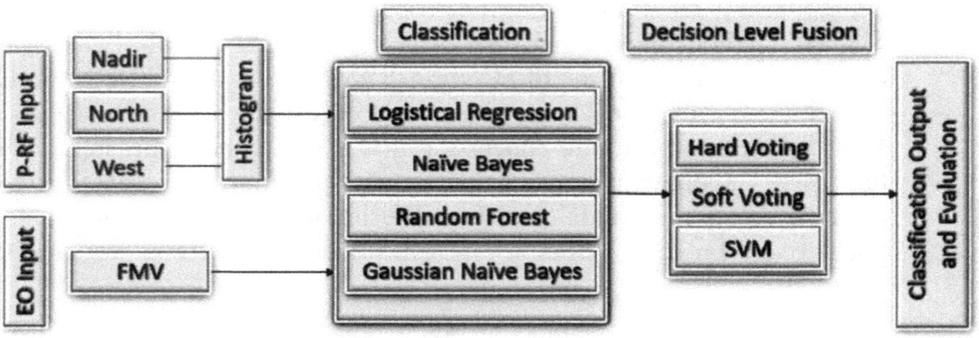

Figure 8.18 Fusion of EO and RF neural network (FERNN) architecture.

possible [42]. The collected data and simulations provide a variety of situations including a variety of signals, 2G, 3G, and 4G. When trained under these constraints, the accuracy of the stand-alone P-RF neural network is 83%, as shown in Figure 8.19. When the EO imagery is utilized (72% accuracy), the fusion of the methods for enhanced classification rises to 95% [47].

8.4 DL Communications Research Discussion

Deep learning techniques applied and tailored for RF signals and communications should progress as methods are demonstrated to enhance improvement over standard ML methods. Determining the benefits requires robustness analysis through performance evaluation including system testing [43], measures of effectiveness [44], and autonomy assessment [50]. Future considerations are the extent to which simulated data can be generated with sufficient fidelity in order to augment the volume of training data available for training ML architectures. However, it is uncertain that such fidelity can be achieved without the details of the RF fabrication

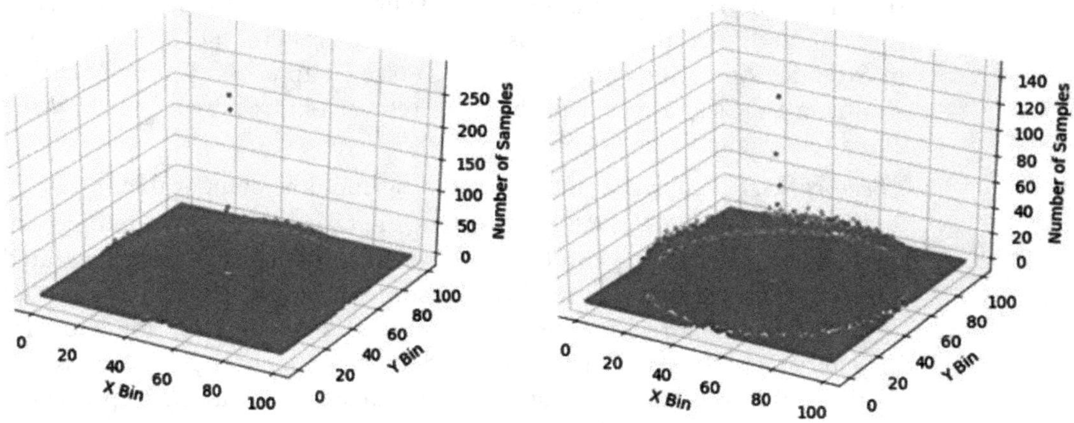

Figure 8.19 Comparison of I/Q histograms collected by SIGINT in simulation 4 (left) and simulation 11 (right). (Simulation 4 is 2 vehicles w Tx (2G, None); simulation 11 is 3 vehicles (2G, 3G, 4G)).

process. Approaches for RF feature learning or fingerprinting for WiFi devices should be robust to the sensor, environment, and target operating conditions. Some of the examples to consider are:

1. *Experiments design and data collections:* Data collections should be in various conditions.
   - *Indoor and outdoor environments*. Within outdoor settings, it is important to include congested RF devices such as that of urban conditions. For indoor settings, it could be as pristine as an anechoic chamber. For training and testing data sets, divide the data into (a) training on indoor and testing on outdoor data, (b) training on outdoor data and testing on indoor data, (c) training in mix data (i.e., indoor and outdoor) and testing on mix data.
   - *Types of WiFi*. With different protocols, then the data should be collected for 802.11a, 802.11g, and 802.11n, and so forth. The data should be divided into (a) training on 802.11g and testing on 802.11a or vice versa, and (b) training on mixed types of devices and testing on mixed types of devices.
   - *Signal-to-noise-ratio (SNR)*. SNR could be low, medium, and high. Then training and testing could be performed on these three combinations of SNRs.
   - *Weather conditions*. Data collection can be done on multiple days. Then training and testing data can be prepared as (a) training on day 1 data and testing on day 1 data, (b) training on day 1 data and testing on day 2 data, (c), training on mixed day data and testing on mixed day data.
   - *Signals transmitted/received*. RF signals can be collected by varying the polarization of the transmitted and received signals as well as changing the bandwidth of the signals.

The types of design of experiments (DOEs) and data collections should be thorough and extensive to demonstrate robust performance. One can say that if the ML algorithms perform well on the worst scenarios of the data sets, they will help to guarantee good results in operational scenarios. These types of DOEs are also known as performance evaluation of ML algorithms in extended operating conditions (EOC) (see Chapter 9). Testing ML algorithms in EOCs is a recent push in DL research known as *adversarial machine learning* utilizing GANs.

2. *Performance analysis:* The data analytics should be comparable.
    - *Preprocessing of the RF signals.* Once data has been collected, ground truth has been done, and training and testing data have been separated, all the data must be normalized to some standard. Data normalization (as part of data registration) is very critical. If the distribution and characteristic of the data (on training and testing) are not the same, classification accuracy would decrease. For RF image classification, preprocessing may include forming images from the raw phase history data and ensuring similar distribution of the pixels in training and testing data. For RF signal classification, spectrograms are formed from the signal data by applying short-time Fourier transform (SFTF) on the data (on a few milliseconds of the data). Forming spectrogram from the raw signal data is similar to creating images. If there is a need to create hundreds to thousands of images for training and testing, it is possible to use the actual SFTF spectrograms. The SFTF spectrograms are typically generated with the frequency on the vertical axis and time on the horizontal, as input images into a DL architecture. In this fashion, it is possible to use a CNN, which has been trained previously using optical images, for the current problem of classifying and identifying a transmitter based upon their emitted signal waveforms.
    - *Training and testing on deep neural networks.* Once data preprocessing have been conducted, a suitable DNNs model should be tailored to train and test for classification accuracy. Researchers have used VGG and Resnet or other DNN models. These models have been tested to provide good classification results.
    - *Classification results analysis*: Performance of classification algorithms are visualized using a confusion matrix. Presentations on the metrics should be understandable to designers, operators, and users.
3. *Hardware/software variations.*
    - *Access control.* There is a major challenge in performing fair and unbiased training of DL architectures for WiFi and related communications signals. In particular, these protocols often use TCP/IP data link layer Medium Access Control (MAC) addresses that effectively serve as the unique serial numbers for devices. These MAC addresses are included in the Data Link Layer 2 protocol format for the TCP/IP protocol suite for standard transmissions within a given WiFi subnet. Thus, it is important to prevent the DL from training on each of these MAC addresses, which gives a unique sequence of symbols for the data frame. Instead, it is important for the DL system to train the small subtle waveform variations that follow from the particular circuit elements implemented in a given manufacturers fabrication process. These unintended variations of the signal waveforms give weaker variation in the signal parameters (e.g.,

amplitude $A(t)$, frequency $\omega_c(t)$, and phase $\alpha(t)$) for the MAC portion of the data frame that is sent from one WiFi device to another. Great care must be taken to ensure that the DL architecture is not training, or figuratively memorizing, the source MAC address itself in attempting to identify the unique device that transmitted the 802.11 data link frame of interest.

- *Waveform design.* In addition, it is theoretically possible, depending on the specific circuit elements fabricated within a given transmitter device, for there to be nontrivial contributions to the signal waveform that are present for some nonzero short time after a given symbol transition has occurred at a previous instant in time. That is, the temporal duration for which one symbol transition affects the shape of other symbol transitions later in the data stream depends on the details of the device impulse response, and thus are not typically known a priori. These seemingly subtle issues can further confound the ability to accurately train DL systems for communications signal classification and identification and thus must be addressed fully in order to generate a viable DL system.
- *Bit checks.* Another point of interest with regard to WiFi on other similar RF communications waveforms is that they include a set of cyclic redundancy check (CRC) frame check sequence (FCS) bits, which are added as trailer at the Data Link Layer 2 frame. These bits are almost guaranteed to be fully random from one communications packet to the next, which could be useful in training DL architectures fairly.
- *Multipath fading.* One of the major challenges in classifying and identifying transmitting devices is due to the variations in the multipath and fading characteristics of the communications channel that can exist from one day to the next. Specifically, it is possible that during the training of a DL system the architecture effectively learns or trains on the specifics of the channel itself rather than on the unintended variations in the transmitted waveform. Then, if the system is tested on a different day for which the channel characteristics have changed, then it is possible that the DL system will incur degraded performance. Therefore, fair and unbiased training in the presence of channel effects that can vary from day to day or even hour to hour constitute an important component of DL research.

8.5 Summary

This chapter highlighted DL methods for RF communication signals classification. The key communication signals design elements include the RF signal, modulation scheme, and analog-to-digital keying approaches. Various modulation schemes of signals present features that can be detected by DL algorithms, and thus classifying that particular signal/waveform disambiguates transmitters. DL algorithms can be used to address the issue of spectrum sharing and enhancing security of RF-based communications. The future of DL approaches to RF communications and signals analysis should produce new ideas in the coming decades.

References

[1] Liang, P., G, Teodoro, H. Ling, E. Blasch, G. Chen, and L. Bai, "Multiple Kernel Learning for Vehicle Detection in Wide Area Motion Imagery," *Int. Conf. on Info Fusion*, 2012.

[2] MCconaghy, T., H. Leung, E. Bosse, and V. Varadan, "Classification of Audio Radar Signals Using Radial Basis Function Neural Networks," *IEEE Transactions on Instrumentation and Measurement*, Vol. 52, No. 6, 2003, pp. 1771–1779.

[3] Falcon, R., R. Abielmona, and E. Blasch, "Behavioral Learning of Vessel Types with Fuzzy-Rough Decision Trees," *International Conference on Information Fusion*, 2014.

[4] Malec, M., T. Khot, J. Nagy, E. Blasch, and S. Natarajan, "Inductive Logic Programming meets Relational Databases: An Application to Statistical Relational Learning," *Statistical Relational AI (StarAI) Workshop*, 2016.

[5] Falcon, R., R. Abielmona, and E. Blasch, "Behavioral Learning of Vessel Types with Fuzzy-Rough Decision Trees," *International Conference. on Information Fusion*, 2014.

[6] Lu, J., L. Li, G. Chen, et al.,"Machine Learning Based Intelligent Cognitive Network Using Fog Computing," *Proc. SPIE*, Vol. 10196, 2017.

[7] Blasch, E., T. Busch, S. Kumar, and K. Pham, "Trends in Survivable/Secure Cognitive Networks," *IEEE International Conference on Computing, Networking, and Communications*, Jan. 2013.

[8] Guerci, J. R., R. M. Guerci, M. Rangaswamy, J. S. Bergin, and M. C. Wicks, "CoFAR: Cognitive Fully Adaptive Radar," *IEEE Radar Conference*, 2014.

[9] Xiong, W., J. Lu, X. Tian, G. Chen, K. Pham, and E. Blasch, "Cognitive Radio Testbed for Digital Beamforming of Satellite Communication," *AIAA Cognitive Communications in Aerospace Applications*, 2017.

[10] Johnson, J., J. Li, E. Culpepper, E. P. Blasch, and H. Klopf , "Learning Algorithms for Suppressing Motion Clutter in Airborne Array Radar," *IEEE National Aerospace and Electronics Conference*, 1997, pp. 840–845.

[11] Wang, R., J. Lu, Y. Xu, et al.,"Intelligent Path Loss Prediction Engine Design Using Machine Learning in the Urban Outdoor Environment," *Proc. SPIE,* Vol. 10641, 2018.

[12] Zhang, Q., Y. Shao, S. Guo, L. Sun, and W. Chen, "Novel Method for Sea Clutter Suppression and Target Detection via Deep Convolutional Autoencoder," *International Journal of Signal Processing*, Vol. 2, 2017, pp. 35–40.

[13] https://www.darpa.mil/program/radio-frequency-machine-learning-systems.

[14] Yang, Y., T. Nguyen, and E. Blasch, "Mobile Positioning via Fusion of Mixed Signals of Opportunity," *IEEE Aerospace and Electronic Systems Magazine,* Vol. 29, No. 4, pp. 2014, pp. 34–46.

[15] Shafiee, E., M. R. Mosavi, and M. Moazedi, "Detection of Spoofing Attack Using Machine Learning Based on Multi-Layer Neural Network in Single-Frequency, GPS Receivers," *The Journal of Navigation,* Vol. 71, No. 1, 2018, pp. 169–188.

[16] Nagothu, D., Y. Chen, E. Blasch, A. Aved, and S. Zhu, "Detecting Malicious False Frame Injection Attacks on Surveillance Systems at the Edge Using Electrical Network Frequency Signals," *Sensors*, Vol. 1, No. 11, 2019, p. 2424.

[17] Roy, D., T. Mukherjee, M. Chatterjee, E. Blasch, and E. Pasiliao, "RFAL: Adversarial Learning for RF Transmitter Identification and Classification," *IEEE Transactions on Cognitive Communications and Networking*, Oct. 2019.

[18] Mukherjee, T., P. Kumar, D. Pati, E. Blasch, E. Pasiliao, and L. Xu, "LoSI: Large Scale Location Inference Through FM Signal Integration and Estimation," *Big Data Mining and Analytics,* Vol. 2, No. 4, 2019, pp. 319–348.

[19] Sira, S. P., Y. Li, A. Papandreou-Suppappola, D. Morrell, D. Cochran, and M. Rangaswamy, "Waveform-Agile Sensing for Tracking," *IEEE Signal Processing Magazine*, Vol. 26, No. 1, 2009, pp. 53–64.

[20] Chaudhari, Q., "Wireless Pi," https://wirelesspi.com/quadrature-amplitude-modulation-qam/.

[21] Guerci, J. R., "Optimum Matched Illumination-Reception Radar for Target Classification," U. S. Patent No. 5,381,154, 1995.

[22] Pillai, S. U., H. S. Oh, D. C. Youla, and J. R. Guerci, "Optimum Transmit-Receiver Design in the Presence of Signal-Dependent Interference and Channel Noise," *IEEE Transactions on Information Theory*, Vol. 46, No. 2, March 2000.

[23] Garren, D. A., M. K. Osborn, A. C. Odom, J. S. Goldstein, S. U. Pillai, and J. R. Guerci, "Optimal Transmission Pulse Shape for Detection and Identification with Uncertain Target Aspect," *IEEE Radar Conference*, 2001, pp. 123–128.

[24] Garren, D. A., M. K. Osborn, A. C. Odom, J. S. Goldstein, S. U. Pillai, and J. R. Guerci, "Enhanced Target Detection and Identification via Optimized Radar Transmission Pulse Shape," Proceedings of Radar, Sonar & Navigation, Vol. 148, No. 3, 2001, pp. 130–138.

[25] Garren, D. A., M. K. Osborn, A. C. Odom, J. S. Goldstein, S. U. Pillai, and J. R. Guerci, "Full-Polarization Matched-Illumination for Target Detection and Identification," IEEE Transactions on Aerospace and Electronic Systems, Vol. 38, No. 3, 2002, pp. 824–837.

[26] Vanhi, Y., A. Damnjanovic, and B. Vojcic, The cdma2000 System for Mobile Communications, Prentice Hall, 2004.

[27] Chen, Y., A. Bayesteh, Y. Wu, et al., "Towards the Standardization of Non-orthogonal Multiple Access for Next Generation Wireless Networks," IEEE Communications Magazine, Vol. 56, No. 3, 2018, pp. 19–27.

[28] Wang, G., Z. Shu, G. Chen, et al., "Performance Evaluation of SATCOM Link in the Presence of Radio Frequency Interference," IEEE Aerospace, 2016.

[29] Wang, G., G. Chen, D. Shen, X. Tian, et al. "Spread Spectrum Design for Aeronautical Communication Systems with Radio Frequency Interference," IEEE/AIAA Digital Avionics Systems Conference, 2015.

[30] Xiong, W., Z. Mo, G. Chen, et al., "Distributed Coherent MIMO System with Advanced CDMA Channel Estimation," IEEE Aerospace, 2017.

[31] Hemp, B., and S. Hsiao, "Low Power IQ Modulator for Digital Communications," Analog Devices, https://www.analog.com/en/technical-articles/low-power-iq-modulator-for-digital-communications.html.

[32] Iannacci, J., RF-MEMS Technology for High-Performance Passives, Ch. 3, Bristol, UK: IOP Publishing Ltd., 2017.

[33] Sharma, R., Y. A. Vershnin, G. Kadambi K. and N. Mukundan, "A Survey of MAC Layer Protocols to Avoid Deafness in Wireless Networks Using Directional Antenna," Ch. 19 in Mobile Computing and Wireless Networks: Concepts, Methodologies, Tools, and Applications (Information Resources Management Association, eds.), Harrisburg, PA: IGI Global, 2014.

[34] Chandrasekhar, V., R. Livingston, and J. R. Cavallaro, "Reducing Dynamic Power Consumption in Next Generation DS-CDMA Mobile Communication Receivers," International Conference on Application-Specific Systems, Architectures and Processors, Proceedings, Vol. 3 No. 3, 2003,pp. 260–270.

[35] Rohwer, J. A., "Machine Learning Methods for CDMA Power Control and Direction of Arrival Estimation," PhD dissertation, University of New Mexico, 2003.

[36] Fasad, N., and A. Goldsmith, "Detection Algorithms for Communication Systems Using Deep Learning," arXiv:1705.08044, 2017.

[37] Gui, G., H. Huang, Y. Song, and H. Sari, "Deep Learning for an Effective Nonorthogonal Multiple Access Scheme," IEEE Transactions on Vehicular Technology, Vol. 67, No. 9, 2018, pp. 8440–8450.

[38] Kim, M., N. Kim, W. Lee, and D. Cho, "Deep Learning-Aided SCMA," IEEE Communications Letters, Vol. 22, No. 4, 2018, pp. 720–723.

[39] Zha, X., H. Peng, X. Qin, G. Li, and S. Yang, "A Deep Learning Framework for Signal Detection and Modulation Classification," Sensors, Vol. 19, No. 18, 2019, pp. 40–42.

[40] Sankhe, K., F. Restuccia, S. D'Oro, et al., "Impairment Shift Keying: Covert Signaling by Deep Learning of Controlled Radio Imperfections," IEEE MILCOM, 2019, pp. 598–603.

[41] Sankhe, K., M. Belgiovine, F. Zhou, S. Riyaz, S. Ioannidis, and K. Chowdhury, "Oracle: Optimized Radio Classification Through Convolutional Neural Networks," Conference on Computer Communications (INFOCOM), 2019, 370–378.

[42] Sankhe, K., M. Belgiovine, F. Zhou, et al., "No Radio Left Behind: Radio Fingerprinting through Deep Learning of Physical-Layer Hardware Impairments," IEEE Transactions on Cognitive Communications and Networking, Vol. 6, No. 1, 2020, pp. 165–178.

[43] Debashri, R., T. Mukherjee, M. Chatterjee, E. Blasch, and E. Pasiliao, "RFAL: Adversarial Learning for RF Transmitter Identification and Classification," IEEE Transactions on Cognitive Communications and Networking, Oct. 2019.

[43] Zulch,P., M. Distasio, T. Cushman, B. Wilson, B. Hart, and E. Blasch, "ESCAPE Data Collection for Multi-Modal Data Fusion Research," IEEE Aerospace, March 2019.

[44] Shen, D., E. Blasch, P. Zulch, et al., "A Joint Manifold Leaning-Based Framework for Heterogeneous Upstream Data Fusion," Journal of Algorithms and Computational Technology (JACT), Vol. 1, No. 4, 2018, pp. 311–332.

[45] Shen, D., J. Lu, P. Zulch, et al., "Joint Manifold Learning Based Distributed Sensor Fusion of Image and Radio-Frequency Data," IEEE Aerospace, March 2019.

[46] Vakil A., J. Liu, P. Zulch, E. Blasch, R. Ewing, and J. Li, "Feature Level Sensor Fusion for Passive RF and EO Information Integration," IEEE Aerospace Conference, 2020.

[47] Bergin, J., D. Kirk, J. Studer, J. Guerci, and M. Rangaswamy, "A New Approach for Testing Autonomous and Fully Adaptive Radars." IEEE Radar Conference, 2017.

[48] Blasch, E., S. Liu, Z. Liu, and Y. Zheng, "Deep Learning Measures of Effectiveness," IEEE NAECON, 2018.

[49] Blasch, E., and B. Pokines, "Analytical Science for Autonomy Evaluation," IEEE National Aerospace and Electronics Systems Conference, 2019.

CHAPTER 9

Radio Frequency ATR Performance Evaluation

9.1 Introduction

Performance evaluation results from a test plan to determine the evaluation methods. Over the years, there have been many research studies published focusing on ATR performance using a standard data set (e.g., MSTAR). While the standard data set and the corresponding challenge problem support development and advances in machine learning approaches for ATR, these results are generally studied in laboratory conditions. From the beginning of the MSTAR program, various publications discussed the need to consider test and evaluation in the laboratory and for real world analysis. Some of the attributes to consider for performance evaluation beyond the laboratory setting include credibility of performance assessments, realistic performance assessment in conditions, affordable performance assessment, comprehensive performance assessment, and analytical understanding of ATR performance [1]. Within the context of ATR performance assessment, this chapter focuses on key attributes of performance evaluation. Key issues not covered in a related book, *High-Level Information Fusion Management and Systems Design* [17], are included here. That book covered evaluation and highlights system design, software management, metrics, and scenario design, all with a focus on increasing situation awareness (for an operator) as well as situational assessment (for a machine). In fact, [17] highlights elements of the MSTAR program in ATR operating conditions (OC) modeling with standard and extended operating conditions leading toward a Bayesian network example, where an OC condition increases ATR in accuracy, timeliness, and confidence (or credibility) [1].

9.2 Information Fusion

A key issue of utilizing context relates to the original distinction of low-level information fusion (LLIF) and high-level information fusion (HLIF) using the Joint Directors of the Lab (JDL) model. In the *Data Fusion Information Group* (DFIG) model (see Figure 9.1), LLIF (L0-1) composes explicit object assessment. HLIF (L2-6) includes much of the discussion and changes in the last decade for the DFIG/JDL model.

In the DFIG model, the goal is to utilize the data fusion and resource management (DF/RM) functions proposed in the JDL 1998 revisions while highlighting the user's involvement [2]. Humans are active observers with the ability to reason over contextual information [3]. RM is divided into fusion process control (L4), user refinement (L5), platform placement/resource collection (L6), with the goal to meet mission objectives. L2 situation assessment (SA) includes structures and relationships that are inferred from L1 object assessment of individual entities from L0 data alignment (i.e., registered features). Since unobserved SA events are difficult for a computer to assess, user knowledge and reasoning are necessary. L3 (impact assessment) includes scenarios, outcomes, sense-making of threats, courses of action, optimal decisions obtained using game theory [4], and estimation of intent, to help refine the SA estimation and information needs for different actions. Table 9.1 shows a listing of general types of information fusion methods applied to the multilook fusion for SAR ATR. Many of the preliminary results aggregate detection and classification results for enhanced performance and reduced uncertainty. While each method has benefits and limitations, they are baseline methods when compared to emerging deep learning methods.

For additional comparisons and analysis, a review was conducted on books and journal articles for SAR ATR [13].

Information fusion provides a systems-level analysis of the implementation of ATR systems, including situation, context, and user contributions. Since ATR has historically been a manual function, the 1980s brought the introduction of machine analytics to ATR tradecraft. One example is the ATR assessment on analytic tradecraft standards outlined in Intelligence Community Directive (ICD) 203 [14], which can provide a framework for assessing how well an AI/ML/ATR system can explain the basis for its decisions (i.e., analysis). To guide the design, development, and deployment of ATR systems, a multisource AI scorecard table (MAST) was created to bring together the elements surrounding the methods [15]. Figure

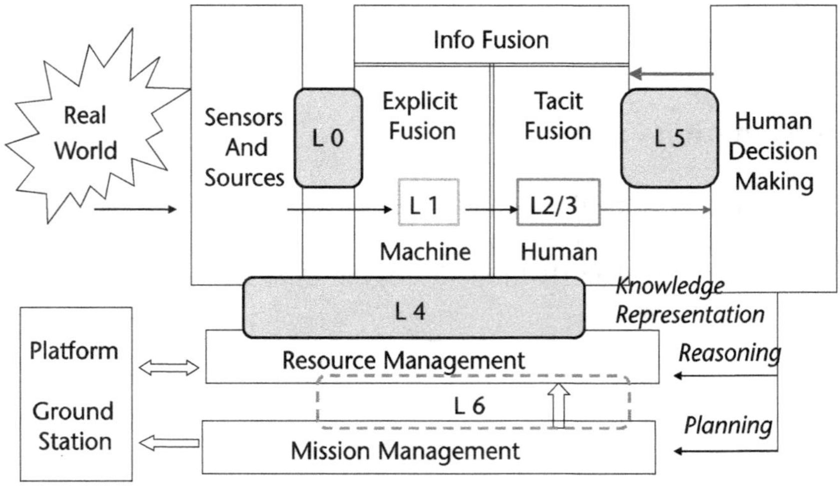

Figure 9.1 DFIG information fusion model (L = level) [6].

Table 9.1 Data Fusion Methods Applied to the MSTAR SAR ATR Problem

Fusion Type	*Description*	*Benefit*	*Limitation*	*Reference*
Boolean logic	Mathematical (Crisp) rules of AND /OR / NOT/NOR	Well-defined Simple to implement Expandable for large systems	High uncertainty Requires many rules Reduces meaning and interpretation	[5]
Fuzzy logic	Provides overlapping (soft) many-valued sets to process partial truth decision rules	Handles imprecise information Defines bounds for classification Amendable to human analysis	Not accurate Decision rules are arbitrary Does not scale to complex situations	[6]
Markov chains	Probabilistic modeling of dynamic systems comprised of defined finite states and transitions between states	Incorporate uncertainty of both the states and the model Computational complexity is minimal Offers scalabilty in complex scenarios	Combinatoric explosion of states can reduce utility Requires aggregation of states and pruning of superfluous stales Calls for knowledge engineering to determine the meaning	[7]
Bayesian networks	Probabilistic reasoning network for partial information providing a system of beliefs in given states that impact all other states in the network	Models a priori conditional situations Propagates information through the network methodically Incorporates partial knowledge for uncertainty analysis	Structure is static and cannot be changed adaptively Approximates the interconnection functions (e.g., Gaussian sums) Numerical approximations can have low bounds on the induced error	[8] [9]
Entropy	Measures data information contained in a signal	Monitors and predicts measurement uncertainty Handles discrete and continuous disparate sources useful for data fusion Widely adopted	Entropy is unitless and can easily misinterpreted Provides a relative assessment versus and absolute analysis Uncertainty level is not conducive to high credibility	[10] [11]
Dempster Shafer	Bayesian evidence accrual system that determines possibility (belief and plausibility) based on observations	Incorporates positive, negative, and conflicting information Does not need a fully specified model for processing Captures multidecision reasoning	Complexity of mathematical implementation does not scale Requires approximation techniques with many fusion rules Challenged with partial information	[12]

9.2 showcases the elements necessary for ATR systems using AI/ML to provide analytic support to users. Three important elements are shown in Figure 9.2: (1) before utilizing ML, much of the process includes support functions of collecting, moving, and transforming the data, (2) ML requires some level of knowledge input of labeling the data towards optimizing learning, and (3) the presentation of results for users should augment decision making versus being the autonomous machine decision maker.

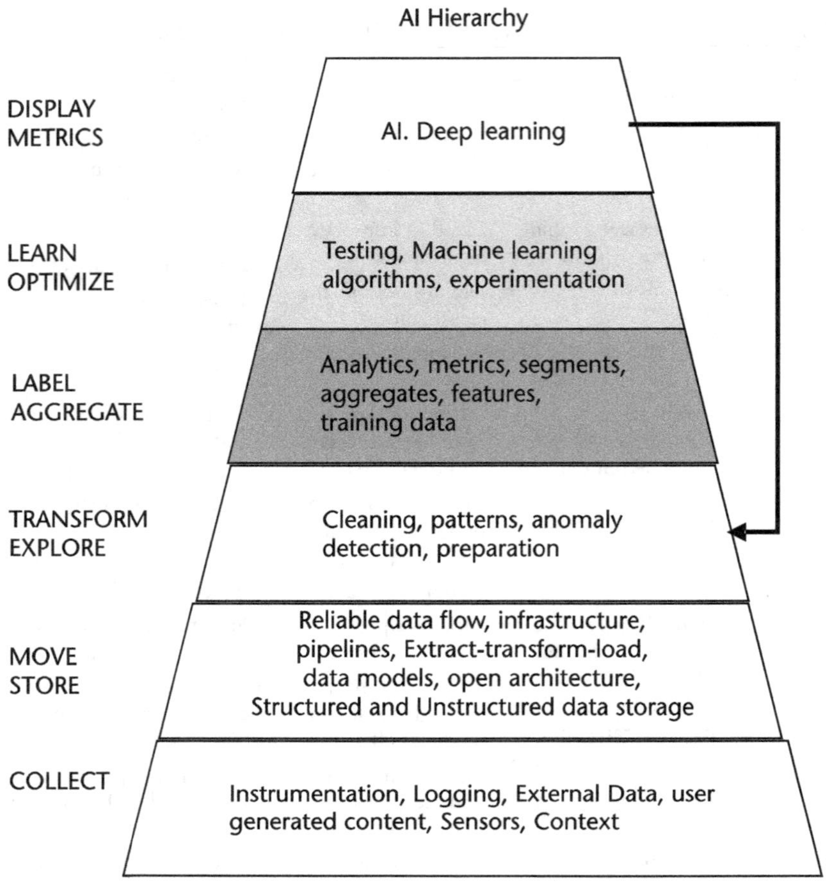

Figure 9.2 AI/ATR hierarchy from data to decisions.

The nine standards are shown below. These definitions provide a meaningful discussion of concerns surrounding the ATR performance assessment from the acknowledgment of data sources and the meaningful presentation of results.

Standard 1. Sourcing: Properly describes quality and credibility of underlying sources, data, and methodologies

Provide statements on factors affecting source quality and credibility. Such factors can include accuracy and completeness, possible denial and deception, timeliness of information, technical elements of collection, source pedigree and priority, evidence analysis, and assumptions.

Standard 2. Uncertainty: Properly expresses and explains uncertainties associated with major analytic judgments

Output assessments should indicate data and decision uncertainties. One common method is to support source, analytic, and user reporting. Analysts' confidence is based on individual experience, topic understanding, and quantity and quality of source material.

Standard 3. Distinguishing: Properly distinguishes between underlying intelligence information and analysts' assumptions and judgments
Assumptions are defined as suppositions used to frame or support an argument and affect analytic interpretation. *Judgments* are defined as conclusions based on underlying intelligence information, analysis, and assumptions. Product reports should explicitly state assumptions of an argument.

Standard 4. Analysis of alternatives: Incorporates analysis of alternatives
Present alternatives that address uncertainties, complexity, or low probability/high impact situations. The standard also looks for explanation of the reasoning and evidence that underpin the alternatives and the alternative's likelihood or implications.

Standard 5. Customer relevance: Demonstrates customer relevance and addresses implications.
Provide information beyond what is generally known, addressing near-term, direct, or first-order implications based on customer requirements. Address prospects, context, threats, or factors affecting opportunity for action.

Standard 6. Logical argumentation: Uses clear and logical argumentation
Present a clear analytic message with clear reasoning with no flaw in logic, combining evidence, context, and assumptions effectively.

Standard 7. Consistency: Explains change to or consistency of analytic judgments
Note how major analytic judgments compare to previous production. Explain how new information or reasoning supports changing or maintaining analytic line.

Standard 8. Accuracy: Makes accurate judgments and assessments
Express absolute probabilities (likely, very likely, and so on) in assessments—not just relative probabilities (more likely, increases the likelihood), or possibilities. Outputs should express judgments clearly and precisely as possible, reducing ambiguity by addressing the likelihood, timing, and nature of the outcome or development (see Section 9.4.1).

Standard 9. Visualization: Incorporates effective visual information where appropriate
Present visuals that are pertinent to the analysis, using visual information to clarify, complement, or augment data or analytic points (see Section 9.4.2).

With these definitions, consideration of the test and evaluation design reflect these attributes as separate functions or together as a system.

9.3 Test and Evaluation

Current directions in AI/ML systems focus on collecting large volumes of data to verify, validate, test, and evaluate system operations. For uncertain scenarios, data sampling may not be adequate to fully explore and represent the entire trade space for verification and validation (V&V). However, leveraging methods from test and evaluation (T&E), a hierarchy of analytics can be developed to narrow the trade space, keeping the opportunity cost of the remaining space as a risk-mitigated deployment strategy. Issues in V&V/T&E employ statistics but could benefit from theoretical analytics, such as the ability to augment data for testing using simulated models or to define tests to minimize operational risk. The use of modeling is not new, however, as analytics of AI/ML are designed to exploit data. These methods are then independent of the data developed from the first-principles physics models.

As machines and products become more sophisticated, there is interest in developing analytical and autonomous systems for evaluation [16, 17]. An autonomous system typically is associated with cognition and adaptation. As a starting point, the OODA loop contains adaptation and cognition. Other paradigms, such as the sense-learn-adapt (for radar processing), sense-learn-avoid (for UAV collision avoidance), and sensemaking (for data fusion) are popular paradigms. These methods all include *closed-loop processing* that utilizes real-time, online evaluation toward mission solutions of (1) *cognition,* the mental action or process of acquiring knowledge and understanding through thought, experience, and the senses, and (2) *adaptation,* making (something) suitable for a new use or purpose (i.e., to modify or become adjusted to new conditions).

As a closed-loop paradigm, elements of cognition and adaption result from intelligent software analysis with recent advances in SDR, SDNs, and software-defined radar, which are associated with cognitive radios and cognitive radar. An example of this is the environmentally aware cognitive fully adaptive radar (CoFAR) [18] with adaptive waveform transmission, reception, and controller/scheduler functions. By learning/understanding the complete multidimensional radar channel (i.e., targets, clutter, interference), and coordinating over the operating conditions, a sense-learn-adapt (SLA) approach supports an OODA loop concept. Results demonstrate a multiple-input/multiple-output (MIMO) probing approach for learning the radar parameters for nonstationary environments [19–21]. Evaluation of performance gains [22] demonstrated optimal waveform as a function of range and radar position. Recent results in using the RFView [23, 24] allow for various comparisons for MIMO multiaccess methods (e.g., time-division multiple Access (TDMA)) for ground moving target indicator analysis as a benchmark for challenge problems.

As with the CoFAR, test and evaluation considerations include data in rest, use, motion, collect, and transmit as shown in Figure 9.3.

A multi-intelligence sensor data system (e.g., radar, electro-optical, text) performs robustly in a SLA approach [3]. To confirm the ability of the system to operate in the real world, there is a need for V&V, such as for radar analysis [25–28], communication [29, 30], and target tracking evaluation [31–39]. The V&V process is part of T&E, of which the development process includes theory (e.g., models), systems design (e.g., data flow), and experiments (e.g., measurements). Training the models requires theory and data that assist in determining the filtering, estimation, and prediction methods. The data processing supports the analysis for prognostics, diagnostic, and predictive elements of the systems uncertainty [40–43]. Finally, experiments prove out performance: as in robustness to operating conditions, failure due to operational control and computing power, and safety and security concerns.

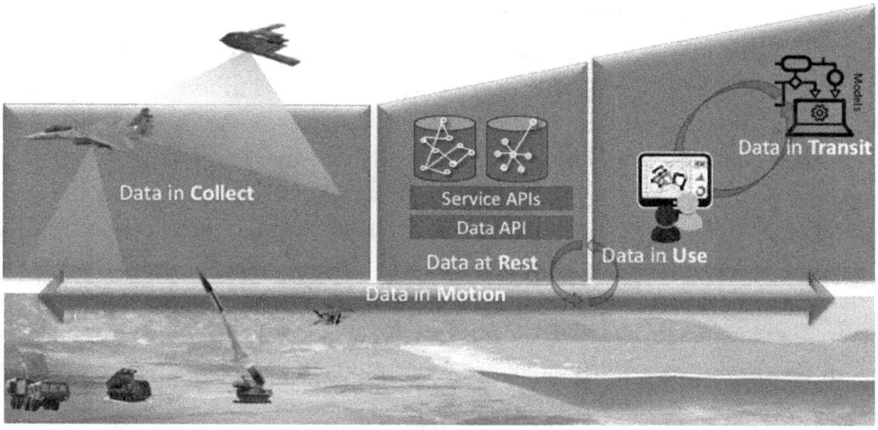

Figure 9.3 Complex autonomous systems environment.

T&E from V&V determines operations and monitoring (O&M) strategies for trust, explainability, robustness, and effectiveness of systems development.

9.3.1 Experiment Design

For DL-based objects classification, experiment design is critical for the expected outcome of the algorithms for worst-case performance analysis and for understanding limitations of the models. This involves developing various (level of complex data representing operating environments) data collections and defining performance metrics. Often researchers design a DL model with simple or toy data sets. However, until a deep architecture has been tested on complex data sets, it will not be prudent to field the system. Table 9.2 represents a typical experiment design with a performance metric. Suppose there are several variations of panda and gibbon species and noise is added to these animals' imagery to generate adversarial samples and add small patches of clutter. These thought experiments allow development of robust DL-based classifiers with expected outcomes.

Table 9.2 Experiment Design

Metric	*Experiment 1 (Easy)*	*Experiment 2 (Not So Easy)*	*Experiment 3 (Difficult)*	...	*Experiment 10 (Most Difficult)*
Number of Animals	5 known	5 known and 1 unknown	10 known and 4 unknown	...	20 known, 5 unknown
Confusers (i.e., similar/dissimilar or modified animals)	2 confusers	2 confusers and 1 unknown	Define	...	5 confusers and 3 unknown
PD/PID	92%/95%	90%/90%	Define	...	Define
Learning Rate	No constraints	Limited training data	Define	...	Define
Run-time	Unlimited	1 minute	30 second	...	Define
Compute Energy	Unlimited	50 Watts, 2 lbs	10 Watts, 1 Lbs	...	Define

9.3.2 System Development

Systems engineering and development [44] (Figure 9.4) highlights the importance of data feeding models for product development. Inherent in any T&E analytics is data quality, which has to be prepared for system use. The empirical data feeds the models, which includes theoretical and data-driven models. Optimizing empirical data is important for analyzing DL systems models. Together, the models support the product for the intended use (whether by a machine or system).

The future deployment of DL systems will require standard (or best) practices to gain widespread acceptance. What are the standards that support each stage (e.g., metrics), toward increasing user understanding? Inherent in the discussion of autonomous deployment are trust metrics [45, 46] that involve security; while transparency supports safety [47].

Figure 9.5 highlights that, after model construction, there is a need for continual V&V, as well as operations and monitoring (O&M). Automation/autonomy DL can provide methods of self-monitoring of operational performance.

Information is planned for applications such as natural language processing (NLP) [48] for target labelling, computer vision (CV) for target assessment [49], as well as information fusion for target tracking [50] to support a decision maker, especially for situations involving huge, multimodal, and novel situations [51].

Product development incorporates many key elements including design, test and evaluation, and operational relevance. Typically, a technical community sees an explosion of methods, determines a set of prominent ideas, and develops standards.

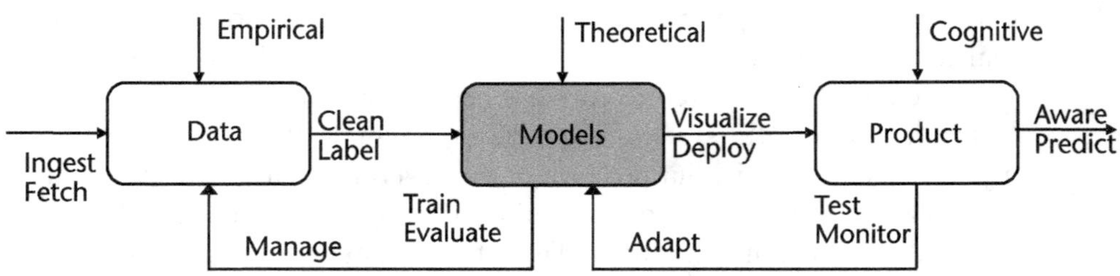

Figure 9.4 Product development.

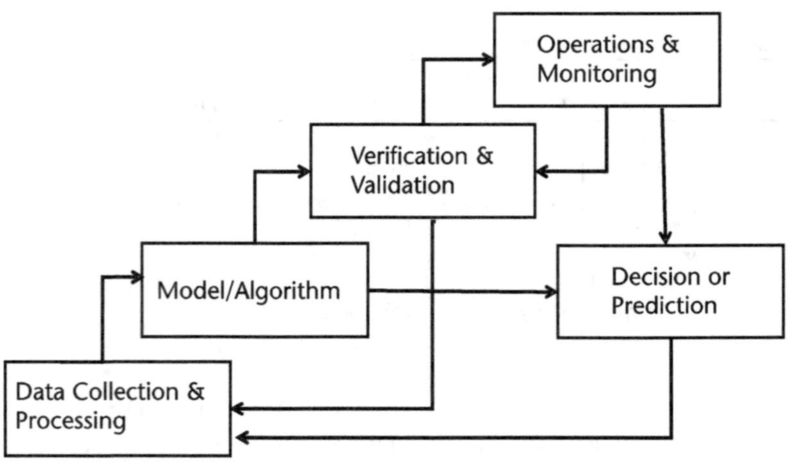

Figure 9.5 ATR product assessment.

Current DL methods have reached the need of developing a set of T&E standards or best practices for the development and deployment of the products [52] to support user performance needs [53, 54].

The deployment of products requires common practices for data collection and processing, algorithm and model development, V&V, as well as O&M. Examples in ATR [55–57] and image fusion [58, 59] demonstrated the value of T&E.

9.3.3 Systems Analysis

The execution of V&V requires experimentation, testing, and evaluation [60]. One example is high-level information fusion evaluation (Figure 9.6) [61], which presents a continual T&E loop.

T&E follows a continual progression of innovation, design, operational, and software methods—each of which expands the theory and model development, as shown in Figure 9.7. These elements are shown in Table 9.3.

9.4 ATR Performance Evaluation

ATR performance evaluation pursues data quality goals of accuracy, precision, and veracity among the big data issues of value, volume, velocity, and variety. Evaluation for ATR systems first determines meaningful MOPs, MOEs, as well as MOFEs. The use of machines to augment human performance seeks to reduce uncertainty in decision making. A current taxonomy of uncertainty analysis objectives is shown using the uncertainty representation and reasoning evaluation framework (URREF) ontology (Figure 9.8).

In the analysis, this book highlights examples of quality (accuracy, precision/recall, and veracity).

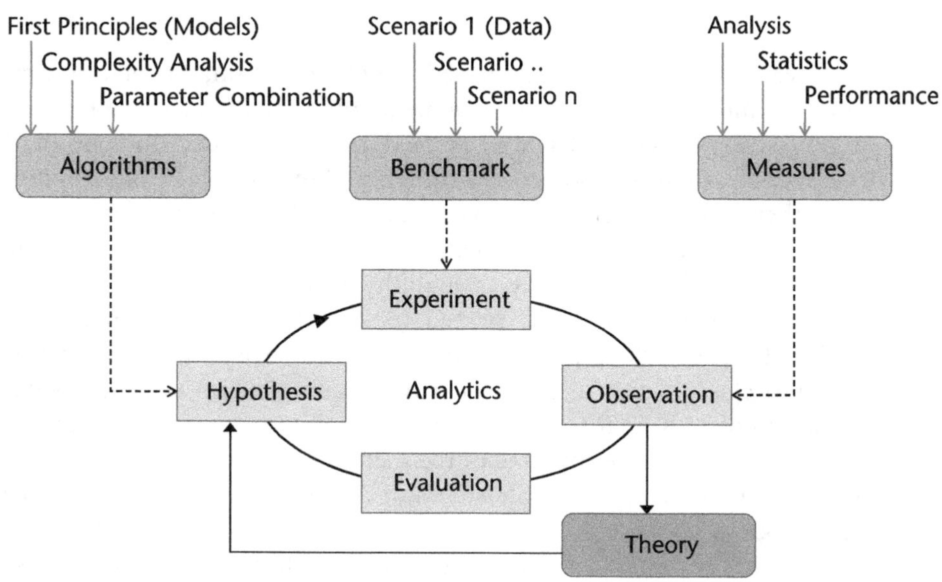

Figure 9.6 Testing and evaluation.

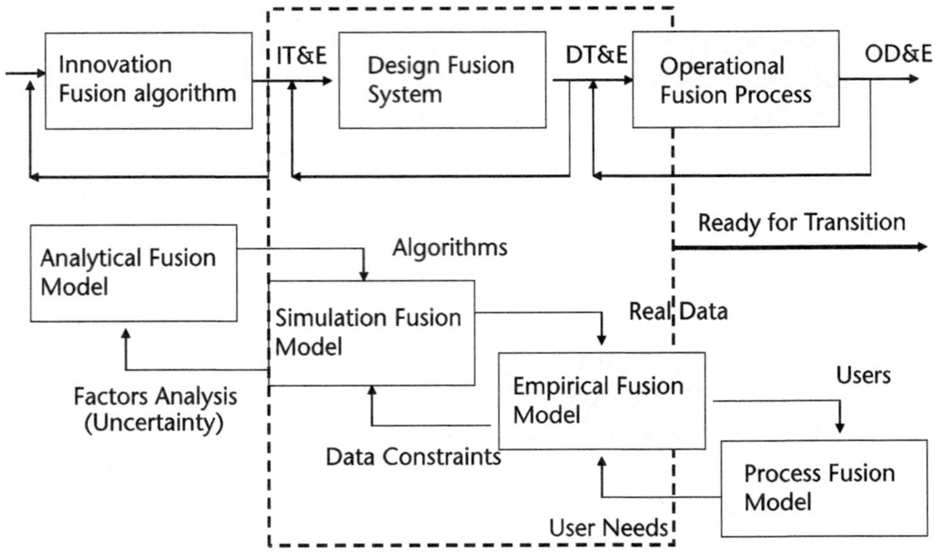

Figure 9.7 Methods for testing and evaluation.

Table 9.3 Test and Evaluation Considerations

Goals (Test and Evaluation)	Innovative test and evaluation—exploratory (IT&E)
	Design test and evaluation—development (DT&E)
	Operational test and evaluation—utilization (OT&E)
	Software test and evaluation—capability (ST&E)
Data (Model Instantiation)	Analytical model—mathematical performance analysis
	Simulation model—robust parameter exploration
	Empirical model—real data collected from sensors
	Process model—observations of humans w/ systems

Given a scenario of interest for SAR ATR collection, the analysis assumes some understanding of the sensor, environment, and target (e.g., scene resolution, region of interest, set of vehicles) as well as the data input and decision output. The system level measures (among others) are shown in Table 9.4. It is important to note that while there are many metrics, this chapter will focus on the first two rows of the key elements of the metrics in detection and classification that support the standard metrics of credibility, accuracy, trainability, robustness, and cost.

The probability of ID is related to identification of friend, foe, or neutral (IFFN). Typically, the pipeline of analysis is the probability of detection, recognition, and classification determined from a comparative test and evaluation. The evaluation of SAR ATR focuses on measures of performance while the effectiveness is determined from the run time of the methods for a given machine. Future efforts are defining efficiency and robustness of AI/ML systems from which calculations are emerging.

A useful example from the MSTAR data is shown in Figure 9.9. Assume there are 10 targets. If there are eight detections (○), then $P_D = 8/10$, while for the one false alarm (□), $P_{FA} = 1/10$, as shown in Figure 9.9(a). Using the extraction location

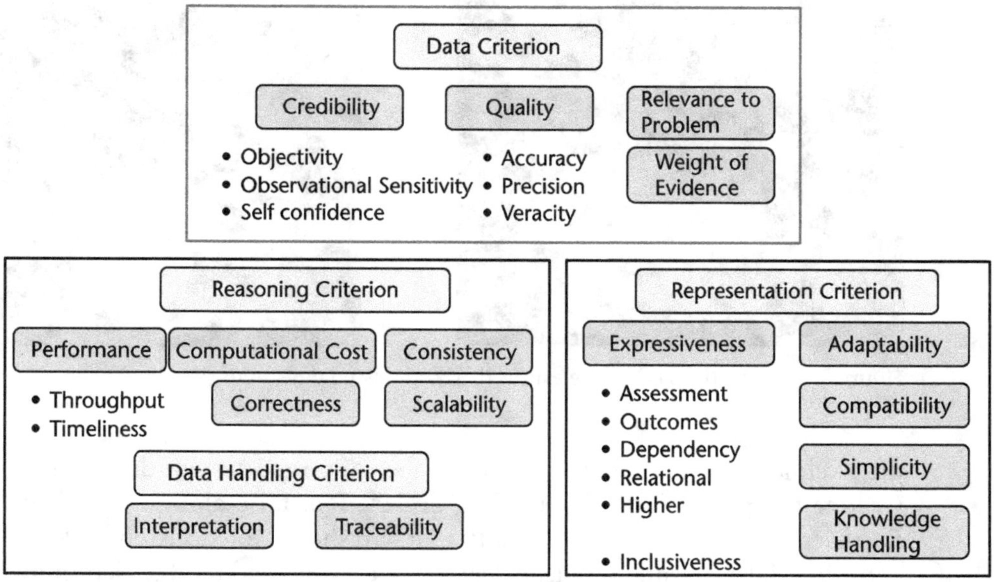

Figure 9.8 URREF categories.

Table 9.4 Measures of Performance and Measures of Evaluation

	Measures of Performance		*Measures of Effectiveness*		
	Credibility/Confidence	*Accuracy*	*Timeliness*	*Throughput*	*Cost*
Detection	Probability of false alarm (PFA) False Alarm Rate (FAR)	Probability of detection (PD)	Trainability	Robustness	Database size
Classification (Recognition)	PCC conditioned on detection PCC conditioned on target database	Probability of correct classification (PCC)	Time to acquisition	Complexity	Energy
Identification	Probability of correct association Fingerprinting	Probability of identification (PID)	Track lifetime	Security	Assurance
Tracking	Track length gap time	Track purity	Revisit rate	Bandwidth	Number of sensors

of only the five targets and the ability to classify the target in Figure 9.9(b), then the conditional analysis of P (classification | detection) = 4/5. Once the detections are determined, then the target type classifications are obtained via a confusion matrix [62].

9.4.1 Confusion Matrix

To conduct an ATR assessment, one needs to identify the truth targets from the test objects in real-world operation. In the real world, there are known and unknown targets. ATR decisions are based on the target labels (target 1, 2, etc.). Each of these targets can be assigned a semantic label of the target type. A confusion matrix (CM) determines the truth and test objects versus ATR decisions (e.g., detection, classification, and identification). The CM determines the summary metrics by assessing

Figure 9.9 Detections (○), false alarms (□), and missed detections (◇).

the likelihood (numerator) over the number of targets tested (denominator) to determine the probability of detection, declaration, false alarm, and identification. Furthermore, the assessment of different thresholds for decision making leads to detection, false alarm, and identification rate. A method used to determine the rate is the receiver operating characteristics (ROC) curves that moves the threshold to yield a summary of performance. Finally, the area under the curve (AUC) and F-score are comprehensive metrics from the ROC analysis.

In a confusion matrix, each row is a test object (or target type) from truth, while the column is a decision (test). When processing, the goal is to determine the target match (truth target type 1 to decision target type 1) as down the diagonal. A robust analysis of real-world testing contains the known targets (including others) and has the ability to reject targets below a given threshold, as well as appreciate that some unknown targets (not in database) can result. Figure 9.10 highlights that the detection acceptance and detection rejection are included in the results for classification [63].

The analysis proceeds by comparing the diagonal information in the confusion matrix over the classifier results for a probability likelihood of classification (P_{CL}):

$$P\left(c_i \middle| t_j\right) = \frac{\sum_{i=\{1,2,3,\ldots,0\}} M_{ii}}{\sum_{i=\{1,2,3,\ldots 0\}} \left(\sum_{j=\{1,2,3,\ldots,0\}} M_{ij}\right)} \tag{9.1}$$

		Decisions						
Test Objects		Detection Accept						Detection Reject
		Target					Non-Target	
		Class Accept				Class Reject		
		Type 1	**Type 2**	**Type 3**	**OTHER**			
	Type 1	**$M_{1,1}$**	**$M_{1,2}$**	**$M_{1,3}$**	**$M_{1,O}$**	**$M_{1,R\text{-}CL}$**	**$M_{1,NT}$**	$M_{1,R\text{-}Det}$
Target	**Type 2**	**$M_{2,1}$**	**$M_{2,2}$**	**$M_{2,3}$**	**$M_{2,O}$**	**$M_{2,R\text{-}CL}$**	**$M_{2,NT}$**	$M_{2,R\text{-}Det}$
	Type 3	**$M_{3,1}$**	**$M_{3,2}$**	**$M_{3,3}$**	**$M_{3,O}$**	**$M_{3,R\text{-}CL}$**	**$M_{3,NT}$**	$M_{3,R\text{-}Det}$
	Other	**$M_{O,1}$**	**$M_{O,2}$**	**$M_{O,3}$**	**$M_{O,O}$**	**$M_{O,R\text{-}CL}$**	**$M_{O,NT}$**	$M_{O,R\text{-}Det}$
Non-Target	**Type 4**	**$M_{4,1}$**	**$M_{4,2}$**	**$M_{4,3}$**	**$M_{4,O}$**	**$M_{4,R\text{-}CL}$**	**$M_{4,NT}$**	$M_{4,R\text{-}Det}$
	Type 5	**$M_{5,1}$**	**$M_{5,2}$**	**$M_{5,3}$**	**$M_{5,O}$**	**$M_{5,R\text{-}CL}$**	**$M_{5,NT}$**	$M_{5,R\text{-}Det}$

Numerator	
Denominator	

Figure 9.10 Detection likelihoods from the ATR.

An interesting situation arises from the confusion matrix analysis (Figure 9.11) as the designers of an ATR system determine the performance with respect to the truth (across a row), whereas operators are concerned with performance given a report (down the columns) [64]. The goal is to define measure credibility relevant to providing a user with improved information.

Figure 9.12 clarifies many of the primary challenges [65]. Assume that the ATR performance is reasonable with respect to the training and testing. However, the ATR must be pragmatic in acknowledging when the analysis rejects some results as well as providing information from which the decision is unknown. Additionally, there might be some targets that are not available but are known to exist, such as unknown newly designed foe targets not available in the training environment. Figure 9.12 captures both an unlikely case (i.e., target 4) as well as a potential target (i.e., target 5). Also, the reader should notice that both targets have a known classification rejection as well as an acknowledgment of an unknown case during ATR testing.

9.4.2 Object Assessment from Confusion Matrix

We use the SAR target identification approach with the likelihood vectors compiled into a confusion matrix (CM). Thus, each single look provides a full analysis of the classifier, presented as a CM, for all target comparisons. The likelihood vectors of the confusion matrix allow for a more thorough analysis with such performance criteria as declaration, P_D, and false alarm, P_{FA}, probabilities. The confusion matrix lists a set of likelihood values with the real targets as the rows $\{T_1, \ldots, T_N\}$, and the

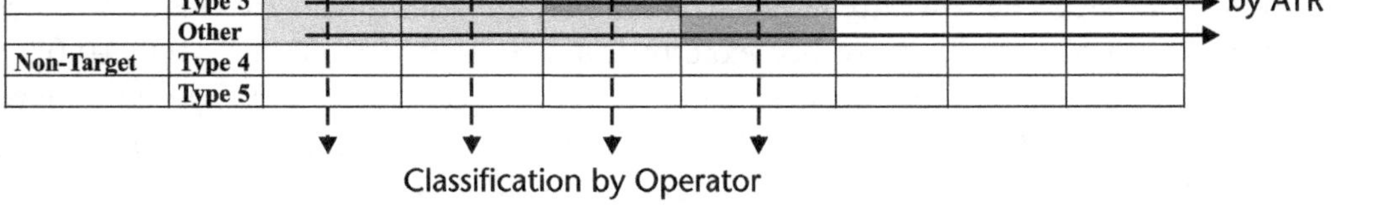

Figure 9.11 Analysis of confusion matrix.

			Target					(Test)
			Class Accept				Class Reject	Unknown
(Truth)			**Type 1**	**Type 2**	**Type 3**	**OTHER**		
	Type 1	Friend	**0.80**	0.03	0.01	0.01	0.02	0.13
Target	**Type 2**	Friend	0.01	**0.85**	0.01	0.03	0.01	0.09
	Type 3	Neutral	0.01	0.01	**0.82**	0.01	0.03	0.12
	Other	Neutral	0.03	0.01	0.01	**0.90**	0.03	0.02
Non-Target	**Type 4**	Foe	0.01	0.02	0.04	0.21	0.30	0.42
	Type 5	Foe	0.20	0.20	0.20	0.02	0.14	0.24

Figure 9.12 Analysis of confusion matrix.

testable hypothesis as the columns $\{T_1, \ldots, T_N, T_O, T_R, \text{unknown}\}$. For example, if the true target is T_1, the CM is

$$\text{CM} = \begin{array}{c|ccccccc|} & \text{T1} & \text{T2} & & \text{TN} & \text{TO} & \text{TR} & \text{Unk} \\ \hline \text{T1} & \text{A} & \text{B} & \ldots & \text{B} & \text{B} & \text{B} & \text{U} \\ \text{T2} & \text{E} & \text{D} & \ldots & \text{D} & \text{D} & \text{D} & \text{U} \\ & \vdots & \vdots & & \vdots & \vdots & \vdots & \vdots \\ \text{TN} & \text{E} & \text{D} & \text{D} & \text{D} & \text{D} & \text{D} & \text{U} \\ \hline \end{array} \tag{9.2}$$

Selecting the likelihood values in the CM, one can compare the performance metrics for different size CMs. From the CM and a defined target-to-confusion ratio given as m (as set by the operational ATR requirements), a set of metrics can be identified to support analysis, including:

$$P_{\text{Declaration}} = \frac{A}{A+B} \tag{9.3}$$

$$P_{\text{FalseAlarm}} = \frac{E}{E+D} \tag{9.4}$$

$$P_{\text{Correct Classification}} = \frac{m \cdot P_{\text{D}}}{\left(m \cdot P_{\text{D}}\right) + P_{\text{FA}}} \tag{9.5}$$

Using Figure 9.12 as an example, let A be the fact that target 1 is the choice (row) and that target 1 is the declared (column) by the ATR to produce a normalized likelihood of 0.80. The rows are normalized to one, but rounding errors lead to values that are close to and yet are not exactly equal to 1. B = 0.03+0.01+0.01+0.02 = 0.07. E = 0.01+0.01+0.01+0.03+0.01+0.20 = 0.26. U is the entire right column of 013+0.09+0.12+0.02+0.42+0.24=1.02. Finally, D is the remaining value D = 4.65. Using the results from Figure 9.12, then P_{Dec} = 0.8/ [0.8+0.07] = 0.92. P_{FA} = 0.26 / [0.26+4.65] = 0.053. If we let m = 1, then P_{CC} = 0.92/[0.92 + 0.05] = 0.95. These metrics are important in that the system is not forced to make a target declaration, and also provide an ability to discern whether there is enough evidence to declare a target in the library.

By choosing a different target truth (row) to selection (column) as defined by the diagonal, this process is repeated for each true target and the resulting metrics are summed and normalized for the number of targets. For example, the second true target is:

$$\mathrm{CM} = \begin{array}{c} \\ \mathrm{T1} \\ \mathrm{T2} \\ \\ \mathrm{TN} \end{array} \begin{array}{|ccccccc|} \mathrm{T1} & \mathrm{T2} & & \mathrm{TN} & \mathrm{TO} & \mathrm{TR} & \mathrm{Unk} \\ \hline \mathrm{D} & \mathrm{E} & \cdots & \mathrm{D} & \mathrm{D} & \mathrm{D} & \mathrm{U} \\ \mathrm{B} & \mathrm{A} & \cdots & \mathrm{B} & \mathrm{B} & \mathrm{B} & \mathrm{U} \\ \vdots & \vdots & & \vdots & \vdots & \vdots & \vdots \\ \mathrm{D} & \mathrm{E} & \mathrm{D} & \mathrm{D} & \mathrm{D} & \mathrm{D} & \mathrm{U} \\ \hline \end{array} \tag{9.6}$$

Thus, Target 2 is $P_{DEC} = 0.93$, $P_{FA} = 0.055$, and $P_{CC} = 0.94$. The analysis of the target types is dependent on the situation, which provides context for a threat assessment. A threat assessment is an element of information fusion (Figure 9.1).

9.4.3 Threat Assessment from Confusion Matrix

The row analysis (for the ATR machine) yields a Type 1 classification of 0.8/(0.8+0.03+0.01+0.01+0.02) = 92%; with an uncertainty of 15%. In contrast, column analysis (for the user) gives Type 1 as 0.8/(0.8+0.01+0.01+0.03+0.01+0.02) = 75%. From the analysis, it is important to note the traditional probability assessment is 80%, while the ATR provides overconfidence with an average of 92%. The user results are conservative around 75%, as is depicted in Figure 9.13. Hence, caution is warranted when the training data do not cover data sampling over the entire space.

In information fusion, there is situation and threat analysis [66]. Using the above example, ATR includes both classification and identification. Typically, IFFN sensors deal with the potential threat analysis. Since the CM analysis assumes each true target is equally likely, then Bayesian probabilities can be used to assess the likelihood from a priori estimation. For example, the known target prior probabilities can be different. For example, let P_F be the prior probability of friendly targets (1, 2), P_N be the prior probability of neutral targets (3, 4), and P_H is the prior probability of hostile (foe) targets (5, 6). Using the total law of probability, the column analysis can utilize the a priori probabilities:

$$\mathrm{P}_{ID}\left(c_j \mid t_i\right) = \frac{P\left(type_i \mid classification_j\right) P\left(type_i\right)}{\sum_{i=\{1,2,3,\ldots,n\}} P\left(type_i \mid classification_j\right) P\left(type_i\right)} \tag{9.7}$$

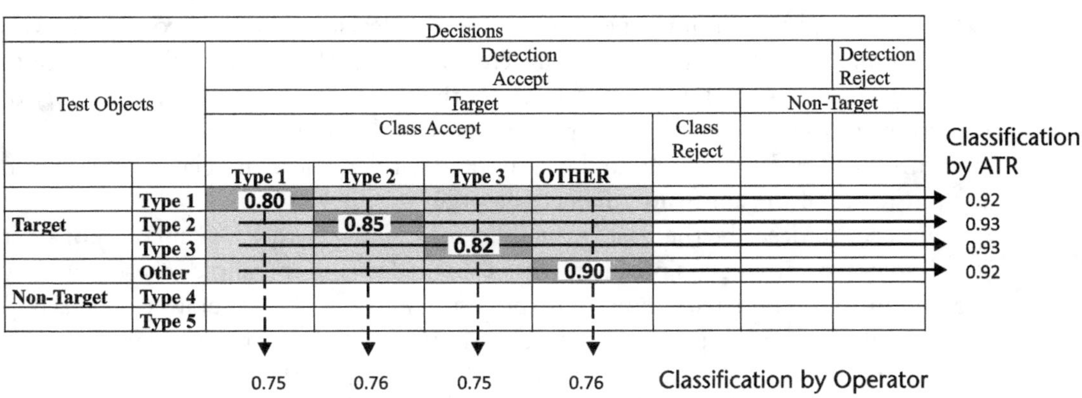

Figure 9.13 Classification by ATR and classification results of interest to the operator.

With a threat environment of $P_H = 4P_F$, the priors for each true target types are (0.1, 0.1, 0.00, 0.00, 0.40 and 0.40). Then, for target type 1:

$$\mathbf{P}_{ID}\left(c_j \middle| t_i\right) = \frac{(0.8)(0.1)}{(0.8+0.01)(0.1)+(0.01+0.03)(0.0)+(0.01+0.20)(0.4)} = \frac{0.08}{0.081+0.084} = 48\%$$

For a nonthreat environment $P_F = 4P_H$, (0.4, 0.4, 0.0, 0.0, 0.1 and 0.1), so that:

$$P_{ID}\left(c_j \middle| t_i\right) = \frac{(0.8)(0.4)}{(0.8+0.01)(0.1)+(0.01+0.03)(0.0)+(0.01+0.20)(0.4)} = \frac{0.32}{0.09+0.0602} = 93\%$$

Hence, operators acting on an ATR decision system should realize that, without using a prior knowledge, the results can be significantly inaccurate (e.g., 48% versus 93%). Hence, the ATR analysis from the machine presents both 93% for classification and 93% for known status quo engagement. However, with the unknown, the user should be aware that the ATR results are 75% credible, and then as a threat (e.g., hostile) situation unfolds, the ATR results would be only 50% credible.

The combination of metrics determines the situation and the thresholds that are used to make decisions. A comprehensive analysis with different thresholds is called a receiver operating characteristic (ROC) curve.

9.5 Receiver Operating Characteristic Curve

9.5.1 Receiver Operating Characteristic Curve from Confusion Matrix

The most common evaluation tool used in ATR is the ROC curve that compares the results to truth. A ROC curve is based in statistical decision theory and was originally developed as a tool for electronic signal detection. The ROC analysis has been extensively applied to human perception and the decision making process. A ROC curve summarizes the performance of detecting a signal (target) in the presence of noise (false alarm objects). The curve relates the detection or *hit* rate (probability of detection) to the false alarm rate (probability of false alarm) while varying a decision threshold. A ROC curve can be computed for detection and classification, and each instance of the confusion matrix is a point on the ROC.

In order to describe the basic principles behind standard ROC curves, consider the simple two-target class problem with a single decision threshold, represented by the two probability density functions (pdfs) in Figure 9.14. The clutter pdf $p(x|C_1)$ is the conditional pdf representing the distribution of clutter objects (class 1), while the target pdf $p(x|C_2)$ is the conditional pdf representing the distribution of target objects (class 2). Both distributions are in a given feature space, where x is the

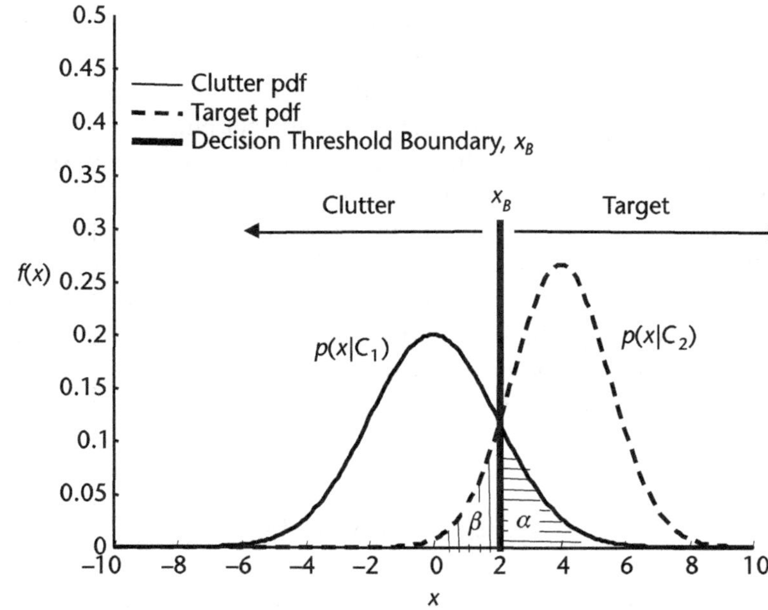

Figure 9.14 Clutter and target pdfs for a two-class problem. Clutter: N(0,2.0); target: N(4,1.5).

feature measurement of the test object. A given decision threshold boundary x_B partitions the feature axis into two regions, clutter and target, resulting in two types of errors:

1. Type I error (α): Misclassifying an actual clutter object as a target (false positive (FP), or false alarm (FA));
2. Type II error (β): Misclassifying an actual target object as clutter (false negative (FN)).

The conditional probabilities, P_{FP} and P_{FN}, of these two types of errors are shown as horizontal (α) and vertical (β) hatched areas under the two distribution curves in Figure 9.14. Associated with these two probability errors are their complementary probabilities of correct classification.

Note that a ROC analysis is used in many other classification tasks for human perception, medical diagnosis, and image processing. Other terms are related such as specificity, confidence, significance, sensitivity, and power.

Using the CM (Figure 9.8), the detection comes from the correct results, while the false alarms are the errors from the columns (T). The notation is P(declare|truth) and hence P(T|c) is declare T when truth is clutter.

False alarm = P(T|c) = 0.26+0.27+0.27+0.73 / (4.98) = 0.13

Detection = P(T|t) = 0.8+0.85+0.82+0.90+0.44 / (4.98) = 0.77

For a typical ROC curve, the decision threshold is varied from a conservative value (i.e., high decision threshold required to declare an image as a target) to an aggressive value (i.e., low decision threshold required to declare an image as a target).

Because of interrelationships among the probabilities, it is only necessary to indicate a single (specificity, sensitivity) pair to describe the performance of an ATR system for a specific decision threshold. These (specificity, sensitivity) probability pairs vary as the decision boundary shifts along the feature axis. For example, if the decision boundary is shifted to the left in Figure 9.14, then β decreases while α increases. A functional relationship exists between these two probability errors or one of the equivalent (specificity, sensitivity) pairs. In a ROC analysis, the decision threshold is purposefully varied to show this relationship in the form of a ROC curve. Figure 9.15 illustrates the generation of a ROC curve for the case of one-dimensional normal distributions for both the clutter and target distributions.

For a given finite data set, a standard ROC curve $f(x)$ is mathematically a function of the estimated performance measures of interest: the probability of false alarm $\hat{P}_{FA}$ and the probability of detection $\hat{P}_D$. These two performance measures are in turn functions of the varying decision threshold boundary x_B as $\hat{P}_{FA} = \hat{P}_{FA}(x_B)$ and $\hat{P}_D = \hat{P}_D(x_B)$. A standard ROC curve $f(x)$ is then empirically generated by varying x_B over its entire range X

$$f = f(x_B) = \left\{ \left(\hat{P}_{FA}(x_B), \hat{P}_d(x_B) \right) \middle| x_B \in X \right\} \tag{9.8}$$

An interesting result is noted in Figure 9.15. If a classifier is listed with a detection of $P_D = 80\%$, what does this mean? From Figure 9.15, the P_D of 80% is at P_{FA} of 10%. While the result is not especially poor, the operator would prefer the lowest number of false alarms. Hence, at a 5% P_{FA}, then the P_D is actually somewhere near 65%. Thus, the ATR designer would want the operator to believe it is PD = 80%, but it is actually at 65% (which is the similar story as presented in Figure 9.13).

A 3D ROC trajectory [67] captures additional information. Using another function, g, is then empirically generated by varying z over its entire range Z of situations:

$$g = g(z) = \left\{ \left(\hat{P}_{FA}(z), \hat{P}_D(z), \hat{P}_3(z) \right) \middle| z \in Z \right\} \tag{9.9}$$

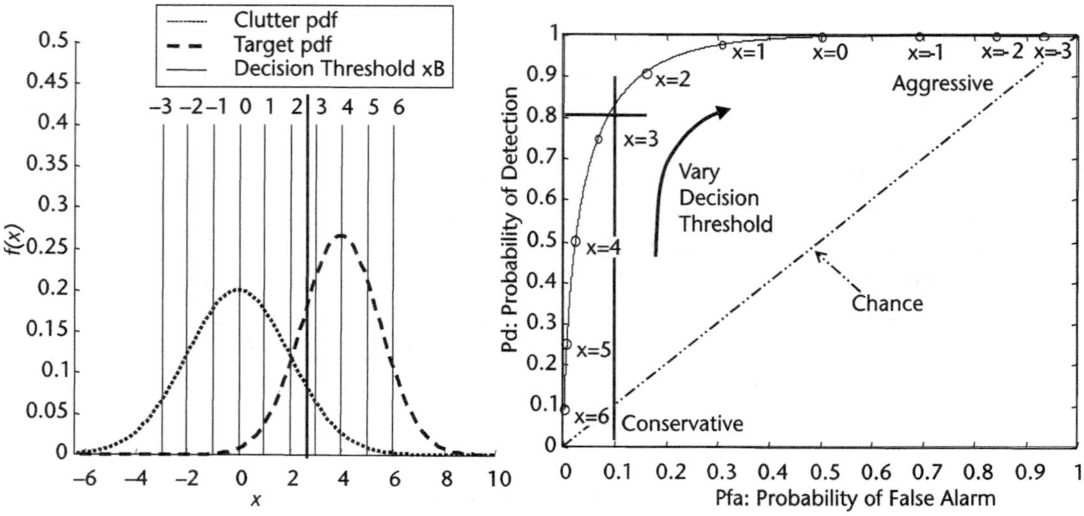

Figure 9.15 Generation of ROC Curve (right) using two 1-D normal distributions (left). Clutter: N(2,0); target: N(4,1.5).

One example is the probability of rejection $\hat{P}_r$ as a third performance measure (see Figure 9.10). The probability of rejection is used in the pattern recognition literature when employing the doubt/rejection option. Exemplars are rejected from consideration when their classification falls in a grey or uncertain region. Our motivation for using this doubt option is to improve ATR performance by rejecting images that are difficult to classify and perhaps that are unknown. Classification can be delayed until further measurements can be made or these difficult exemplars are passed onto a second classifier specifically designed to deal with the gray area of classification. The 3D ROC trajectory can be a useful performance tool for understanding the trade-offs between the probability of rejection and the two performance measures commonly used in detection problems: the probability of detection and the probability of false alarm.

After setting the three performance measures, z is the size or width of the rejection zone because of the definition of the probability of rejection. To illustrate the use of this varying rejection zone, consider a generic two-class detection problem wherein class 1 is the nontarget or confuser class and class 2 is the target class. Let the outputs o_1 and o_2 for each exemplar for the competing classifiers be the estimated posterior probabilities for each class, normalized such that:

$$\sigma_1 + \sigma_2 = 1 \tag{9.10}$$

Since the normalized estimated posterior probabilities sum to one, class decision can be based on just o_2:

$$class = \begin{cases} class\ 2 \text{ if } o_2 > t_u \\ class\ 1 \text{ if } o_2 < t_l \\ reject \text{ if } t_l \le o_2 \le t_u \end{cases} \tag{9.11}$$

where t_u and t_l are the upper and lower thresholds that are functions of the center threshold t_c and the rejection zone width z:

$$\begin{aligned} t_u &= t_c + \frac{1}{2}z \\ t_l &= t_c - \frac{1}{2}z \end{aligned} \tag{9.12}$$

To generate the 3D ROC trajectory, set $t_c = 0.5$ and vary z from 0 (no rejections) to 1 (all exemplars rejected). Hence, probability of rejection ($\hat{P}_r$) = the probability that the image is rejected as unknown or is too difficult to classify (Figure 9.16 sets $P_R < 0.4$), the 3D ROC trajectories indicate that an ATR (from a traditional ML) has fewer false alarms and greater detection rates.

As indicated, there are two comprehensive metrics for analyzing a ROC, which include the area under the curve (AUC) and the F-metric. The F-measure is deter-

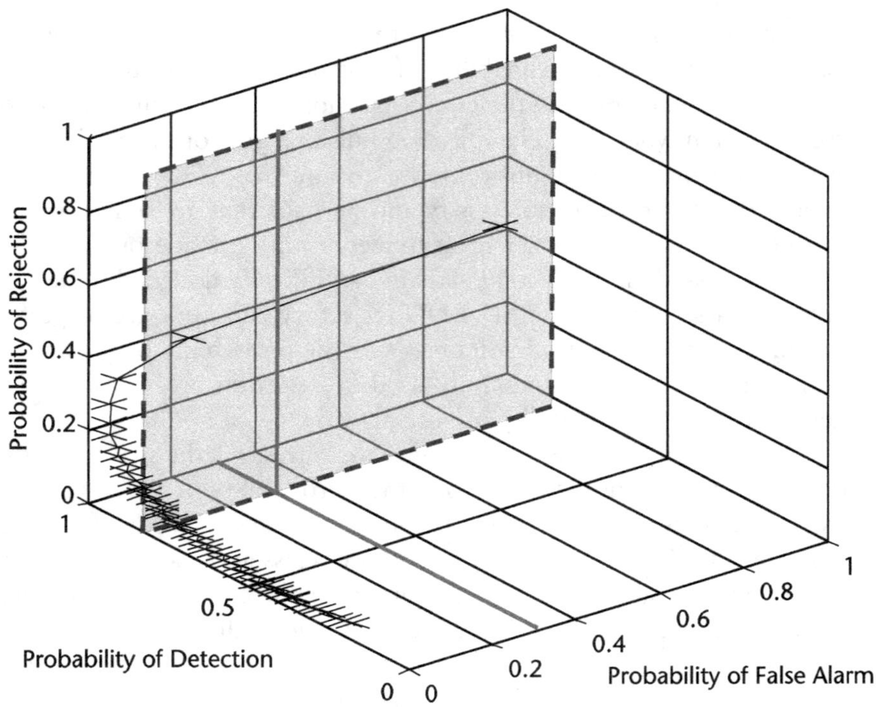

Figure 9.16 3D ROC.

mined from a precision-recall (curve) using the confusion matrix and the number of tests.

9.5.2 Precision-Recall from Confusion Matrix

To compare to truth, related metrics presented in Table 9.5 provide corresponding information in terms of true positive (TP), true negative (TN), false positive (FP), and false negative (FN). The metrics for precision and recall are determined from the CM as a comprehensive F-metric.

From the example in Figure 9.12, assume there are 100 tests for each of 6 targets (n = 600) and that the unknown situations are potential elements of FNs or TNs, as a lack of knowledge of the classification is representative of the potential uncertainty. Hence, the unknown is split between FN/TN. In Figure 9.17, the analysis of the confusion matrix is denoted with the TP, FP, TN, and FN.

	ATR Classification	
	Yes	*No*
Truth(Yes)	TP = 337	FN = 51
Truth(No)	FP = 108	TN = 104

Hence, the metrics are:

$$\text{Precision}(P) = \frac{\text{TP}}{\text{TP} + \text{FP}} = \frac{337}{337 + 108} = 76\%$$

Table 9.5 Measures Associated with a ROC

		Classified As (Reported)	
		Clutter (C) *Normal (N)* *Null hypothesis H_0*	*Target (T)* *Abnormal (A)* *Alt hypothesis H_1*
Actual (truth)	*Clutter (c)* *Normal (n)* *Null hypothesis H_0*	True negative (P_{TN}) Correct rejection (P_{CR}) Specificity Confidence (1-α) P(C|c) P(N|n)	False positive (P_{FP}) **False alarm** (P_{FA}) 1- Specificity Level of significance (α) P(T|c) P(A|n)
	Target (t) *Abnormal (a)* *Alt hypothesis H_1*	False negative (P_{FN}) Missed detection (P_M) 1 - Sensitivity 1 - power (β) P(C|t) P(N|a)	True positive (P_{TP}) Detection (P_D) Sensitivity Power (1 - β) P(T|t) P(A|a)

Table 9.6 Measures Associated with an F-Metric

		Classified As (Reported)	
		Clutter (C) *Normal (N)* *Null hypothesis H_0*	*Target (T)* *Abnormal (A)* *Alt hypothesis H_1*
Actual (truth)	*Clutter (c)* *Normal (n)* *Null hypothesis H_0*	True negative (P_{TN}) Correct rejection (P_{CR}) P(C|c)	False positive (P_{FP}) False alarm (P_{FA}) P(T|c)
	Target (t) *Abnormal (a)* *Alt hypothesis H_1*	False negative (P_{FN}) Missed detection (P_M) P(C|t)	True positive (P_{TP}) Detection (P_D) P(T|t)

$$\text{Recall}(R) = \frac{\text{TP}}{\text{TP}+\text{FN}} = \frac{337}{337+51} = 87\%$$

$$\text{Error}(E) = \frac{\text{FP}+\text{FN}}{\text{n}} = \frac{108+51}{600} = 27\%$$

$$\text{Accuracy}(A) = \frac{\text{TP}+\text{TN}}{\text{n}} = \frac{337+104}{600} = 74\%$$

Methods for relevance include the F-measure or balanced F-score (F_1 score), which is the harmonic mean of precision and recall, where an F_1 score reaches its best value at 1 (perfect precision and recall) and worst at 0. *Precision* is a measure of how many selected items are relevant, while *recall* is how many relevant items are selected:

			Target					(Test)
			Class Accept				Class Reject	Unknown
(Truth)			**Type 1**	**Type 2**	**Type 3**	**OTHER**		
	Type1	Friend	**0.80 = TP**	0.03 = FP	0.01 = FP	0.01 = FP	0.02 = TN	0.13 = FN/TN
Target	**Type2**	Friend	0.01 = FP	**0.85 = TP**	0.01 = FP	0.03 = FP	0.01 = TN	0.09 = FN/TN
	Type3	Neutral	0.01 = FP	0.01 = FP	**0.82 = TP**	0.01= FP	0.03 = TN	0.12 = FN/TN
	Other	Neutral	0.03 = FP	0.01 = FP	0.01 = FP	**0.90 = TP**	0.03 = TN	0.02 = FN/TN
NonTarget	**Type4**	Foe	0.01 = FP	0.02 = FP	0.04 = FP	0.21 = FP	0.30 = TN	0.42 = FN/TN
	Type5	Foe	0.20 = FP	0.20 = FP	0.20 = FP	0.02 = FP	0.14 = TN	0.24 = FN/TN

Figure 9.17 CM analysis for precision and recall.

$$F_1 = 2\frac{\text{Precision} * \text{Recall}}{\text{Precision} + \text{Recall}} \tag{9.13}$$

The general formula for positive real β modifies the weights of precision and recall:

$$F_\beta = \left(1+\beta^2\right)\frac{\text{Precision} * \text{Recall}}{\left(\beta^2 * \text{Precision}\right) + \text{Recall}} \tag{9.14}$$

where $F_{0.5}$ weighs recall lower than precision (by attenuating the influence of false negatives) and F_2 weighs recall higher than precision (by placing more emphasis on false negatives). Hence, $F_{0.5} = 0.78$, $F_1 = 0.81$, and $F_2 = 0.85$.

An additional method of increasing accuracy and robustness is obtaining multiple looks (e.g., circular SAR). With the additional information, a confusion matrix fusion approach can lead to better results.

9.5.3 Confusion Matrix Fusion

The decisions from an ATR are often stored in a CM, which is an estimate of likelihoods. For the single-look ATR performance, these estimates are treated as priors. Decisions from multiple ATRs or from multiple looks from different geometric perspectives are fused using the decision level fusion (DLF) technique. Researchers are encouraged to read our paper [68] for detailed information on DLF. With respect to the DLF, the CMs represent the historical performances of the ATR system.

For DLF the data sets from the MSTAR 15° depression and the 17° depression were used. In this case, Run 1 trained on 15 and tested on 17°; while Run 2 trained on the 17° and tested on the 15° depression data sets. Using the PCNN confusion matrix results from each result, DLF using the confusion matrix fusion (CMF) resulted in higher accuracy, as shown in the confusion matrix of Figure 4.15. As an average comparison, the PCC for Run 1 was 74% and Run 2 was 79%, while the CM fusion resulted in 87%. Additional results were conducted over the years in numerous SAR and HRR data sets, showing improved performance using CMF.

9.6 Metric Presentation

From the various SAR ATR methods presented, the standard theoretical methods perform analytical analysis of the targets through an experiment using a look-up table (LUT) (e.g., collected MSTAR imagery) [69]. However, it is important for the reader to notice that the difference between the machine test and the operator interest (see Figure 9.11). Figure 9.18 lists the various types of theoretical and operational tests. For the theoretical case, the standard experiment is applied over collected imagery and supports a score-based ATR performance analysis. As with enhancements of the standard operating conditions, real analysis affords considerations for differences. Likewise, if extended operating condition tests are needed, then synthetic imagery can be generated, and coordination with an operator determines the final performance of the system.

The other scenario is the determination of whether the human is outside, on, or in the loop. Typically, for cases in which the operator is not part of the experiment and the data is static, it is a constructive test. A virtual test also does not include the operator, but the incoming data is presented in real time. Finally, when the streaming data analysis exists, then it is a live experiment. For a live experiment, the operator requires some data quality in the analysis as defined by the National Imagery Interpretability Rating Scale (NIIRS) as well as a meaningful presentation of the ATR results.

9.6.1 National Imagery Interpretability Rating Scale

An image quality model was developed using the NIIRS to assess the quality of SAR data. Given a SAR image with a NIIRS rating (see Table 9.7), then the discernibility

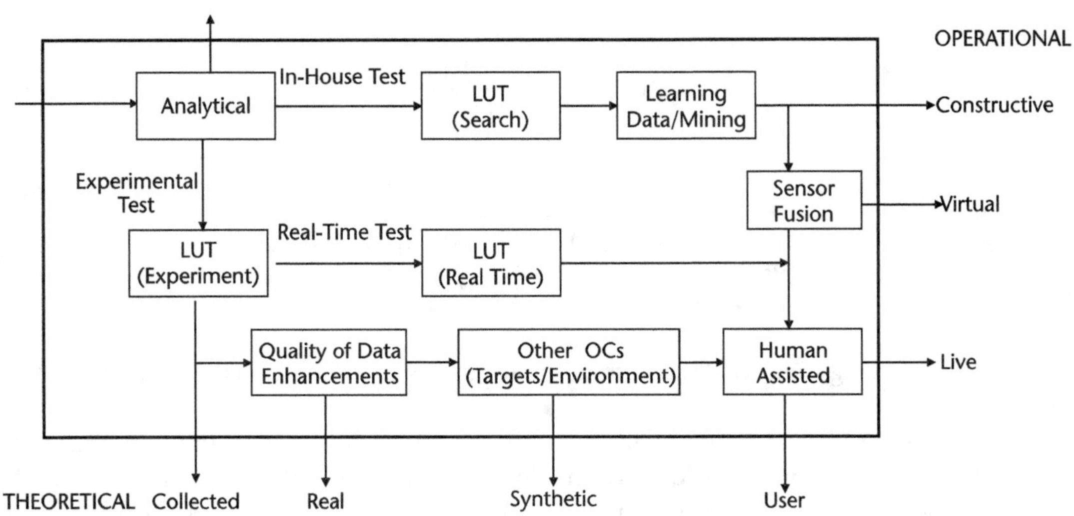

Figure 9.18 Evaluation of SAR system for live user analysis.

Table 9.7 NIIRS Dentitions and Examples

NIIRS	*1*	*2*	*3*	*4*	*5*	*6*	*7*	*8*	*9*
Resolution	> 9m	4.5–9.0m	2.5–4.5m	1.2–2.5m	0.75-1.2m	0.4–0.75m	0.2–0.4m	0.1–0.2 m	< 0.1m
Detect	Road networks	Defense area	Buildings	Convoy	Semi truck	Wheeled vs. tracked tanks	Medium tank versus car	Turrets	Guns

of certain features is possible. For example, in many of the MSTAR images, the ground sampling distance (GSD) resolution is 1 ft (0.3 m), which corresponds to a NIIRS of 7.

The SAR NIIRS and resolution (GSD) [70] is then determined by sensor analogy to a general image quality equation (GIQE-4):

$$\begin{aligned}\text{NIIRS}_{\text{IR}} &= 10.751 - A * \log_{10}(GSD) + B * \log 10(RER) \\ &-0.656H - [0.344(G/SNR)]\end{aligned} \tag{9.15}$$

where *GSD* is the geometric mean of the ground sample distance, *H* is the geometric mean height due to edge overshoot, *RER* is the geometric mean of the normalized relative edge response, *G* is the noise gain, *SNR* is the signal to noise ratio, *A* is constant (3.32 if $RER = 0.9$, 3.16 if $RER<0.9$), and *B* is constant (1.559 if $RER = 0.9$, 2.817 if $RER < 0.9$). Typically, the SAR GSD is related to the NIIRS_{IR} via:

$$GSD = 10^{[(10.751 - \text{NIIRS}_{IR})/A]} \tag{9.16}$$

Through experiment, an improved empirically fit result was determined to be:

$$\text{NIIRS}_{\text{IR}} = 1.14 + 0.18 * \text{NIIRS}_{\text{SAR}} + 0.08 * \text{NIIRS}_{\text{SAR}}^2 \tag{9.17}$$

Using $y = ax^2 + bx + c$, then the solution is:

$$x = \frac{-\text{b} \pm \sqrt{\text{b}^2 - 4\,\text{a}\,\text{c}}}{(2a)} \tag{9.18}$$

$$NIIRS_{SAR} = \frac{-0.18 \pm \sqrt{(0.18)^2 - 4(0.08)(1.14NIIRS_{IR})}}{(2 * 0.08)} \tag{9.19}$$

Knowing that NIIRS > 1, then the following are determined as found in Table 9.8.

The predicted image quality [71] is shown in Figure 9.19 (and compared to other sensors for image quality) in that, if a SAR probability of correct classification or correct identification (P_{CC}) of 80% is desired, then a NIIRS of 7 also

Table 9.8 NIIRS Values for IR (I) and SAR (S)

NIIRS-I	*NIIRS-S*	*GSD (m)*	*2*GSD(m)*
2	2.3	5.88	11.76
3	3.8	2.84	5.67
4	5.0	1.37	2.74
5	5.9	0.66	1.32
6	6.8	0.32	0.64
7	7.5	0.15	0.31
8	8.2	0.07	0.15
9	8.9	0.04	0.07

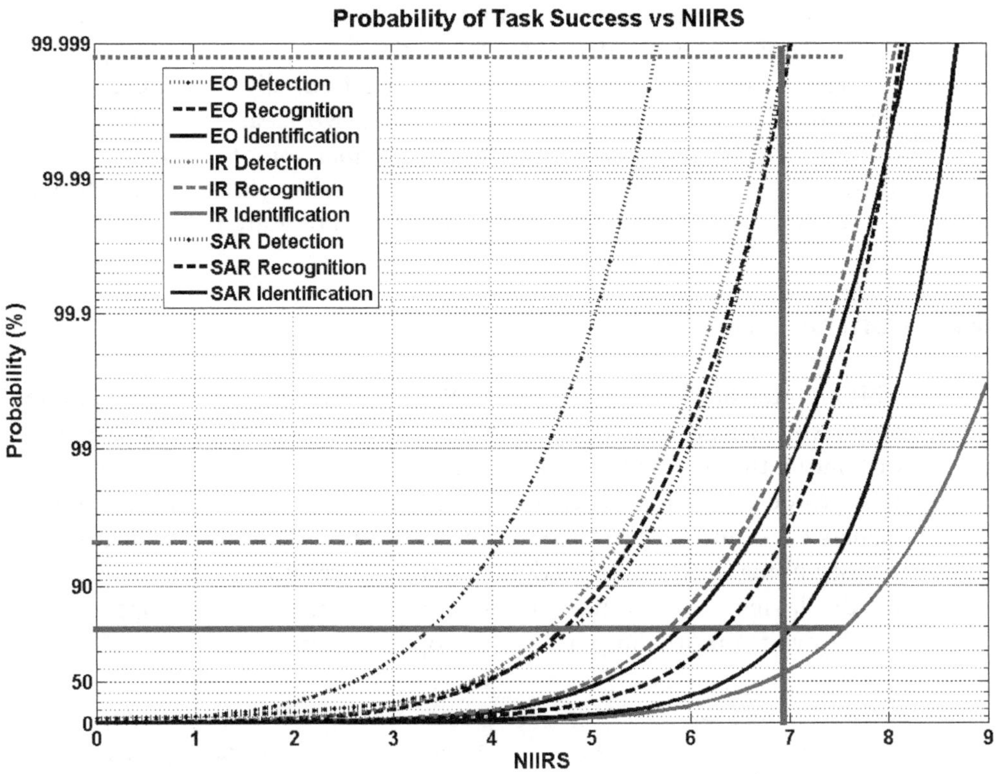

Figure 9.19 SAR NIIRS analysis.

indicates that $P_D = 99.9$ and $P_R = 93\%$. Hence, the results in the book demonstrate the improvements from ML to achieve classification rates previously seen as detection rates.

It is noted that caution exists for real-world situations in which streaming SAR imagery is compressed for easy of transport, since these analysis results could degrade from loss of NIIRS through compression [72]. Thus, an appropriate compression rate should be considered as being dependent on the needs of the ATR system as well as that of the user.

9.6.2 Display of Results

The final step in the processing is presenting results to a user. The previous results are presented to an ATR developer; however, pragmatic methods are sought to display to the operator. One idea is to present the complete results in a perspective, such as from the aspect angle. Using the MSTAR data, a cockpit design could include just the metrics with the data and the target type. The display is not a complete picture, and as with ATR methods, might only present one results from a single look. Other results include the ROC approach by giving the operator the top choices with a measure of confidence.

As a potential upgrade, the entire performance is presented to an operator where multiple looks can be provided. From this example, complete confidence/credibility (P_{CC} = 100%) is the outside circle and the chance (P_{CC} = 50%) drawn to show the relative performance [73]. In Figure 9.20, the declared target has a relative performance at any angle as the target choice is around 80% and exceeded the second choice from the data set.

The presentation of ATR results was traditionally only a decision score (e.g., 80% as show in Figure 9.21). For example, two ideas can be a radar plot of the comparative results along with a feature comparison of the SAR imagery with the known classification. As operators can continually refine the user display, the best practices determine the style of the ATR results that are preferred.

9.7 Conclusions

This chapter has provided some ATR performance evaluation examples with an emphasis on ATR as part of the following: (a) an information fusion system (coordinating the ATR machine and the user), (b) the fundamental analysis from the ROC to derive ATR metrics, (c) confusion matrix fusion, and (d) initial ideas on display of results. The data coordination is part of a systematic hierarchy of analysis, beginning with the collection of data and resulting in the choice on strategies for displaying the information. Since the ATR machine and operator are part of a cohesive team, it is important to adopt a continual approach of train, test, and

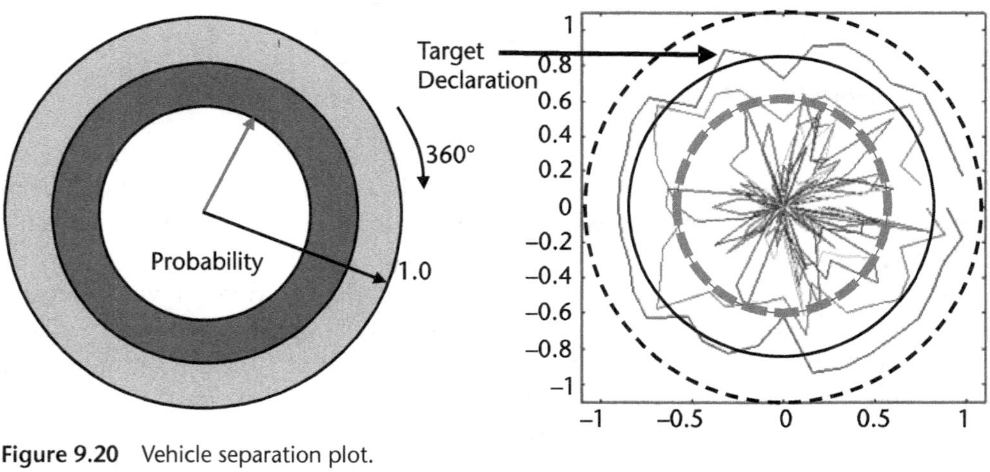

Figure 9.20 Vehicle separation plot.

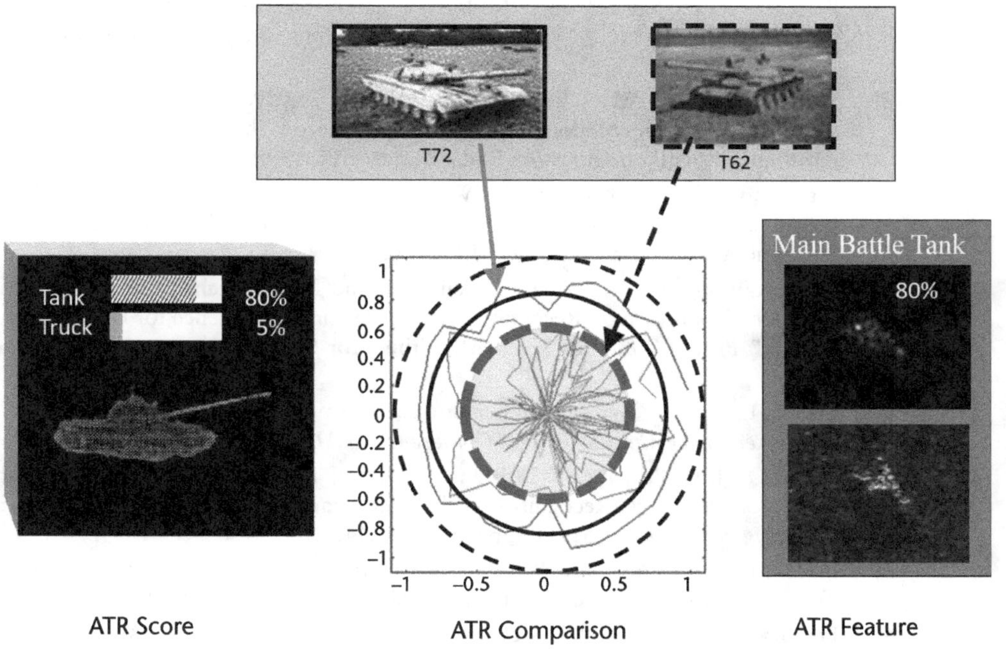

Figure 9.21 ATR presentation for operators.

deploy technology that the consumer desires. Inherent in the analysis is the fact that the data collection and generation can be instrumental in the robustness of the ATR decision. Thus, measures and metrics of analysis, user interaction protocols, and managed expectations from test and evaluation are expected to render the ATR systems as de facto methods for autonomous SAR exploitation. As deep learning (Chapter 4) replaces machine learning (Chapter 3) approaches, the T&E will ensure that the systems are accurate, robust, and usable.

References

[1] Ross, T. D., and L. C. Goodwon, "Improved Automatic Target Recognition (ATR) Value through Enhancements and Accommodations," *Proc. SPIE,* Vol. 6237, Algorithms for Synthetic Aperture Radar Imagery XIII, 2006.

[2] Blasch, E., M. Pribilski, B. Daughtery, B. Roscoe, and J. Gunsett, "Fusion Metrics for Dynamic Situation Analysis," *Proc. of SPIE*, Vol. 5429, 2004, pp. 428–438.

[3] Blasch, E.,"Situation, Impact, and User Refinement," *Proc. of SPIE,* Vol. 5096, 2003.

[4] Blasch, E., D. A. Lambert, P. Valin, et al., "High Level Information Fusion (HLIF) Survey of Models, Issues, and Grand Challenges," *IEEE Aerospace and Electronic Systems Magazine,* Vol. 27, No. 9, Sept. 2012.

[5] Chen, G., D. Shen, C. Kwan, J. Cruz, M. Kruger, and E. Blasch, "Game Theoretic Approach to Threat Prediction and Situation Awareness," *J. of Advances in Information Fusion,* Vol. 2, No. 1, 2007, pp. 1–14.

[6] Kaufman, V. I., T. D. Ross, E. M. Lavely, and E. P. Blasch, "Score-Based SAR ATR Performance Model with Operating Condition Dependencies," *Proc. SPIE,* Vol. 6568, Algorithms for Synthetic Aperture Radar Imagery XIV, 2007.

[7] Blasch, E. P., and S. H. Huang, "Multilevel Feature-Based Fuzzy Fusion for Target Recognition," *Proc. SPIE,* Vol. 4051, Sensor Fusion: Architectures, Algorithms, and Applications IV, 2000.

[8] Weisenseel, R. A., W. C. Karl, D. A. Castanon, G. J. Power, and P. Douville, "Markov Random Field Segmentation Methods for SAR Target Chips," *Proc. SPIE,* Vol. 3721, Algorithms for Synthetic Aperture Radar Imagery VI, 1999.

[9] Ettinger, G. J., G. A. Klanderman, W. M. Wells III, and W. E. L. Grimson, "Probabilistic Optimization Approach to SAR feature matching," *Proc. SPIE,* Vol. 2757, Algorithms for Synthetic Aperture Radar Imagery III, 1996.

[10] Westerkamp, J. J., T. Fister, R. L. Williams, and R. A. Mitchell, "Robust Feature-Based Bayesian Ground Target Recognition Using Decision Confidence for Unknown Target Rejection," *Proc. SPIE,* Vol. 3721, Algorithms for Synthetic Aperture Radar Imagery VI, 1999.

[11] Blasch, E., and M. Byrant, "Information Assessment of SAR Data for ATR," *IEEE National Aerospace and Electronics Conference (NAECON)*, 1998.

[12] Zhao, Q., J. C. Principe, V. L. Brennan, D. Xu, and Z. Wang, "Synthetic Aperture Radar Automatic Target Recognition with Three Strategies of Learning and Representation," *Opt. Eng.,* Vol. 39 No. 5, 2000, pp. 1230–1244.

[13] Blasch, E., *Derivation of a Belief Filter for Simultaneous High Range Resolution Radar Tracking and Identification,* Ph.D. thesis, Wright State University, 1999.

[14] El-Darymli, K., E. W. Gill, P. McGuire, D. Power, and C. Moloney, "Automatic Target Recognition in Synthetic Aperture Radar Imagery: A State-of-the-Art Review," *IEEE Access*, Vol. 4, 2016, pp. 6014–6058.

[15] "Intelligence Community Directive 203," 02 Jan 2015; https://www.dni.gov/files/documents/ICD

[16] Blasch, E., J. Sung, T. Nguyen, C. P. Daniel, A. P. Mason, "Artificial Intelligence Strategies for National Security and Safety Standards," *AAAI Fall Symposium Series*, Nov. 2019.

[17] Blasch, E., E. Bosse, and D. A. Lambert, *High-Level Information Fusion Management and Systems Design*, Norwood, MA: Artech House, 2012.

[18] Snidaro, L., J. Garcia, J. Llinas, et al. (eds.), *Context-Enhanced Information Fusion: Boosting Real-World Performance with Domain Knowledge*, Springer, 2016.

[19] Guerci, J. R., R. M. Guerci, M. Rangaswamy, J. S. Bergin, and M. C. Wicks, "CoFAR: Cognitive Fully Adaptive Radar," *IEEE Radar Conference*, 2014.

[20] Bergin, J. S., J. R. Guerci. R. M. Guerci, and M. Rangaswamy, "MIMO Clutter Discrete Probing for Cognitive Radar," *IEEE Radar Conference*, 2015.

[21] Guerci, J. R., J. S. Bergin, R. J. Guerci, M. Khanin, and M. Rangaswamy, "A New MIMO Clutter Model for Cognitive Radar," *IEEE Radar Conference*, 2016.

[22] Bergin, J., D. Kirk, J. Studer, J. Guerci, and M. Rangaswamy, "A New Approach for Testing Autonomous and Fully Adaptive Radars." *IEEE Radar Conference*, 2017.

[23] Bergin, J., J. Guerci, D. Kirk, and M. Rangaswamy, "Site-Specific Performance Gain of Optimal MIMO Radar in Heterogeneous Clutter," *IEEE Radar Conference*, 2018.

[24] *RFView(TM)*, http://rfview.islinc.com.

[25] Guerci, J. R., J. S. Bergin, S. Gogineni, and M. Rangaswamy, "Non-Orthogonal Radar Probing for MIMO Channel Estimation," *IEEE Radar Conference*, 2019.

[26] Xu, X., R. Zheng, et al., "Performance Analysis of Order Statistic Constant False Alarm Rate (CFAR) Detectors in Generalized Rayleigh," *Proc SPIE*, Vol. 6699, 2007.

[27] Yang, C., T. Nguyen, et al., "Mobile Positioning via Fusion Mixed Signals of Opportunity," *IEEE Aerospace and Electronic Systems Magazine,* Vol. 29, No. 4, 2014, pp. 34–46.

[28] de Villiers, J. P., W. D. van Eeden, et al., "A Comparative Cepstral-Based Analysis of Simulated and Measured S-band and X-band Radar Doppler Spectra of Human Motion," *IEEE Radar Conf.*, 2015.

[29] Van Eeden, W. D., J. P. de Villiers, et al., "Micro-Doppler Radar Classification of Humans and Animals in an Operational Environment," *Expert Systems with Applications*, No. 102, 2018, pp. 1–11.

[30] Wang, G., G. Chen, D. Shen, Z. Wang, K. Pham, and E. Blasch, "Performance Evaluation of Avionics Communication Systems with Radio Frequency Interference," *IEEE/AIAA Digital Avionics Systems Conference*, 2014.

[31] Wang, G., Z. Shu, G. Chen, et al., "Performance Evaluation of SATCOM Link in the Presence of Radio Frequency Interference," *IEEE Aerospace Conference*, 2016.

[32] Yang, C., et al., "Optimality Self Online Monitoring (OSOM) for Performance Evaluation and Adaptive Sensor Fusion," *International Conference on Information Fusion*, 2008.

[33] Kadar, I., E. Blasch, C. Yang, et al., "Panel Discussion: Issues and Challenges in Performance Assessment of Multitarget Tracking Algorithms with Applications to Real-World Problems," *Proc. SPIE*, Vol. 6968, 2008.

[34] Blasch, E., C. Yang, I. Kadar., et al., "Net-Centric Layered Sensing Issues in Distributed Target Tracking and Identification Performance Evaluation," *International Conference on Information Fusion,* 2008.

[35] Chen, H., and G. Chen, et al., "Information Theoretic Measures for Performance Evaluation and Comparison," *International Conference on Information Fusion*, 2009.

[36] Blasch, E., and P. Valin, "Track Purity and Current Assignment Ratio for Target Tracking and Identification Evaluation," *International Conference. on Information Fusion*, 2011.

[37] Straka, O., J. Dunık, M. Šimandl, et al., "Randomized Unscented Transform in State Estimation of non-Gaussian Systems: Algorithms and Performance," *International Conference on Information Fusion*, 2012.

[38] Blasch, E., C. Yang, and I. Kadar, "Summary of Tracking and Identification Methods," *Proc. SPIE*, Vol. 9119, 2014.

[39] Dunik, J., O. Straka, et al., "Random-Point-Based Filters: Analysis and Comparison in Target Tracking," *IEEE Transactions on Aerospace and Electronic Systems,* Vol. 51, No. 2, 2015, pp. 1403–1421.

[40] Dunik, J., O. Straka, et al., "Survey of Nonlinearity and Non-Gaussianity Measures for State Estimation," *International Conference on Information Fusion*, 2016.

[41] Costa, P. C. G., K. B. Laskey, E. Blasch, and A.- L. Jousselme, "Towards Unbiased Evaluation of Uncertainty Reasoning: The URREF Ontology," *International Conference on Information Fusion*, 2012.

[42] Costa,P. C. G., E. P. Blasch, K. B. Laskey, et al., "Uncertainty Evaluation: Current Status and Major Challenges—Fusion2012 Panel Discussion," *International Conference on Information Fusion*, 2012.

[43] de Villiers,J. P., R. Focke, G. Pavlin, A.- L. Jousselme, V. Dragos, K. B. Laskey, P. C. Costa, E. Blasch, "Evaluation mMtrics for the Practical et al., "Application of URREF Ontology: An Illustration on Data Criteria," *International Conference on Information Fusion*, 2017.

[44] de Villiers, J. P., A. -L. Jousselme, A. de Waal, et al., "Uncertainty Evaluation of Data and Information Fusion within the Context of the Decision Loop," *International Conference on Information Fusion*, 2016.

[45] Oliva, R., E. Blasch, and R. Ogan, "Bringing Solutions to Light through Systems Engineering," *IEEE Southeast Conference,* 2012.

[46] Blasch, E., "Trust Metrics in Information Fusion," *Proc. SPIE*, Vol. 9091, 2014.

[47] Cho, J.- H., K. Chan, and S. Adali, "A Survey on Trust Modeling," *ACM Computing Surveys*, Vol. 48, No. 2, Article 28, Oct. 2015.

[48] Yu, W., H. Xu, et al., "Public Safety Communications: Survey of User-Side and Network-Side Solutions and Future Directions," *IEEE Access*, Vol. 6, No. 1, 2018, pp. 70397–70425.

[49] Malec, M., T. Khot, J. Nagy, et al., "Inductive Logic Programming Meets Relational Databases: An Application to Statistical Relational Learning," *Inductive Logic Programming (ILP),* 2016.

[50] Zuo, H., H. Fan, et al., "Combining Convolutional and Recurrent Neural Networks for Human Skin Detection," *IEEE Signal Processing Letters*, Jan. 2017.

[51] Blasch, E., and A. J. Aved, "Physics-Based and Human-Derived Information Fusion Video Activity Analysis," *International Conference on Information Fusion*, 2018

[52] Blasch, E., R. Cruise, A. Aved, et al., "Methods of AI for Multi-modal Sensing and Action for Complex Situations," *AI Magazine*, July 2019.

[53] "Intelligence Community Directive 203," Jan. 2, 2015, https://www.dni.gov/files/documents/ICD.

[54] Blasch, E., "Multi-Intelligence Critical Rating Assessment of Fusion Techniques (Mi-CRAFT) Method," *IEEE Nat. Aerospace and Electronics Conf.*, 2015.

[55] Chen, H.- M., E. Blasch, K. Pham, Z. Wang, and G. Chen, "An Investigation of Image Compression on NIIRS Rating Degradation through Automated Image Analysis," *Proc. SPIE*, Vol. 9838, 2016.

[56] Blasch, E., E. Lavely and T. Ross "Fidelity Score for ATR Performance Modeling" *Proc. SPIE,* Vol. 5808, 2005, pp. 383–394.

[57] Blasch, E., D. Pikas, and B. Kahler, "EO/IR ATR Performance Modeling to Support Fusion Experimentation," *Proc. SPIE,* Vol. 6566, 2007.

[58] Blasch, E., Y. Chen, M. Wie, H. Chen, and G. Chen, "Performance Evaluation Modeling for Multi-Sensor ATR and Track Fusion System," *International Conference on Info Fusion,* 2008.

[59] Blasch, E., X. Li, G. Chen, and W. Li, "Image Quality Assessment for Performance Evaluation of Image fusion," *International Conference on Info Fusion,* 2008.

[60] Zheng, Y., E. Blasch, and Z. Liu, *Multispectral Image Fusion and Colorization*, SPIE Press, 2018.

[61] Blasch, E., R. Moses, D. Castanon, A. Willsky, and A. Hero, "Integrated Fusion, Performance prediction, and Sensor Management for Automatic Target Recognition," *International Conference on Information Fusion*, 2008.

[62] Laudy, C., and N. Museux, "How to Evaluate High Level Fusion Algorithms?" *International Conference on Information Fusion*, 2019.

[63] Mossing, J. C., and T. D. Ross, "Evaluation of SAR ATR Algorithm Performance Sensitivity to MSTAR Extended Operating Conditions," *Proc. SPIE,* Vol. 3370, Algorithms for Synthetic Aperture Radar Imagery V, 1998.

[64] Ross, T. D., L. A. Westerkamp, R. L. Dilsavor, and J. C. Mossing, "Performance Measures for Summarizing Confusion Matrices: The AFRL COMPASE Approach," *Proc. SPIE,* Vol. 4727, Algorithms for Synthetic Aperture Radar Imagery IX, 2002.

[65] Sadowski, C., "Measuring Combat Identification", *69th MORS Symposium,* Annapolis, MD, 2001.

[66] Blasch, E., "Fundamentals of Information Fusion and Applications Tutorial," *International Conference on Information Fusion,* 2002.

[67] Blasch, E., "Fusion Evaluation Tutorial," *International Conference on Information Fusion*, 2004.

[68] Alsing, S., E. P. Blasch, and R. Bauer, "Three-Dimensional Receiver Operating Characteristic (ROC) Trajectory Concepts for the Evaluation of Target Recognition Algorithms Faced with the Unknown Target DetectionPproblem," *Proc. SPIE,* Vol. 3718, 1999.

[69] Kahler, B., and E. Blasch, "Decision-Level Fusion Performance Improvement from Enhanced HRR Radar Clutter Suppression," *J. Advances in Information Fusion,* Vol. 6, No. 2, Dec. 2011, pp. 101–118.

[70] Lavely, E. M., and P. B. Weichman, "Model-Based and Data-Based Approaches for ATR Performance Prediction," *Proc. SPIE,* Vol. 5095, Algorithms for Synthetic Aperture Radar Imagery X, 2003.

[71] Driggers, R., J. Ratches, J. Leachtenauer, and R. Kistner, "Synthetic Aperture Radar Target Acquisition Model Based on a National Imagery Interpretability Rating Scale to Probability of Discrimination Conversion", *Optical Engineering*, Vol. 42, No. 7, July 2003.

[72] Kahler B., and E. Blasch, "Predicted Radar/Optical Feature Fusion Gains for Target Identification," *Proc. IEEE Nat. Aerospace Electronics Conf (NAECON)*, 2010.

[73] Blasch, E., H,- M. Chen, J. M. Irvine, et al., "Prediction of Compression Induced Image Interpretability Degradation," *Opt. Eng.*, Vol.57, No. 4, 043108, 2018.

[74] Blasch, E. P., and P. Svenmarck, "Target Recognition Using Vehicle Separation Plots (VSP) for Human Assessment," *5th World Multi-Cnference on Systems, Cybernetics, and Informatics (SCI 2001)*, July 2001.

CHAPTER 10

Recent Topics in Machine Learning for Radio Frequency ATR

10.1 Introduction

This chapter highlights recent ML research topics impacting RF image and signal classification for ATR. These advances include adversarial machine learning (AML), transfer learning (TL), energy efficient-learning (EEL), and near-real-time data-efficient learning. For more details on the topics, the reader can review the original manuscripts for AML [1–2], TL [3–4], EEL [5], and near-real-time data-efficient training [6].

AML is a contemporary topic highlighted for deep learning (Chapter 4), signals analysis (Chapter 8), and imagery. AML presents how ML algorithms/models/data are perturbed (intentional or unintentional), so that the ML methods deliver unintended results (i.e., inaccurate classification). The concerns raised from AML are important as inaccuracies in the desired outcome may lead to detrimental effects in decision making. For example, declaring a school bus as a tank could have huge consequences if action is taken. AML RF includes classification when the radar or return signals are modified with noise. Future classification methods could then use AML to determine if the data has been altered, the model has been manipulated, or decisions have been corrupted.

TL is useful in situations of classifying objects when measured data are very limited. Sometimes there are few snapshots of measured RF data of an object; however, a machine may generate representative synthetic RF signature data to facilitate learning. Then the research centers can conduct pragmatic combinations of the measured and synthetic data to achieve enhanced classification performance. In the ML literature TL, also known as domain adaptation (DA), aims to fill in the gaps of measured and synthetic data mismatch or adapting data from one domain to another (one sensor or environment to another), so that classification performance results. This chapter provides an example of current TL trends in the RF domain.

The third direction in ML for RF data includes *efficiency of data and energy learning*. For certain ML-based target classification, size, weight, and power (SWaP) computation is necessary to achieve energy-efficient computing for ML applications. Several computing architectures are evolving for SWaP constraint computation. This chapter discusses advances in adversarial learning (Section 10.2), transfer learning (Section 10.3), and data efficient near-real-time training for developing deep learning models (Section 10.4).

10.2 Adversarial Machine Learning

AML attacks have been demonstrated in multifaceted scenarios [2, 7 ,8]. The first type of AML is *noise insertion* into otherwise standard imagery [1]. Here, the adversary intentionally tries to modify signal/image/data for misclassification to happen. A classic example of a noise-insertion attack has been shown as an image of a panda (57% confidence) being classified to a gibbon (99.3% confidence) after adversarial noise insertion [2]. A second type of AML is *data poisoning* such as small patches being inserted into a STOP sign and thus causing the ML algorithm to be misclassified [9]. A more sophisticated AML type has been presented wherein an adversary can hack a computer system and *modify a learned model.*

Formal definitions exist to describe various types of adversarial attacks. In a *white box attack*, the attacker can modify the model parameters to cause inaccurate classification [10]. In a *black box attack*, the attacker may not have access to the model parameters; however, it can generate adversarial images for misclassification to happen [11, 11, 12]. AML also has been defined as a nontargeted or targeted attack. In a nontargeted attack, the adversary tries to misclassify an object (it could be any class of object); however, in a targeted attack, the adversary tries to modify the model or imagery so that the output of the classifier will be a particular target class.

As an emerging research topic, AML algorithms have been demonstrated in non-RF (i.e., EO/video imagery) imagery [12, 13] and RF (i.e., SAR imagery) [14]. Recently, Inkawhich et al. [1] evaluated AML and presented a comprehensive treatment of mitigating adversarial issues in SAR imagery/signal. The technical approach consists of developing advanced deep learning techniques known as *adversarial training* (AT) to mitigate the detrimental effects of sophisticated noise and phase errors. It was demonstrated that AT improves performance under extended operating conditions (EOCs)—in some cases with a 10% improvement over models without AT. Furthermore, AT improves performance when sinusoidal or wideband phase noise is present, gaining 40% in accuracy that typically would be lost in the presence of noise. Also, it was noted that model architectures have significant impact on AML algorithm robustness, with more complex networks showing a greater improvement in adversarial settings. Finally, research reveals that the availability of multipolarization data is always advantageous to mitigate adversarial threats.

10.2.1 AML for SAR ATR

Due to AML issues, secure, robust, and reliable ATR for SAR/RF data under extended operating conditions is crucial. In ATR, SOCs refer to the execution of the algorithms under a controlled/scheduled plan. In the ATR community, ML algorithms' performance evaluation under adversarial settings represents robustness for EOCs, as discussed in Chapter 9. EOCs enable execution of the ML algorithms under various DL models and data usage (intentional or unintentional). The various EOC parameters include elevation and azimuth angles, radar polarization, frequency, bandwidth, signal errors, and so forth. In general, ATR performance evaluation under EOCs mitigates significant risks associated with AML issues.

Traditionally, SAR-ATR corresponds to a signal processing task, accomplished with algorithms such as constant false alarm rate (CFAR) detectors and template-based matching. With recent breakthroughs in DL, the SAR-ATR community has begun to consider the application of DNNs for ATR with AT. In the ML community, a significant research effort is currently focused on AML, which is the study of how worst-case noise impacts DL models. By adopting the AT method from AML [15], a model is trained to be robust to worst-case noise by including it in the training process. Intuitively, if a model is trained to be robust to worst-case noise, it may increase robustness to other forms of noise. Through extensive experimentation of both normally trained and adversarially trained DL models for SAR-ATR, it has been shown that (1) using an adversarial objective increases the robustness of the ATR model in noisy/corrupted environments and EOCs, (2) model architecture and capacity has a significant impact on quality of the adversarially trained model, and (3) using information from multiple polarizations is beneficial to ATR performance and robustness.

10.2.2 AML for SAR Training

Various mathematical models have been developed to understand adversarial attacks in video imagery. These models provide an impact analysis and support reverse engineering techniques to achieve an optimum outcome. Adversarial training is one of the most prominent ways to guard against the classifier attacks. Among the attack models, the well-known approaches [3, 16] were applied on SAR phase history data.

Fast Gradient Sign Method

The fast gradient sign method (FGSM) is a simple, one-step attack to input data point/image, x. For an input image x, the FGSM method uses the gradients of the loss with respect to the input image to create a new image that maximizes the loss. This new image is called the *adversarial image*. The FGSM adversarial example computation is:

$$x_{adv} = x + \varepsilon * sgn\left(\nabla_x L(x, y; \theta)\right) \tag{10.1}$$

In the above equation, *x_adv* is adversarial image, x is the original input image, ε is the perturbation/adversarial noise level, L is the loss, y is the ground truth label for x, and θ is the model parameter.

Projected Gradient Descent

Projected gradient descent (PGD) is a more powerful, multistep attack to input data point, x. To compute a PGD adversarial example:

$$x_{adv}^{t+1} = \prod_{x+s} (x^t + \varepsilon * sgn(\nabla_x L(x, y; \theta))) \tag{10.2}$$

In (10.2), x_{adv}^{t+1} is adversarial image sample at $t + 1$ iteration, x is the original input image, ε is the perturbation, L is the loss, y is the ground truth label for x, θ is the model parameter, s is the set of all samples in the data set, and Π is the projection operator.

DeepFool

DeepFool has been presented by Moosavi-Dezfooli et al. [17]. DeepFool is an accurate approach for constructing adversarial examples based on the distance of samples to the closest decision boundary. DeepFool tries to find the minimum perturbation r that changes the output of function $f(\cdot)$ if added to input x. The calculated perturbation r is multiplied by a constant $(1 + \zeta)$ to ensure that x crosses the decision boundary, resulting in a net perturbation of δ.

$$\begin{aligned} \delta &= r.(1 + \zeta) \\ x_{adv} &= x + \delta \end{aligned} \tag{10.3}$$

Other (RF/radar-specific) adversarial attack models include random noise per pixel, signal phase errors, interference/jamming, and data collection geometry mismatch in training and testing (elevation, azimuth, etc.).

In general, when generating an adversarial image from the original image, the parameter, ε (i.e., adversarial perturbation) determines the classification accuracy. The higher the value of ε, the more noticeable the noise in the imagery and thus more inaccurate classification. Figure 10.1 plots the value of ε vs. classification accuracy and conveys the impact of higher adversarial noise. Plotting the classification accuracy results versus the adversarial perturbation exists with various data sets, such as MSTAR public release or MNIST database.

Figure 10.2 demonstrates perturbed SAR images at different values of ε and different adversarial noise generation methods. The first column shows an original SAR image of an object. The second column shows an SAR image generated by perturbing it with $\varepsilon = 4$ using random, FGSM, and PGD adversarial attacks. As the value of ε increases, noises in the imagery become more visible, and the greatest image degradation is at the value of $\varepsilon = 32$. One thing to notice is that, unlike advanced adversarial attacks models (FGSM or PGD), random noise is not causing a significant degradation to the SAR image.

In an adversarial setting, training a network involves optimizing network parameters to minimize an adversarial loss, as shown in Figure 10.3. Extending Figure 3.7, the nonlinear decision boundary is created to mitigate adversarial impact.

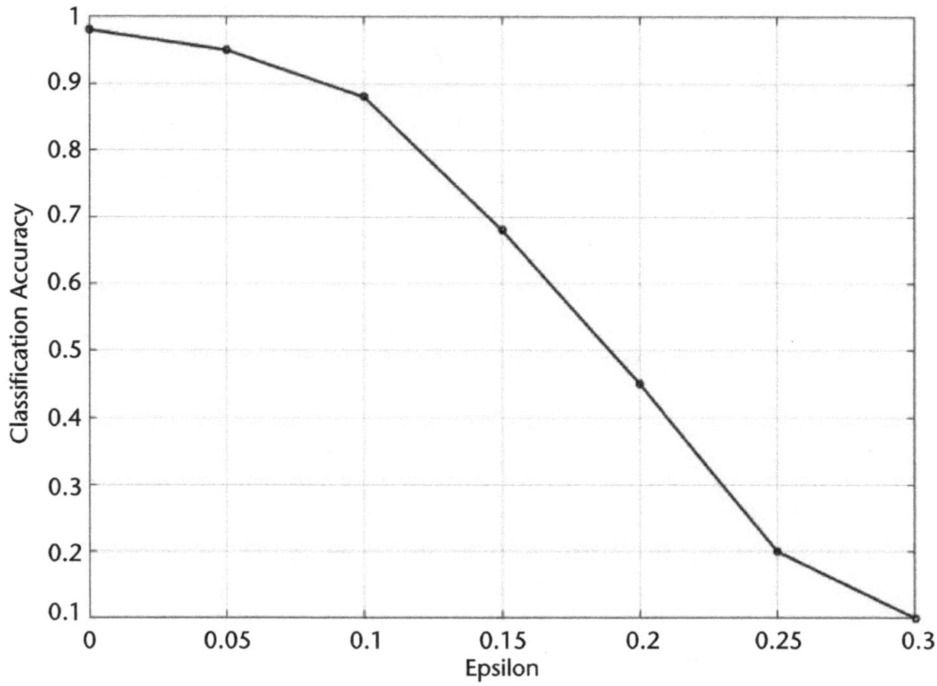

Figure 10.1 Classification accuracy vs. epsilon (ε). For a given image, as the value of ε (i.e., adversarial perturbation) increases, the classification accuracy falls.

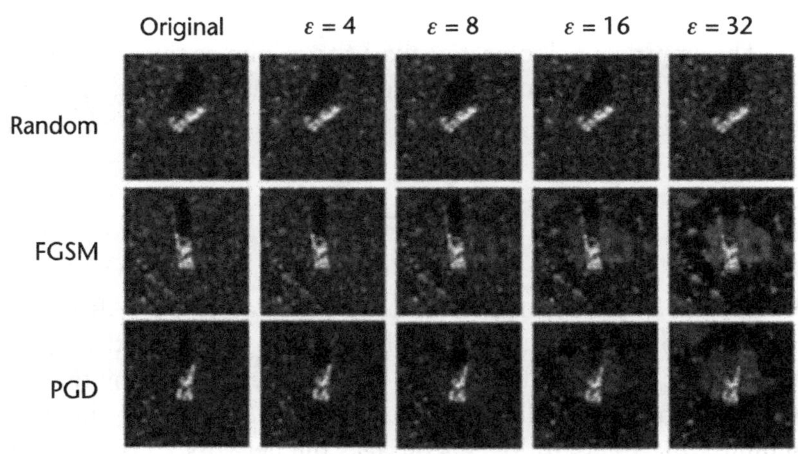

Figure10.2 SAR imagery (MSTAR public released) perturbed by different adversarial attacks models and the degree of attack (i.e., $\varepsilon = 4,8,16,32$.) Imagery from the left column to the right shows the most significant impact/degradation as the value of epsilon increases.

Decision boundaries respect L_∞-norm (i.e., provides the largest magnitude among each element of a vector) balls around the training data [3].

For standard training, optimization involves a minimizing the single parameter θ:

$$\theta^{*} = \arg\min_{\theta} \underset{(x,y)\sim p_{true}}{\mathbb{E}} \left[L(x,y;\theta) \right] \tag{10.4}$$

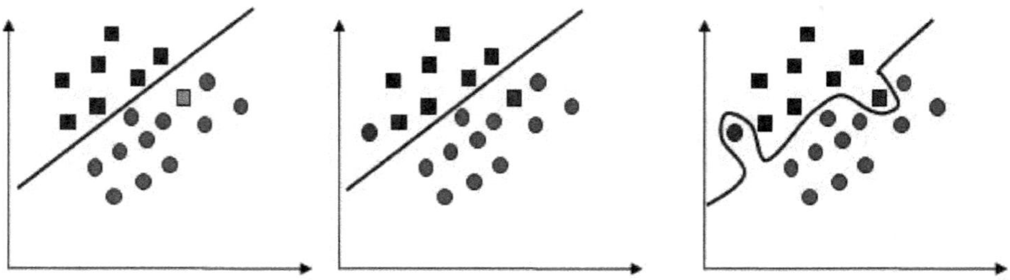

Figure 10.3 Intuition with L_∞ norm ball: (a) normal decision boundary, (b) adversarial attack, and (c) decision boundary accounting for adversarial challenges.

where θ^* is the classifier that minimizes the expected classification loss L for the samples drawn from p_{true}, with x as the SAR image and y as the true classification label. However, it is not possible to train over the entirety of p_{true} as the systems does not have access to an infinite set of data (e.g., SAR chips). Rather, a strategy is to determine a finite dataset $D = \{x_n, y_n\}$ for $n = 1, 2, \ldots, N \sim p_{true}$, of N samples from p_{true} (e.g., from the MSTAR data set). The learning objective becomes an empirical risk minimization over samples in D,

$$\min_{\theta} \frac{1}{N}\sum_{i=1}^{N}\left[L(x,y;\theta)\right) \tag{10.5}$$

The empirical risk objective works to minimize the expected loss of samples in D. Both the MSTAR and CVDome are 10-class problems, so L is a cross-entropy loss over 10 classes (C), defined as:

$$L(x,y;\theta) = -\sum_{c\in C} y_c \log\left[\left(f(x;\theta)_c\right]\right. \tag{10.6}$$

where f is a DNN, parameterized by θ, which accepts a SAR image chip x and outputs a vector of 10 logit values. The function $f(x;\theta)c$ is the log-odds that input x belongs to class c. Similarly, y is a one-hot vector encoding the ground truth class, with y_c equal to the value corresponding to class c. Thus, the training objective to minimize the standard loss encourages the network to output a high value corresponding to the ground truth class via updates to θ. Note that the objective does not include a term to maximize the distance to a decision boundary, and further, the loss signal is very small for adaptive predictions. Thus, the network does not learn much from correct predictions.

For adversarial training, optimization involves two parameters with:

$$\min_{\theta} \mathop{\mathbb{E}}_{(x,y)\sim D}\left[\max_{\delta\in S} L(x+\delta,y;\theta)\right] \tag{10.7}$$

where S is the set of allowable perturbations and δ is a perturbation taken from that set. Here, the inner maximization term works to maximize the loss on the current $(x; y)$ sample, given the current state of θ, by adding a perturbation (δ) to x. The outer minimization term works to minimize the loss on the perturbed sample through an update of θ. Note that the goal of maximizing a DNN loss through constrained perturbations of the input is the generation of an adversarial example, so the maximized loss term is referred to as an adversarial loss.

Besides adversarial training, it was found that most robust deep neural networks models, such as ResNet18, can withstand adversarial attacks to some degree. The following are the most commonly used DNN models, and their performance in adversarial settings varies, where the most complex model performs better:

- SAR ATR community DNN models:
  - A-ConvNets or aconv (all convolutional networks) sparsely connected layers
  - ConvNetB or convb, fully connected layers
- Computer vision community DNN models:
  - ResNet18 (rn18)
  - VGG11bn (vgg11)
  - ShuffleNetv2 (shuf)

Table 10.1 shows the learning rate, number of parameters, and classification accuracies achieved by these models on the MSTAR and CVDome data set. The robust model such as ResNet18 (rn18) utilizes a huge number of parameters and performs very well on adversarial attacks.

Figure 10.4 highlights performance of various standard DNN models using the MSTAR data. With epsilon value zero (i.e., no adversarial perturbation), all the models perform very well at 95%–98% accuracy. As the value of ε increases, performance degradation is observed. Simple models, such as aconv and convb, suffer the most when significant attacks happen whereas ResNet18 stays competitive.

Figure 10.5 shows the test accuracy versus training ε against FGSM, and PGD attacks. Note that the random attack causes little accuracy degradation. This is because small random noise is unlikely to move a sample over a decision boundary. The FGSM and PGD attacks cause significant performance decrease, especially for $\varepsilon = 0, 2, 4$ trained models. For PGD attacked $\varepsilon = 0$ models, the accuracy is less than 10%, meaning the performance of the classifier is worse than that of random guessing. Finally, FGSM

Table 10.1 DNN Model Parameters

			Accuracy	
Model	*Learning Rate*	*No. of Parameters*	*MSTAR*	*CVDome*
aconv	0.001	373,898	98.39	91.91
convb	0.001	9,512,970	98.54	90.96
vgg11	0.01	598,698	98.94	
rn18	0.1	111,753,370	97.57	95.67
shuf	0.1	115,863	96.89	

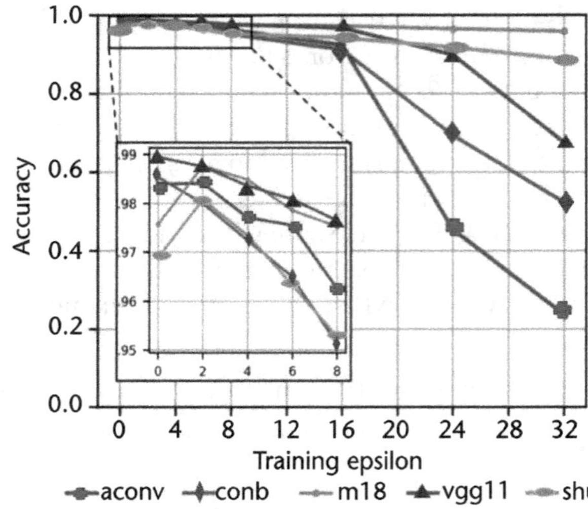

Figure10.4 Performance of various standard DNN models with adversarial perturbation [1].

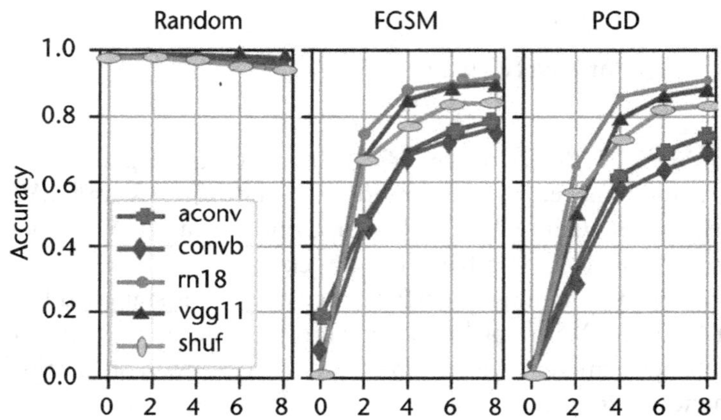

Figure 10.5 Accuracy versus training ε. Robustness to random noise and adversarial attacks at different training epsilons [1].

and PGD attacks cause the most performance impact on aconv and convb, and the least on ResNet18 model.

10.3 Transfer Learning

TL includes (1) homogeneous TL, (2) heterogeneous TL, and (3) negative TL. Within these categories, additions subtopics are identified [4].

One challenge for SAR target recognition is the balance between the measured and synthetic data for training of which bootstrap approaches are common [18]. Efforts include the determination of the sample size, resolution balance, and model-based support for the dynamic data-driven interaction [19]. Recent results [31, 32] on TL have been applied to SAR imagery support situations in which there are few measured SAR images of targets, but there is enough synthetic SAR imagery of these targets. Utilizing both data sets to train a DNN enhances opportunities to achieve good classification accuracy. The technical challenges include the following:

1. *Measured data* is limited and it does not provide sufficient information (the sensor did not collect the entire 360-degree azimuth to capture critical salient features) about the target for good classification performance.
2. *Synthetic data* provides sufficient information of the target. However, this data is pristine (less clutter, noise, etc.) and there is mismatch between synthetic and measured data. Figure 10.6 shows RF measured and synthetic data of BMP2 tank.

Recently, the RF transfer learning data set SAMPLE was published by the Air Force Research Lab [5]. The SAMPLE data set provides an AI/ML researcher much needed data, technical challenges, and an innovative algorithm development opportunity to advance the state-of-the-art TL research. The SAMPLE paper used the DenseNet [20] architecture for the initial TL results. The experiments consist of training the DenseNet with a combination of measured and synthetic SAR data, then test with the measured data. Figure 10.7 shows that as the measured fraction of the data is decreased, the classification accuracy degrades.

Additionally, SAR TL results utilize the Wide ResNet (WRN) architecture [21], the domain adversarial neural network (DANN) [22], and the attention augmented (AA) convolutional network [23]. Figure 10.8 shows that WRN and DANN networks performed better over the DenseNet architecture used in the SAMPLE baseline example at high synthetic fractions until 100% synthetic data is reached for each. The improved performance of the WRN over the DenseNet was noteworthy, since it does not do anything in particular to improve transfer performance. However, the WRN was included in the experiment as the baseline. The increased performance at high synthetic fractions shows the architecture is more efficient at training on fewer measured images than the DenseNet, but breaks down when no measured examples are present. The DANN networks showed similar performance to the WRN when the synthetic fraction is close to 50%, but it ends up outperforming both the WRN and DenseNet architectures at very high synthetic

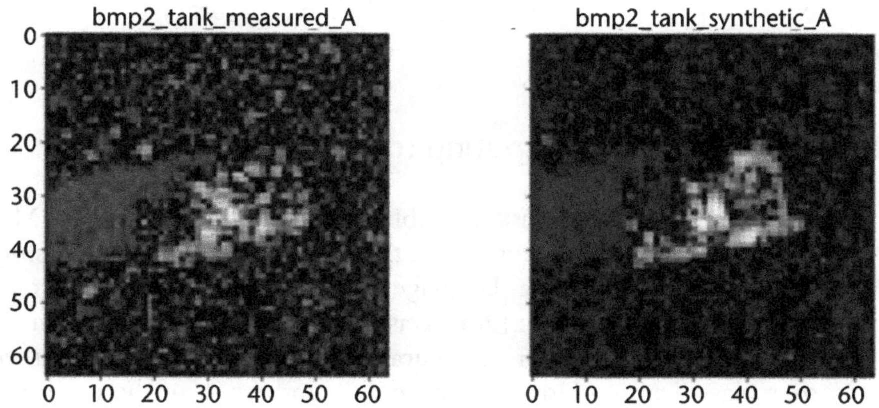

Figure 10.6 Example of (a) measured and (b) synthetic images of a BMP2. Note the differences between the two with respect to the background terrain and the target reflections.

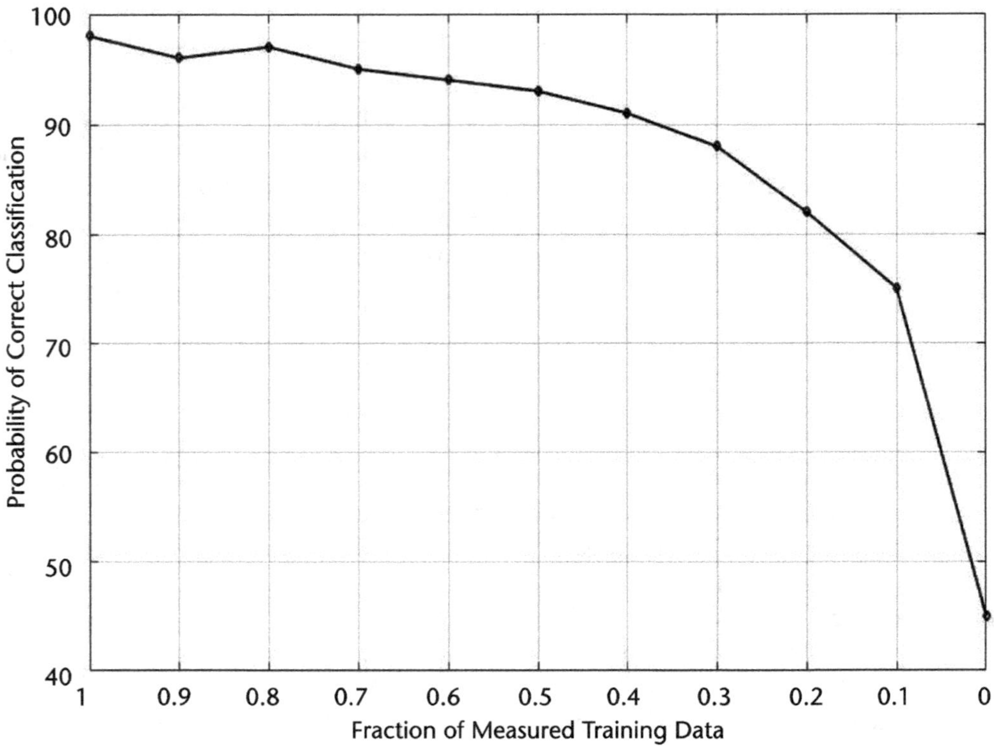

Figure 10.7 DenseNet accuracies versus synthetic fraction for original SAMPLE experiment [5].

fractions. The DANN performance is likely due to the gradient reversal layer on the DANN, which actively adapts the training epoch (i.e., the number of times that the learning algorithm works through the entire training data set) to make the two domains indistinguishable.

Bello et al. [24] presented augmenting CNN with self-attention. The attention map output features were fundamentally different from the convolutional output features, as shown in Figure 10.9.

The attention features show unique segmentation properties between the target and background that are not observed in the convolution output. Overall, the attention network (90.7%) may provide better accuracy than WRN (74.1%), or DANN (88.1%).

10.4 Energy-Efficient Computing for AI/ML

Computing technology is a key enabler for the success of modern ML algorithms (i.e., DNN) that have revolutionized the automatic big data processing (signal and image classification, natural language processing, text analysis, recommender system, medical diagnosis, etc.) for decision making. In general, the more data used for training and the deeper the neural network, the better the learned model for correct classification. However, more data and depth of the networks (i.e., more parameters) entails computational issues. Hence, SWaP-constrained energy-efficient machine learning architectures have emerged as a subfield of ML research. In hardware domain, capabilities (to solve ML problems) of graphics processing

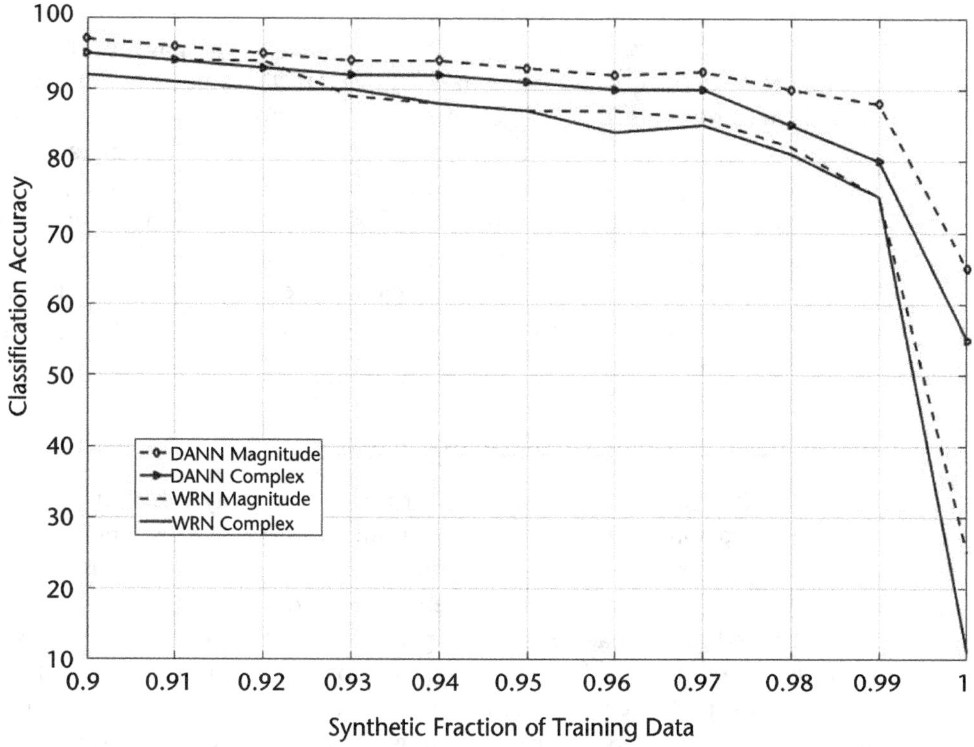

Figure 10.8 WRN and DANN architecture classification performances as a function of the fraction of synthetic targets signature. As more synthetic data is added to the training data, classification accuracy drops.

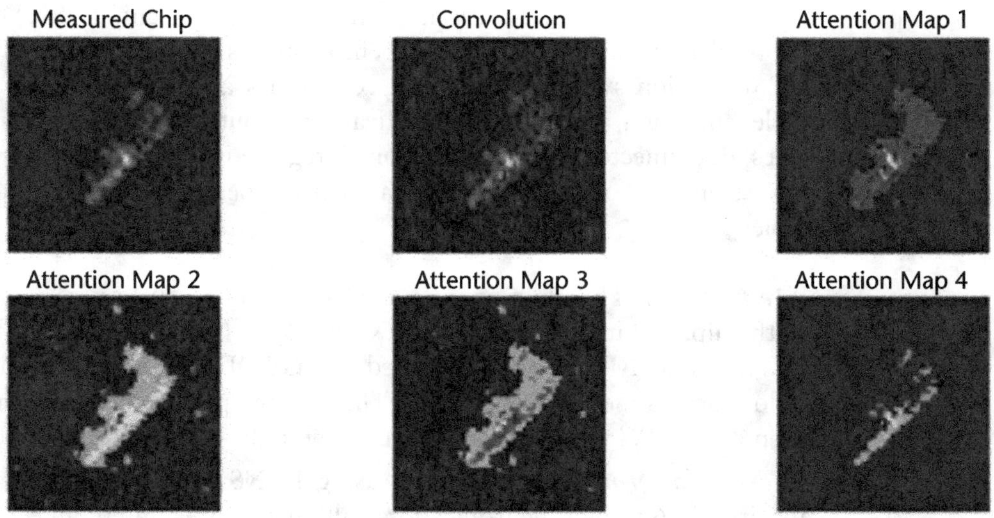

Figure 10.9 Comparison of a measured chip from SAMPLE data set with a convolutional output and four attention output features. Note the difference in the attention features in terms of highlighting/suppressing target attributes and the better overall segmentation of the target area with the background.

units (GPUs), multicore central processing units (CPUs), Google tensor processing units (TPUs), and distributed computing have been explored vigorously. Since these are all *von Neumann*-based computing architectures, there is not much that can be done to improve energy efficiency or make them more compact (in size).

Non-von Neumann based computing architectures seek to advance unprecedented capability of the ML algorithms. Current, non-von Neumann compute architectures (i.e., IBM's TrueNorth) have a huge impact on ML applications in terms of energy consumption and classification performance. SAR image classification on IBM's TrueNorth architecture is possible [24]. Another ML computing technology on the horizon is a memristor for in situ learning [25]. Like IBM's TrueNorth architecture, the memristor is also very efficient for energy consumption. Memristor technology has been demonstrated for MNIST digits classification by implementing CNNs [26].

10.4.1 IBM's TrueNorth Neurosynaptic Processor

Neuromorphic computing is the new frontier of energy-efficient computing for AI/ML applications. There are certain applications/platforms where an unlimited amount of power may not be available for ML algorithm training and testing. Neuromorphic computing has proven to provide unparalleled energy savings. Once a trained model has been developed, the inference stage (i.e., testing/classification) can be done at a fraction of energy costs compared to the traditional von Neumann compute architecture. For example, IBM's TrueNorth demonstrated an ability to classify objects while consuming 100 milliwatts of power, as compared to 3–4 watts needed for von Neumann architecture (CPU/GPU). The TrueNorth architecture is a neuromorphic complementary metal-oxide-semiconductor (CMOS) inspired by the human brain. The TrueNorth employs a parallel, event-driven, non-von Neumann kernel for neural networks that is efficient with respect to computation, memory, and communication. A simple TrueNorth chip (NS1e) consists of 4,096 core tiles as a 64 × 64 array. Each chip consists of over 1 million neurons and over 256 million synapses. Each chip consumes approximately 70 mW of power while running a typical vision application. Input spike activates an axon, which drives all connected neurons. Neurons integrate incoming spikes, weighted by synaptic strength. For the earliest release, three specifications for TrueNorth are available:

- *NS1e platform*: The main processing element of the NS1e is a single TrueNorth chip, and it is coupled with a Xilinx Zynq (FPGA) and two advanced RISC machine (ARM) cores connected to 1-GB DDR3 synchronous dynamic random access memory (SDRAM). The average power consumption is between 2W to 3W with TN NS1e consuming only ~3% of the total power.
- *NS1e-16 platform*: It is constructed using 16 NS1e boards, with aggregate capacity of 16 million neurons and 4 billion synapses, interconnected via a 1-Gig Ethernet packer switched network.
- *NS16e platform*: This architecture integrates 16 TrueNorth chip into a scale-up solution. It can execute neural networks 16 times larger than the NS1e. It can be used to classify large imagery with multiple targets.

10.4.2 Energy-Efficient Deep Networks

Energy-efficient deep networks (EEDN) implement a CNN whose connections, neurons, and weights have been adapted to run inference tasks on the neuromorphic hardware. The TrueNorth uses a 1-bit spike to provide event-based computation and communications and uses low-precision synapses. IBM introduces binary-valued neurons with approximate derivatives for the activation function:

$$Activation\,function, y = \begin{cases} 1\,, & \text{if } r \geq 0 \\ 0\,, & \text{otherwise} \end{cases} \quad r = \text{neuron's filter response} \tag{10.8}$$

The IBM TrueNorth architecture comes with trinary-valued (–1, 0, 1) synapses and weight updates are bounded in the range –1 to +1 by clipping. Training is performed offline on traditional GPUs using a library of custom training layers built on functions from the MatConvNet toolbox.

The TrueNorth also uses lightening memory mapped database (LMDB) technology to store datasets used for training and testing. LMDB is a fast, space-efficient, robust storage mechanism that can support large-scale training. The format supports the storage of labeled, multichannel data. LMDB supports data of type either uint8 or single, where the default is single.

10.4.3 MSTAR SAR Image Classification with TrueNorth

In TrueNorth the entire classification process is divided into five steps: (1) create the data set files (LMDB), (2) perform any necessary preprocessing, (3) compose a network and train it, (4) compile the trained network into a spiking TrueNorth model, and (5) evaluate the classification performance.

By applying the above steps, the MSTAR imagery (Figure 1.2) converted into 10 different MATLAB-compatible .mat files (10 different targets). To fit the images within a single chip of TrueNorth, cropping the images from the center reduces the size. As MSTAR images contain a target in the center, most target information will not get lost. Then a LMDB database is created both for training and testing using those previously created .mat files. Data augmentation (adding noise, transformation, etc.) can also be performed in the preprocessing stage. In typical experiments, the data sets used for training and testing are captured in different angles (15 degrees and 17 degrees).

For the MSTAR data set, the DL energy efficient method achieved 96.66% classification accuracy, which is on par with other computing platforms (GPUs, CPUs, and FPGA) as shown in Chapter 4.

10.5 Near-Real-Time Training Algorithms

It is noted that large data sets for training of DNNs require long training times due to computational backpropagation, numerous weight updates, and memory-intensive weight storage. Exploiting randomness during the training of DNNs can mitigate these concerns by reducing the computational costs without sacrificing network performance. However, a fully randomized network has limitations for real-time target classification, as it leads to poor performance. Therefore, researchers considered *semirandom DNNs* to exploit random fixed weights.

Recent work on semirandom DNNs can achieve near-real-time training with comparable accuracies to conventional DNNs models [8]. Semirandom networks are enhanced by using skip connections and train rapidly at the cost of dense memory usage. With greater memory resources available, these networks can train on larger data sets at a fraction of the training time costs (see Figure 10.10). These semirandom DNN architectures enable an avenue for further research in utilizing random fixed weights in neural networks.

Minimizing the overall training time for conventional (nonrandom) DNNs is essential to ensure feasibility in a wide range of applications. For example, conventional DNNs, such as AlexNet and ResNet, require lengthy training times for complex tasks, such as ImageNet. More recent analysis shows that training AlexNet on ImageNet for 100 epochs and a batch size of 512 on the powerful DGX-1 station requires 6 hours 10 minutes. Training ResNet-50 on ImageNet for 90 epochs and a batch size of 256 on a DGX-1 station requires 21 hours. Such long training times are untenable for several applications. Faster inference models with low latency features are available to accelerate during the postdeployment phase of the DNNs. However, the training time bottleneck for large and small data sets is still an active research problem. With the evolving need to move training on the edge, it is critical to study compute-lite architectures with fast training times.

A *random projection network-based* architecture avoids iterative tuning of parameters and minimizes the gradient computation resources. There are several architecture models that support the universal sampling strategy for any low dimensional data. Examples include extreme learning machines (ELM) [27], no-prop, random kitchen sinks, shallow random networks, and reservoir networks [28].

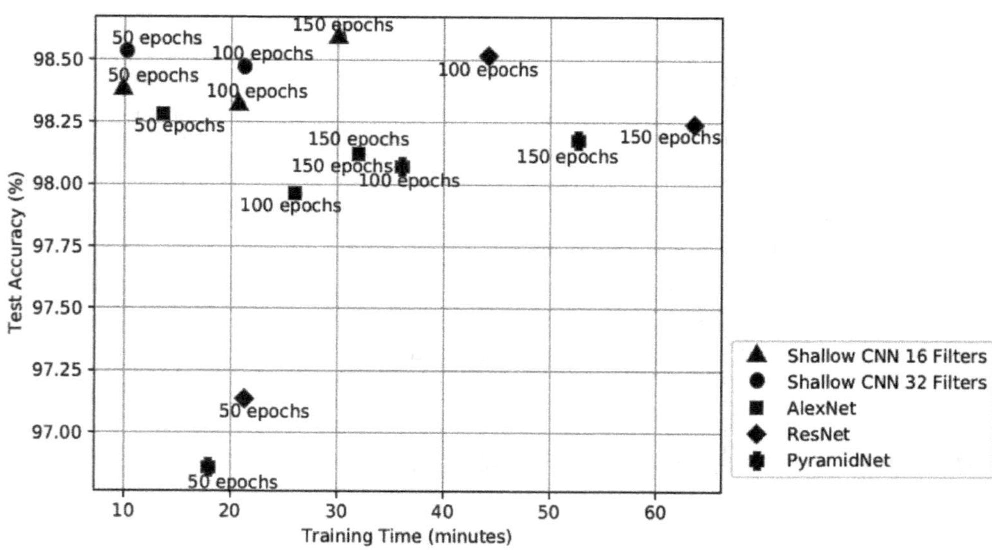

Figure 10.10 Plot of accuracy vs. time of conventional DNN architectures at various training epochs [6].

Figure 10.11 plots the convolutional LEM (CLEM) versus the random vector functional link (RVFL) approaches. Early versions of DNNs relied on the bottom-most layer with randomly connected units that computed random binary functions of their inputs.

Syed et al. [6] focused on using semirandom neural network models for SAR image classification. Specifically, networks such as an ELM with a convolutional layer and RVFL networks with semirandom weights and skipped input-to-output layer connections are analyzed for minimizing training times for the MSTAR data set. It was presented that these semirandom networks can train faster than standard gradient descent approaches by orders of magnitude.

10.6 Summary

This chapter highlighted emerging research topics in machine learning applied to radio frequency signal and image classification. Of the three major concerns, the first introduced the impact of adversarial attacks in SAR imagery and methods to mitigate these issues by incorporating adversarial training. Second, leveraging transfer learning in the context of limited measured data affords the ability to incorporate synthetic data with measured data for training to achieve desired performance accuracy. Third, for energy-efficient computing (e.g., IBM's TrueNorth), SAR image classification SWaP constraints acceptance affords edge computing that can be combined with fog computing for radio frequency target classification. Finally, the chapter introduced an algorithmic approach for near-real-time training for data-efficient approaches.

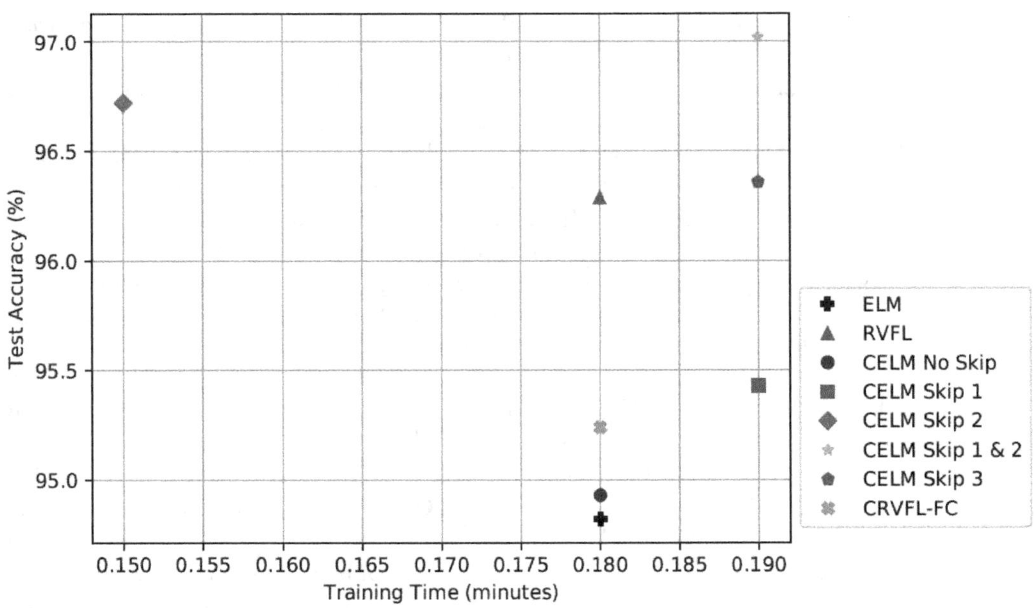

Figure 10.11 Plot of accuracy versus time of semirandom DNNs with varying topologies comparing the ELM, CELM, RVFL, and convolutional RVFL (CRVFL) [6].

References

[1] Inkawhich, N., E. Davis, U. Majumder, C. Capraro, and Y. Chen, "Advanced Techniques for Robust SAR ATR: Mitigating Noise and Phase Errors," *IEEE International Radar Conf.* 2020.

[2] Goodfellow, I. J., J. Shlens, and C. Szegedy, "Explaining and Harnessing Adversarial Examples," *ICLR*, 2015.

[3] Madry, A., A. Makelov, L. Schmidt, D. Tsipras, and A. Vladu, "Towards Deep Learning Models Resistant to Adversarial Attacks," *ICLR*, 2018.

[4] Weiss, K., T. M. Khoshgoftaar, and D. Wang, "A Survey of Transfer Learning," *J. Big Data,* Vol. 3, No. 9, 2016.

[5] Lewis, B., T. Scarnati, E. Sudkamp, J. Nehrbass, S. Rosencrantz, and E. Zelnio, "A SAR Dataset for ATR Development: The Synthetic and Measured Paired Labeled Experiment (SAMPLE)," *Proc. SPIE,* Vol. 10987, 2019.

[6] Syed, H., R. Bryla , U. K. Majumder, and D. Kudithipudi. "Towards Near Real-Time Training with Semi-Random Deep Neural Networks and Tensor-Train Decomposition," 2020, *IEEE AES* (in press).

[7] Merolla, P., J. V. Arthur, R. Alvarex-Icaza, et al., "A Million Spiking-Neuron Integrated Circuit with a Scalable Communication Network and Interface" *Science*, Vol. 345, No. 6197, 2014, pp. 668–673.

[8] Syed, H., R. Bryla, U. K. Majumder, and D. Kudithipudi, "Semi-Random Deep Neural Networks for Near Real-Time Target Classification," *Proc SPIE,* Vol. 10987, 2019.

[9] Eykholt, K., I. Evtimov, E. Fernandes, et al., "Robust Physical-World Attacks on Deep Learning Models," *CVPR,* 2018.

[10] Kurakin, A., I. J. Goodfellow, S. Bengio, et al., "Adversarial Attacks and Defences Competition," in *The NIPS '17 Competition: Building Intelligent Systems*, S. Escalera and M. Weimer (eds), Springer, 2018.

[11] Chen, X., C. Liu, B. Li, K. Lu, and D. Song, "Targeted Backdoor Attacks on Deep Learning Systems Using Data Poisoning," *arXiv:1712.05526*, 2017.

[12] Papernot, N., P. McDaniel, S. Jha, M. Fredrikson, Z. B. Celik, and A. Swami, "The Limitations of Deep Learning in Adversarial Settings," *IEEE European Symposium on Security and Privacy (EuroS&P),* 2016, pp. 372–387.

[13] Nazemi, A., and P. Fieguth, "Potential Adversarial Samples for White-Box Attacks," arXiv:1912.06409.

[14] Papernot, N., I. J Goodfellow, et al., "Practical Black-Box Attacks against Machine Learning," *Asia Conference on Computer and Communications Security*, 2017, pp. 506–519.

[15] Liu, S., V. John, Z. Liu, et al., "IR2VI: Enhanced Night Environmental Perception by Unsupervised Thermal Image Translation," *IEEE CVPRW*, June 2018.

[16] Gao, F., Y. Yang, J. Wang, J. Sun, E. Yang, and H. Zhou, "A Deep Convolutional Generative Adversarial Networks (DCGANs)-Based Semi-Supervised Method for Object Recognition in Synthetic Aperture Radar (SAR) Images," *Remote Sensing*, Vol. 10, No. 6, p. 846, 2018.

[17] Inkawhich, N., W. Wen, H. Li, and Y. Chen, "Feature Space Perturbations Yield More Transferable Adversarial Examples," *IEEE CVPR*, 2019.

[18] Guo, C., J. R. Gardner, et al., "Simple Black-Box Adversarial Attacks," *ICML*, 2019.

[19] Huang, G., Z. Liu, L. van der Maaten, and K. Q. Weinberger, "Densely Connected Convolutional Networks," *IEEE CVPR*, 2017.

[20] Blasch, E. P., S. Alsing, and R. Bauer, "Comparison of Bootstrap and Prior Probability Synthetic Data Balancing Method for SAR Target Recognition," *SPIE Int. Sym. on Aerospace/Defense Sim. & Control,* Vol. 3721, 1999, pp. 740–747.

[21] Blasch, E., S. Ravela, A. Aved (eds.), *Handbook of Dynamic Data Driven Applications Systems*, Springer, 2018.

[22] Moosavi-Dezfooli, S.- M., A. Fawzi, and P. Frossard. "Deepfool: A Simple and Accurate Method to Fool Deep Neural Networks," *IEEE Conference on Computer Vision and Pattern Recognition*, 2016, pp. 2574–2582.

[23] Zagoruyko, S., and N. Komodakis, "Wide Residual Networks," arXiv:1605.07146, 2016.

[24] Ganin, Y., E. Ustinova, et al., "Domain-Adversarial Training of Neural Networks," *Journal of Machine Learning Research*, Vol. 17, 2016, pp. 1–35.

[25] Bello, I., B. Zoph, A. Vaswani, J. Shlens, and Q. V. Le, "Attention Augmented Convolutional Networks," arXiv:1904.09925, 2019.

[26] Spear, J., U. Majumder, et al., "Transfer Learning for SAR Measured and Synthetic Data Mis-match Correction for ATR," *Proc. SPIE, Algorithms for SAR*, 2020.

[27] Yao, P., H. Wu, B. Gao, et al., "Fully Hardware-Implemented Memristor Convolutional Neural Network," *Nature,* Vol. 577, 2020, pp. 641–646.

[28] Huang, G.-B., Q.- Y. Zhu, and C.-K. Siew, "Extreme Learning Machine: A New Learning Scheme of Feedforward Neural Networks," *IEEE International Joint Conference on Neural Networks*, 2004.

[29] Park J., and H. Cho, "Additional Feature CNN-Based Automatic Target Recognition in SAR image," *Fourth Asian Conference on Defence Technology*, 2017.

[30] Tanaka, G., T. Yamane, et al., "Recent Advances in Physical Reservoir Computing: A Review," *Neural Networks*, Vol. 115, 2019, pp. 100–123.

[31] Arnold, Julia M., Linda J. Moore, and Edmund G. Zelnio, "Blending synthetic and measured data using transfer learning for synthetic aperture radar (SAR) target classification," *Algorithms for Synthetic Aperture Radar Imagery XXV,* Vol. 10647, International Society for Optics and Photonics, 2018.

[32] Clum, Charles, Dustin G. Mixon, and Theresa Scarnati, "Matching Component Analysis for Transfer Learning," SIAM *Journal on Mathematics of Data Science 2.2 (2020):* 309-334.

About the Authors

Dr. Uttam K. Majumder is a senior electronics engineer at Air Force Research Laboratory (AFRL), working there since 2003. His research interests include Artificial Intelligence/Machine Learning (AI/ML) for automatic target recognition, high performance computing (HPC), synthetic aperture radar (SAR) algorithms development for surveillance applications, radar waveforms design, and digital image processing. Among various awards, Dr. Majumder received the Air Force's civilian achievement award for distinguished performance, AFRL's science and technology achievement award for radar systems development, and SPIE Rising Researcher award for machine learning algorithms development.

Dr. Majumder earned his Bachelor of Science (BS) degree from the Department of Computer Science, The City College of New York (CCNY), an MS degree from Air Force Institute of Technology, Dayton, Ohio, and a Ph.D degree in Electrical Engineering from Purdue University, West Lafayette, Indiana. He also received an MBA degree from Wright State University, Dayton, Ohio. Dr. Majumder served as an adjunct faculty at Wright State University, Dayton, Ohio, where he developed and taught a graduate course on SAR Signal and Image Processing. Dr. Majumder has published more than 50 technical papers (journal and conference) with IEEE, SPIE, MILCOM, and Tri-Service radar symposium. He submitted two patent applications. Dr. Majumder is a senior member of IEEE. He is a reviewer of IEEE TAES and serves as a technical program committee and session chair for multiple conferences.

Erik Blasch is a program officer at the United States (US) Air Force Research Laboratory (AFRL)–Air Force Office of Scientific Research (AFOSR) in Arlington, VA. Previously he was a principal scientist at the AFRL Information Directorate in Rome, NY, USA (2012-2017), exchange scientist to the Defence Research and Development Canada (DRDC) in Valcartier, Quebec (2010-2012), and Information Fusion Evaluation Tech Lead for the AFRL Sensors Directorate–COMprehensive Performance Assessment of Sensor Exploitation (COMPASE) center in Dayton, OH (2000-2009). Additional assignments include USAF Reserve Officer Colonel supporting intelligence, acquisition, and space technology with 18 medals. Academically, he was an adjunct associate professor in Electrical and Biomedical Engineering (2000-2010) at Wright State University and the Air Force Institute of Technology (AFIT) teaching classes in signal processing, electronics, and information fusion as well as research adjunct appointments (2011-2017) at the Univ. of Dayton, Binghamton University, and Rochester Institute of Technology.

Dr. Blasch was a founding member of the International Society of Information Fusion (ISIF), (www.isif.org), 2007 President, and Board of Governors (2000–2010). He served on the IEEE Aerospace and Electronics Systems Society (AESS) Board of Governors (2011–2016), distinguished lecturer (2012—2019), co-chair of 5 conferences, and associate editor of 3 academic journals. He has focused on information fusion, target tracking, robotics, and pattern recognition research compiling 850+ scientific papers and book chapters. He holds 31 patents, received 33 team-robotics awards, presented 60+ tutorials, and provided 12 plenary talks. His coauthored books include *High-Level Information Fusion Management and Systems Design* (Artech House, 2012), *Context-Enhanced Information Fusion* (Springer, 2016), *Multispectral Image Fusion and Colorization* (SPIE, 2018), and *Handbook of Dynamic Data Driven Applications Systems* (Springer 2018).

Dr. Blasch received his B.S. in Mechanical Engineering from the Massachusetts Institute of Technology (1992) and Masters' Degrees in Mechanical Engineering (1994), Health Science (1995) and Industrial Engineering (Human Factors) (1995) from Georgia Tech and attended the University of Wisconsin for a MD/PhD Neuroscience/Mechanical Engineering until being call to military service in 1996 to the United States Air Force. He completed an MBA (1998), MS Economics (1999), and PhD in Electrical Engineering (1999) from Wright State University and is a graduate of Air War College (2008). He is the recipient of the IEEE Bioengineering Award (Russ-2008), IEEE AESS magazine best paper Award (Mimno-2012), Military Sensing Symposium leadership in Data Fusion Award (Mignogna-2014), Fulbright scholar selection (2017), and 18 research/technical and team awards from AFRL. He is an American Institute of Aeronautics and Astronautics (AIAA) Associate Fellow, Society of Photonics and Industrial Engineers (SPIE) Fellow, and Institute of Electrical and Electronics Engineers (IEEE) Fellow.

David Alan Garren, Ph.D., is a tenured Professor in the Electrical and Computer Engineering Department at the Naval Postgraduate School (NPS)—a position he has held since 2012. He has authored or co-authored 20 refereed journal papers and over 50 conference publications, mostly in the field of radar. He holds 7 U.S. Patents, all pertaining to radar, and is a Senior Member of the IEEE. In addition, he is currently serving as an Associate Editor in the Electronic Warfare area for the refereed journal *IEEE Transactions on Aerospace and Electronic Engineering.* Also, while currently at NPS, he has served as Principal Investigator for numerous research projects sponsored by various federal government agencies.

Professor Garren received the B.S. degree as Salutatorian from Roanoke College in 1986 as a double major in mathematics and physics, with a concentration in computer science. He received the M.S. and Ph.D. degrees in physics from William & Mary in 1988 and 1991, respectively. He was awarded a competitive Office of Naval Research (ONR) Postdoctoral Fellowship at the Naval Research Laboratory (NRL) from 1991 through 1993. From 1994 through 2012, he held positions at two Fortune-500 defense companies while performing research in various technologies, including radar, signal processing, and modeling. His industry service culminated with being awarded the titles of both Technical Fellow and Assistant Vice President at one of these companies. Professor Garren enjoys conveying his experience in applied research and teaching to NPS graduate students as part of their service to the defense of the U.S.A. and allied nations.

Index

D

E

F

L

M

Recent Titles in the Artech House Radar Series

Dr. Joseph R. Guerci, Series Editor

Adaptive Antennas and Phased Arrays for Radar and Communications, Alan J. Fenn

Advanced Techniques for Digital Receivers, Phillip E. Pace

Advances in Direction-of-Arrival Estimation, Sathish Chandran, editor

Airborne Pulsed Doppler Radar, Second Edition, Guy V. Morris and Linda Harkness, editors

Basic Radar Analysis, Mervin C. Budge, Jr. and Shawn R. German

Basic Radar Tracking, Mervin C. Budge, Jr. and Shawn R. German

Bayesian Multiple Target Tracking, Second Edition , Lawrence D. Stone, Roy L. Streit, Thomas L. Corwin, and Kristine L Bell

Beyond the Kalman Filter: Particle Filters for Tracking Applications, Branko Ristic, Sanjeev Arulampalam, and Neil Gordon

Cognitive Radar: The Knowledge-Aided Fully Adaptive Approach, Second Edition, Joseph R. Guerci

Computer Simulation of Aerial Target Radar Scattering, Recognition, Detection, and Tracking, Yakov D. Shirman, editor

Control Engineering in Development Projects, Olis Rubin

Deep Learning for Radar and Communications Automatic Target Recognition, Uttam K. Majumder, Erik P. Blasch, and David A. Garren

Design and Analysis of Modern Tracking Systems, Samuel Blackman and Robert Popoli

Detecting and Classifying Low Probability of Intercept Radar, Second Edition, Phillip E. Pace

Digital Techniques for Wideband Receivers, Second Edition, James Tsui

Electronic Intelligence: The Analysis of Radar Signals, Second Edition, Richard G. Wiley

Electronic Warfare in the Information Age, D. Curtis Schleher

Electronic Warfare Target Location Methods, Second Edition, Richard A. Poisel

ELINT: The Interception and Analysis of Radar Signals, Richard G. Wiley

EW 101: A First Course in Electronic Warfare, David Adamy

Monopulse Principles and Techniques, Second Edition, Samuel M. Sherman and David K. Barton

MTI and Pulsed Doppler Radar with MATLAB®, Second Edition, D. Curtis Schleher

Multitarget-Multisensor Tracking: Applications and Advances Volume III, Yaakov Bar-Shalom and William Dale Blair, editors

Non-Line-of-Sight Radar, Brian C. Watson and Joseph R. Guerci

Precision FMCW Short-Range Radar for Industrial Applications, Boris A. Atayants, Viacheslav M. Davydochkin, Victor V. Ezerskiy, Valery S. Parshin, and Sergey M. Smolskiy

Principles of High-Resolution Radar, August W. Rihaczek

Principles of Radar and Sonar Signal Processing, François Le Chevalier

Radar Cross Section, Second Edition, Eugene F. Knott, et al.

Radar Equations for Modern Radar, David K. Barton

Radar Evaluation Handbook, David K. Barton, et al.

Radar Meteorology, Henri Sauvageot

Radar Reflectivity of Land and Sea, Third Edition, Maurice W. Long

Radar Resolution and Complex-Image Analysis, August W. Rihaczek and Stephen J. Hershkowitz

Radar RF Circuit Design, Nickolas Kingsley and J. R. Guerci

Radar Signal Processing and Adaptive Systems, Ramon Nitzberg

Radar System Analysis, Design, and Simulation, Eyung W. Kang

Radar System Analysis and Modeling, David K. Barton

Radar System Performance Modeling, Second Edition, G. Richard Curry

Radar Technology Encyclopedia, David K. Barton and Sergey A. Leonov, editors

Radio Wave Propagation Fundamentals, Artem Saakian

Range-Doppler Radar Imaging and Motion Compensation, Jae Sok Son, et al.

Robotic Navigation and Mapping with Radar, Martin Adams, John Mullane, Ebi Jose, and Ba-Ngu Vo

Signal Detection and Estimation, Second Edition, Mourad Barkat

Signal Processing in Noise Waveform Radar, Krzysztof Kulpa

Signal Processing for Passive Bistatic Radar, Mateusz Malanowski

Space-Time Adaptive Processing for Radar, Second Edition, Joseph R. Guerci

Special Design Topics in Digital Wideband Receivers, James Tsui

Systems Engineering of Phased Arrays, Rick Sturdivant, Clifton Quan, and Enson Chang

Theory and Practice of Radar Target Identification, August W. Rihaczek and Stephen J. Hershkowitz

Time-Frequency Signal Analysis with Applications, Ljubiša Stanković, Miloš Daković, and Thayananthan Thayaparan

Time-Frequency Transforms for Radar Imaging and Signal Analysis, Victor C. Chen and Hao Ling

Transmit Receive Modules for Radar and Communication Systems, Rick Sturdivant and Mike Harris

For further information on these and other Artech House titles, including previously considered out-of-print books now available through our In-Print-Forever® (IPF®) program, contact:

Artech House	Artech House
685 Canton Street	16 Sussex Street
Norwood, MA 02062	London SW1V HRW UK
Phone: 781-769-9750	Phone: +44 (0)20 7596-8750
Fax: 781-769-6334	Fax: +44 (0)20 7630-0166
e-mail: artech@artechhouse.com	e-mail: artech-uk@artechhouse.com

Find us on the World Wide Web at: www.artechhouse.com